ELECTRONICS:

A Contemporary Approach

ELECTRONICS:
A Contemporary Approach

WILLIAM H. GOTHMANN

Senior Engineer
International Systems Corporation

PRENTICE-HALL, INC., *Englewood Cliffs, New Jersey* 07632

Library of Congress Cataloging in Publication Data

Gothmann, William H
  Electronics: a contemporary approach.

  Includes index.
  1. Electronics. I. Title.
TK7815.G62   621.381   79-1379
ISBN 0-13-252254-3

Editorial production supervision
and interior design by: JAMES M. CHEGE

Page layout by: GAIL COLLIS

Manufacturing buyer: GORDON OSBOURNE

Printed in the United States of America

10 9 8 7 6 5 4 3 2

PRENTICE-HALL INTERNATIONAL, INC., *London*
PRENTICE-HALL OF AUSTRALIA PTY. LIMITED, *Sydney*
PRENTICE-HALL OF CANADA, LTD., *Toronto*
PRENTICE-HALL OF INDIA PRIVATE LIMITED, *New Delhi*
PRENTICE-HALL OF JAPAN, INC., *Tokyo*
PRENTICE-HALL OF SOUTHEAST ASIA PTE. LTD., *Singapore*
WHITEHALL BOOKS LIMITED, *Wellington, New Zealand*

To:
Myrna, Sue,
Bud, and *Mark*

By wisdom a house is built,
And by understanding it is established,
And by knowledge the rooms are filled
With all precious and pleasant riches.

Solomon

CONTENTS

19. FILTERS   **396**

PART III ACTIVE ELEMENTS

20. ACTIVE ELEMENTS   **415**

PREFACE

A popular song once lamented, "The Old, Gray Mare, She Ain't What She Used To Be." Like the mare, electronics is very different now from what it was a scant ten years ago. The new and imaginative areas of citizens'-band radio, microprocessors, and electrooptical techniques have expanded what used to be a purely communications industry into one encompassing such diverse uses as monotoring patients in hospitals and processing data electronically. These revolutionary and exciting developments have been integrated into this beginning electronics text together with the more traditional subjects so that students can be well grounded in all phases of the industry. Hardware such as integrated circuits and light-emitting diodes appear alongside tubes, transistors, and neon lamps. Techniques such as wire wrapping and printed-circuit-card fabrication are presented along with the traditional soldering and cable-wrapping skills. The goal is to introduce the reader to the full range of contemporary basic concepts in both communications and industrial electronics. As such, the book is suitable for a first course in electronic technology programs.

Not only must our subjects be contemporary, but our teaching methods must be updated to use the power of the hand-held calculator. Whereas complex numbers were once avoided because of the lengthy arithmetic required, this text uses them extensively as a tool for greater understanding of alternating current. In addition, RC time constants are treated more mathematically than graphically, reflecting the change in approach used by most instructors. Similarly, determinants are used in the solution of simultaneous equations, recognizing the ease with which they can be worked on the calculator. A section on determinants is included in the Appendix for students who are unfamiliar with them. Finally, four-place accuracy is used in all calculations so that students can check to make certain their methods are correct.

The classroom must also be the governing factor in the order of presentation. Two fundamental arrangements could have been used: the theoretical and the functional.

The theoretical approach ignores the practical realities of the classroom by assuming that no knowledge of a particular subject is needed until a point is reached at which the entire subject can be revealed. It is a fact of life that the functional approach is used within the classroom; material is presented in the order in which it can be learned most easily. This text was written for such a classroom situation. For example, the operation of meters is introduced very early to enable students to start to work immediately in the laboratory. Then, much later, their internal construction is examined. In a similar manner, resistors and their color codes are introduced very early. The attempt has been to order the subjects as the classroom demands rather than as the author finds convenient.

The text is divided into three logical major parts, corresponding to those used in most technology programs:

Part I: Direct Current

Part II: Alternating Current

Part III: Active Elements

This enables the student to step back and view the forest while keeping the trees in sight.

Part I introduces the principles of direct current. Chapter 7 is rather unique in an electronics text, for it is devoted to understanding those parts and pieces of hardware which we all tend to ignore. It was included at the request of instructors who felt their graduates to be unduly weak in this area.

Part II introduces the principles of alternating current. The opening discussion on magnetism is quite detailed, including material about bubble memories. Additionally, this part includes:

1. RMS of rectangular waveforms (Chapter 11).

2. Polar–rectangular methods on the calculator (Chapter 12).

3. An extensive discussion of dielectrics (Chapter 15).

4. Application of network theorems to ac problems (Chapter 17).

5. A discussion of active filters (Chapter 19).

Part III, Active Elements, provides a survey of transistors, integrated circuits, and vacuum tubes. The discussion is more qualitative than quantitative, to enable the student to receive a good overview of these subjects.

This text could not have been written without the help of the outstanding faculty at Spokane Community College and Spokane Falls Community College. Henry Peden's detailed review of the entire manuscript provided many constructive comments for improving its technical accuracy. In addition, the photographs supplied by George Ruple of SCC, Steve Vento of American Sign and Indicator, Robert Gilchrist, and the many manufacturers greatly improve its appeal and understandability. The sketches drawn by Jeanette Kirishian of SFCC add that touch of humor that enables the student to remember the concepts.

Finally, I sincerely appreciate the support, encouragement, and constructive criticism of my wife, Myrna, during the preparation of the manuscript.

Spokane, WA. BILL GOTHMANN

INTRODUCTION

The electronics industry is perhaps the fastest growing, most dynamic industry in the world today, serving us in such diverse areas as electronic ignition in cars, citizens-band radio, and electronic data processing. It is truly a varied field, requiring a sound knowledge of basic principles used in electronics. This text is designed to satisfy this goal. As such it is divided into three major parts:

PART I: Direct Current

PART II: Alternating Current

PART III: Active Elements

PART I discusses a type of circuit whereby electrical charges flow only in one direction along a wire. Examples of such circuits include the automobile electrical system and the portable calculator.

PART II, Alternating Current, introduces a type of circuit whereby the direction of charge flow along a wire reverses direction periodically. Radio and television transmission uses this type of circuit, as does ordinary 120-volt wiring in our houses and businesses.

PART III introduces devices used to amplify, that is, to take a small electronic signal and boost it into a large signal. These devices—transistors, vacuum tubes, and integrated circuits—are used daily in our radios, calculators, and even our watches.

Direct current, alternating current, and active elements form the basis for all electronics. Turn the page and enter this challenging, exciting world.

PART I

DIRECT CURRENT

This part of the text is devoted to those electrical charges that refuse to be deterred in their direction along a wire. Such unidirectional behavior in the world of charges is called *direct current* (dc). In our study of this phenomenon we shall first examine the structure of matter, then the principles and laws governing dc flow. Next, we shall apply these laws to the circuits typically encountered in electronic equipment. We shall then examine the hardware used in building such equipment, some rather sophisticated methods of analyzing circuits, and, finally, the circuits used to form the meters by which we measure direct current. The chapters within this part include:

1. Introduction to Electronics
2. Current, Voltage, and Resistance
3. Ohm's Law
4. Series Circuits
5. Parallel Circuits
6. Series–Parallel Circuits
7. System Hardware
8. Network Analysis
9. Direct-Current Meters

1

INTRODUCTION
TO ELECTRONICS

This chapter reveals the exciting field of electronics: the people who developed the past, those who mold its future, the basic structure of matter, and the mighty electron, without whom this study would be impossible. The sections include:

1-1. Beginnings

1-2. The impact of electronics

1-3. The electronic industry

1-4. People in electronics

1-5. Electrostatic fields

1-6. Atoms and molecules

1-7. Inside the atom

1-1. BEGINNINGS

From the dawn of history, man has been acquainted with the effects of electricity. The caveman observed the fire produced by "the great thunder from above," Fig. 1-1 when it struck a tree and brought this fire into his stony mansion for heat and light, using it to greet the neighborhood saber-tooth tigers and mammoths.

Later, the Greeks observed that, by rubbing a substance called amber, it could be made to attract bits of straw and dust. The Greek name for amber is *elektron*, from which we derive our words electricity and electronics. Many years later, in A.D. 1600, William Gilbert further refined this science of static electricity by defining which substances were positively charged and which were negatively charged. Charles Coulomb further developed the science by specifying the quantity of charge present.

3

FIGURE 1-1: *The power of electricity (photo by Ruple).*

Alessandro Volta (1745–1827) used chemical energy similar to a battery to generate electricity, rather than friction, allowing study of the science under better-controlled conditions, Fig. 1-2. In 1820, Hans Christian Oersted discovered the relationship between magnetism and electricity. Others, such as André Marie Ampère, Michael Faraday, and Georg Simon Ohm, added substantially to the knowledge of electricity.

In 1887, Heinrich Hertz, a German physicist, stirred the interest of the world in radio by demonstrating that electromagnetic (radio) waves can be transmitted through the air, Fig. 1-3(a). In 1901, Guglielmo Marconi, using a long wire antenna, transmitted these radio waves across the Atlantic Ocean. However, it remained for Lee De Forest to introduce the science of electronics with his invention of the vacuum tube in 1906, Fig. 1-3(b). The vacuum tube could amplify signals that had been reduced in strength by distance or circuitry. The term "electronics" infers amplification, whereas the term "electricity" does not.

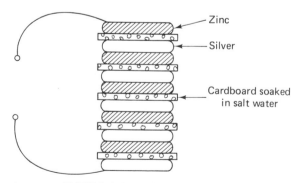

Zinc

Silver

Cardboard soaked
in salt water

FIGURE 1-2: *Volta's electric battery.*

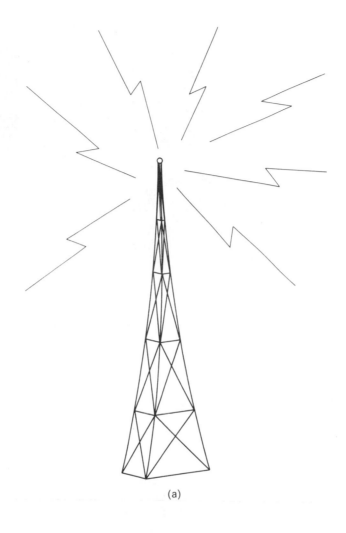

(a)

(b)

FIGURE 1-3: *Radio transmission (photo by Ruple).*

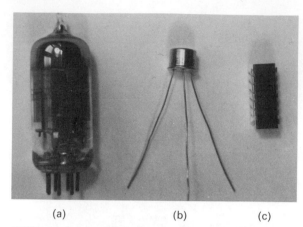

(a)   (b)   (c)

FIGURE 1-4: *Amplifying devices ushered electronics into reality: (a) vacuum tube; (b) transistor; (c) integrated circuit (photo by Ruple).*

In 1948 a second amplifying device was invented by Bell Telephone Laboratories: the transistor. It required far less power to operate than the vacuum tube, was smaller in size, and would not deteriorate with age, Fig. 1-4. Thus, the science of electronics entered the age of miniaturization. Further inventions resulted in many transistors being placed in the same small package producing the integrated circuit.

1-2. THE IMPACT OF ELECTRONICS

The power and variety of electronics has virtually transformed our entire universe. We now depend upon it for our work, our schools, and our home life, relying upon its accuracy, speed, and ease to enrich our lives. Television and radio bring news and entertainment; computers provide accurate data for our industries; the telephone provides instant communication to anywhere on the planet (or outside it). Thus, we live in an electronic age, dependent upon this industry as upon no other. It is no wonder that this industry attracts many of us to learn its secrets and ply its trade.

1-3. THE ELECTRONIC INDUSTRY

With each passing year, more opportunities are opened to us within electronics. We can divide the field into roughly two areas: communications and industrial electronics. As you will observe (you are observing, aren't you?), these areas overlap considerably.

Communications

The communications area employs many people in a diversity of industries. Radio and television stations employ engineers and technicians to design, modify, repair, and operate increasingly complex transmitting equipment, Fig. 1-5. Thousands of

FIGURE 1-5:   *Modern television station (Courtesy, KHQ-TV, Spokane, Wash.).*

television service shops employ technicians to repair radio and television receivers. The advent of citizens' band (CB) radio has provided opportunities for numerous technicians to service such equipment.

However, many of the communication opportunities go unnoticed by the public. Telephone companies employ many people to maintain such diverse equipment as microwave radio links, teletypewriters, data transmission equipment, and television channels. Aircraft use radio navigation and communication equipment that must be maintained to "keep 'em flying." The expanding communications industry will continue to provide opportunities for those who have prepared themselves for this exciting challenge.

Industrial Electronics

Much of this part of the electronics industry remains hidden from the public. Electronics used in signs and display systems increase the "pizazz" of advertising, Fig. 1-6. Computers deal daily with such problems as bank balances, television games, inventory control, airline reservations, college registration, even weighing chicken parts that are packed in boxes for consumers. Hand-held calculators now do for us what once took days to do. Medical electronics now provide faster methods of laboratory analysis, patient monitoring, and assistance to human organs. All these industries need electronic engineers and technicians to design and maintain their complex equipment.

FIGURE 1-6: *Data terminals provide a point of entry into a computing system (photo by Vento).*

1-4. PEOPLE IN ELECTRONICS

As in any other industry, many people are required to keep the industry moving: clerks, presidents, and accountants are all necessary. However, let us direct our attention to those who deal with the hardware we use: the engineer, the engineering aide, the technician, and the assembler.

The Engineer

The *electronic engineer* is the person responsible for the design and modification of an electronic system. He/she is the person who forms the concepts, designs the system on paper, and supervises the manufacturing and installation. He/she relies heavily upon the expertise of technicians and engineering aides for assistance in trouble shooting the design, documenting the system, and carrying out its installation. Most firms require a four-year degree, usually a bachelor of science in electronic engineering. However, many allow any person with the proper credentials through experience and preparation to be engineers.

With the advent of computers, a new field has opened up for engineers and technicians, software engineering. These people are responsible for assuring that the computer does its work in the proper sequence; this sequence is called a program, Fig. 1-7.

FIGURE 1-7: *Software engineer checks out her program (photo by Vento).*

The Engineering Aide

The *engineering aide* is an assistant to the engineer, working under his/her direction. The engineer may ask the aide to design parts of the system, and relies upon his/her knowledge and experience to assure good documentation and schedule adherance. The engineer and the engineering aide form a team to assure that the job is completed on time and according to the customer's specifications, Fig. 1-8. The aide's educational preparation usually consists of a two-year technology degree from a reputable institution.

The Technician

The *technician* is the person responsible for solving malfunctions within a system. He/she is the expert in operation of test equipment, diagnosis of faults within equipment, and the repair of such faults, Fig. 1-9. Thus, both the engineer and the aide rely upon him/her for assistance when equipment must be repaired. His/her educational preparation is similar to that of the aide, usually an associate's degree from a community college.

The Assembler

The *assembler* is the person who "puts it all together" according to the engineer's blueprints, Fig. 1-10. He/she may or may not know what the electronic parts and pieces do, but his/her skill in assembly techniques exceeds that of the aide, technician,

FIGURE 1-8: *Engineer and aide work out a design problem (photo by Vento).*

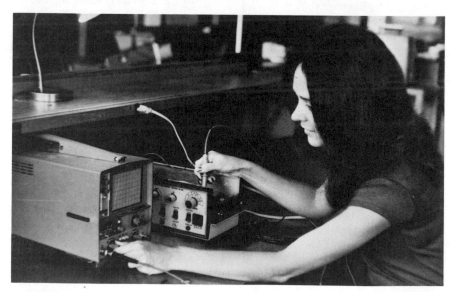

FIGURE 1-9: *Electronic technician solves an equipment malfunction (photo by Ruple).*

FIGURE 1-10: *Assembler at work (photo by Vento).*

and engineer. His/her educational preparation usually consists of fabrication technique classes given by the company. Note that the job requires the teamwork of all these specialists in order to produce the electronic product. In the next section we shall examine the scientific foundations upon which these products are built.

1-5. ELECTROSTATIC FIELDS

How many times have you walked across a carpet on a hot, summer day and touched a metal object, whereupon you heard a sharp "snap" as your teeth rattled from a static shock? This phenomenon of static electricity has been the object of many past experiments, one of the earliest of which consisted of hanging two light pith balls on silk strings, Fig. 1-11(a). Under normal conditions, the balls are pulled toward earth by gravity and the silk strings are parallel. In this condition they are said to be neutral in charge. However, if we were to rub an ebonite rod with fur and touch each of the

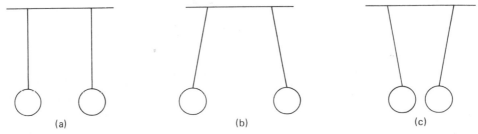

FIGURE 1-11: *Pith-ball experiment:* (a) *neutral charge;* (b) *like charges repel;* (c) *unlike charges attract.*

balls, a force field would develop that would cause the balls to repel one another, Fig. 1-11(b). This field is called an *electrostatic field*: *electro*, because the field is electric, and *static*, because the field does not move. If we were to now rub a glass rod with silk and touch each of the balls, they would behave in a similar manner. However, if we were to touch one of the balls with the stroked ebonite rod and the other with the stroked glass rod, the two balls would be attracted to one another, Fig. 1-11(c). The reason for this is that the ebonite rod, once it has been rubbed with the fur, would become charged in the negative direction; the glass rod, once it has been rubbed with silk, would become charged in the positive direction. Thus, from this experiment two very important principles of static electricity emerge:

1. Unlike charges attract.
2. Like charges repel.

That is, when one charge is negative and the other positive, the two are attracted to one another; when both are negative or both are positive, they repel one another.

Benjamin Franklin (the man with the key and the kite) arbitrarily assigned the charge of two attracting bodies as one being negative and the other positive. This nomenclature has remained with us in designating current flow and polarity of other electrical signals and devices, such as the battery in Fig. 1-12.

FIGURE 1-12: *Battery polarities (Courtesy, Union Carbide, Inc.).*

The basic unit of positive charge is called the *proton* and the unit of negative charge is called the *electron*. Each is equal in its charge. Thus, in our pith-ball experiment, if the number of protons exactly balanced the number of electrons, that ball is said to be electrically neutral. However, if one pith ball had an excess of 1 million electrons and the other pith ball had an excess of 1 million electrons, an electrostatic field of repulsion would exist between the two balls. If, on the other hand, one ball had an excess of 1 million electrons over its number of protons and the second ball had 1 million less electrons than protons, an electrostatic field of attraction would exist between the two. We shall examine the relationship of protons and electrons in the next few paragraphs.

1-6. ATOMS AND MOLECULES

In 460 B.C., Democritus proposed that, if a substance were divided, then divided again, and this were to continue, a point would be reached beyond which division would be impossible. This basic unit of matter he called the *atom*. We know now that the smallest unit of a substance is that of a *molecule*. If, for example, you were to sub-divide a sodium chloride (table salt) crystal, the smallest unit that could still be called sodium chloride would be the molecule, this particular one consisting of one sodium atom and one chlorine atom joined together, Fig. 1-13. The substances sodium and

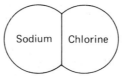

FIGURE 1-13: *Sodium chloride molecule.*

chlorine are called *elements*, for all substances are merely combinations of atoms of these elements. Table 1-1 lists some of the common elements. A molecule of water,

TABLE 1-1: *Some common elements.*

Element	Atomic Number	Atomic Weight[a]
Hydrogen	1	1
Helium	2	4
Carbon	6	12
Nitrogen	7	14
Oxygen	8	16
Sodium	11	23
Aluminum	13	27
Chlorine	17	35
Iron	26	56
Copper	29	64
Germanium	32	73
Silver	47	108
Gold	79	197
Uranium	92	238

[a]Approximate.

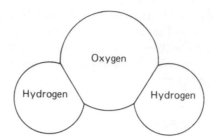

FIGURE 1-14: *Water molecule.*

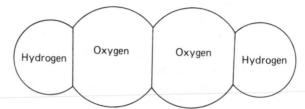

FIGURE 1-15: *Hydrogen peroxide molecule.*

for example, consists of two atoms of hydrogen joined to one atom of oxygen, Fig. 1-14. Peroxide, on the other hand, consists of two atoms of hydrogen joined to two atoms of oxygen, Fig. 1-15. Substances formed by the combination of two or more elements are called *compounds*. Thus, all substances are either compounds or elements.

1-7. INSIDE THE ATOM

Although Democritus believed that the atom was the smallest division of matter, we know now that atoms are primarily composed of three basic particles: protons, electrons, and neutrons. A good approximation of the structure of the atom is attributed to Niels Bohr, a Danish physicist; Fig. 1-16 illustrates the Bohr model of a helium atom. The center of the atom, called the *nucleus*, consists of one proton (labeled P) and one neutron (labeled N). Circling around the nucleus are the helium's two electrons. The proton and the neutron are roughly the same mass, being 1840 times heavier than their puny electron neighbors. However, the electron has the same quantity of charge as the proton but of opposite polarity; the neutron is neutral in charge.

It would seem that, if the electron were negative and the proton were positive, the force of attraction of two opposite charges would cause the electrons to crash into the nucleus. However, this centripetal (center-seeking) force is balanced by the centrifugal (center-fleeing) force created by the orbiting electron. This is similar to swinging a ball on a string around your head. If the string were cut, the ball would flee outward. Thus, this balance of forces, centripetal and centrifugal, maintains the orbit of the electron.

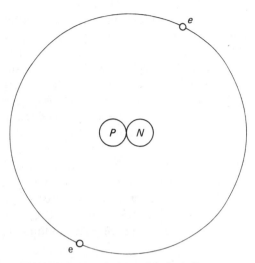

FIGURE 1-16: *Bohr model of a helium atom.*

The Shell Game

The preceding illustration was that of the helium atom. However, a more complex atom is shown in Fig. 1-17. This is a generalized model of any atom showing how the electrons are arranged around the nucleus. Note that they are arranged in *shells*,

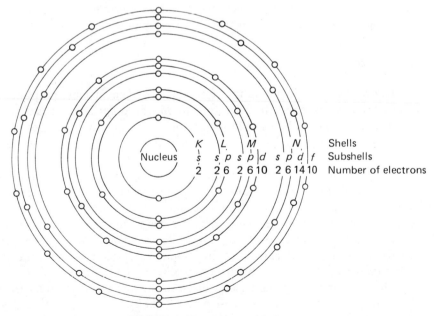

FIGURE 1-17: *Bohr model of an atom.*

labeled K, L, M, and N, and these shells are subdivided into *subshells*, labeled s, p, d, and f. The number of electrons each shell can contain is expressed by the formula $2n^2$, where n is the shell number. Thus, shell one can have a maximum of 2, shell two, 8, shell three, 18. Each of these shells is broken into subshells as shown.

Each shell or subshell has associated with it an energy level. It takes more energy to orbit in an outer orbit than one of the inner ones. They are similar to earth satellites, where it takes a larger booster engine to put the satellite out 100 miles than 50 miles. However, once there, it is easier to remove a satellite from a 100-mile orbit and propel it toward outer space than the 50-mile satellite.

Each element has associated with it a certain number of protons; this number is its *atomic number*. Oxygen has an atomic number of 8, fluorine, 9, iodine, 53, and thorium, 90, for example. The number of electrons in a neutral atom is equal to the number of protons. These electrons fill the lower energy levels first (the inner orbits) and each orbit, once filled, cannot accept any additional electrons at that level.

Valence Electrons

Those electrons in the outermost subshells are called *valence electrons*. By adding energy to the atom in the form of heat, light, or electrical energy, these electrons can be removed from their orbit and freed to drift about the atomic structure. In removing these electrons, the atom has more protons than electrons, and it becomes positively charged. The free electron might then join the orbit of another atom, making that atom negatively charged. This was what happened in the pith-ball experiment: these electrons were removed from their orbits and transferred to the other object, charging that object and the one from which they were removed.

Bonding

In order to form compounds, atoms of one element must join another element. There are three manners in which this union, called *bonding*, takes place. In *covalent bonding*, electrons in the valence shell of one atom share valence electrons of another atom. Thus, these electrons are not associated with any particular atom, but with the molecule as a whole. In *ionic bonding*, the valence electron of one atom is actually given up to fill the shell of a second atom. Thus, both atoms are no longer neutral; these are called *ions*. The donating ion has lost an electron and is therefore a positive ion. The accepting ion has gained an electron and is therefore a negative ion. These two ions, being of opposite charge, are attracted to one another, forming the molecule.

In metals, a third type of bonding, called *metallic bonding*, takes place in which the electrons in the outer shell acquire enough energy at room temperature to become completely free of the atom. Such electrons are called *free electrons* and float throughout the substance, forming a cloud of electrons free to move under the influence of any external force. This allows these metals to be excellent conductors of electricity.

Conductors and Insulators

Materials differ in the amount of energy that must be added to cause a valence electron to become a free electron. In metals such as copper, aluminum, silver, and gold, the valence electrons have already attained sufficient energy to become free at room temperature, whereas in materials such as ceramic, rubber, and plastic, these electrons are tightly held, and considerable energy must be added to cause a valence electron to become a free electron. Those materials with many free electrons are called *conductors*, for they can easily conduct electrons along a path from one point to another. Those that hold their valence electrons tightly are called *insulators*, for these materials can be used to prevent any free-electron flow through the material, thus to insulate one conductor from another.

Atomic Number and Atomic Weight

Whereas the atomic number represents the number of protons in the nucleus, the *atomic weight* represents the sum of neutrons and protons in the nucleus. Thus, oxygen has an atomic weight of 16 and an atomic number of 8. There are, therefore, 16 minus 8 neutrons in its nucleus.

> **EXAMPLE 1-1**: An iron atom has an atomic weight of 56 and an atomic number of 26. How many electrons, protons, and neutrons does a neutral atom of iron contain?
>
> *SOLUTION:* The number of electrons and protons are both 26, the atomic number. The number of neutrons is
>
> $$56 - 26 = 30$$

1-8. SUMMARY

The development of electronics has been a slow, difficult journey. Men such as Gilbert, Coulomb, Volta, Oersted, Hertz, and De Forest have brought us to an era of micro-miniaturization of components. The term "electronic" refers to a circuit or device using an amplifier. Electronics pervades our lives, enriching each of us and extending our five senses. The industry includes the areas of communications and industrial electronics. These industries provide jobs for many, including the engineer who designs the systems, the engineering aide who assists the engineer, the technician who repairs the system, and the assembler who "puts it all together." As a team, these people work to give us the finest in electronics.

Electrostatics provides us with two very important laws: like charges repel and unlike charges attract. A substance is charged if it has an unequal number of protons and electrons. Atoms contain electrons, which have one unit of negative charge; protons, which have one unit of positive charge; and neutrons, which are electrically neutral. The proton and the neutron have about the same weight, whereas the electron weighs $\frac{1}{1840}$ as much. Atoms of elements can join other atoms to form molecules of a compound. Each atom has its electrons in shells, with the outermost shell containing valence electrons. These valence electrons have the highest energy levels without being free of the atom. Insulators have tightly held valence electrons, whereas conductors have many free electrons at room temperature. The atomic number of an atom represents the number of protons, whereas the atomic weight represents the weight of the nucleus.

1-9. REVIEW QUESTIONS

1-1. Our word "electron" came from a name for what substance?

1-2. _____ discovered the relationship between magnetism and electricity.

1-3. _____ invented the vacuum tube.

1-4. The two divisions of the electronics industry are _____ and _____.

1-5. Name three industries that are involved in communications electronics.

1-6. Name three industries that are involved in industrial electronics.

1-7. Name and describe the functions of the four types of jobs that deal with electronic hardware.

1-8. State the two laws of electrostatics.

1-9. The particle of negative charge is the _____.

1-10. The particle of positive charge is the _____.

1-11. The smallest unit of water that can still be called water is the _____.

1-12. All compounds are made up of atoms of _____.

1-13. The electrons in the outer shell of an atom are called _____ electrons.

1-14. The number of protons in an atom is called its _____ _____.

1-15. The sum of protons and neutrons within an atom is called its _____ _____.

1-16. A substance that holds its valence electrons very tightly is called an _____, whereas a substance that has many free electrons is called a _____.

1-10. PROBLEMS

1-1. Compute the number of electrons, protons, and neutrons in a neutral copper atom.

1-2. Compute the number of electrons, protons, and neutrons in a neutral aluminum atom.

1-11. PROJECTS

1-1. Oersted made his famous discovery in a very unusual manner. Describe the incident.

1-2. Tesla, a pioneer in electronics, lead a very unusual life. Write a biographical sketch about him.

1-3. What are the job possibilities, expected salaries, and relocation requirements of the following jobs: electronic engineer, engineering aide, electronic technician, and assembler.

1-4. Although this chapter presumed only three particles of physics (electrons, protons, and neutrons), in reality, others exist. Describe the positron and the antiproton.

2

CURRENT, VOLTAGE, AND RESISTANCE

In the previous chapter we dealt with the electron and its relationship to the atom. In this chapter we shall relate the electron to the fundamental quantities used in electronic circuits: current, voltage, and resistance. We shall also study how we generate a constant flow of electrons and how we measure these little fellows. The sections include:

2-1. The unit of charge

2-2. Current flow

2-3. Voltage

2-4. Resistance and conductance

2-5. Fluid analogy

2-6. Resistors

2-7. The electrical circuit

2-8. Measuring current, voltage, and resistance

2-9. Engineering notation

2-10. Electrical safety

2-1. THE UNIT OF CHARGE

As with any science, it is necessary to quantify a concept: to apply a number to it. In electronics, the fundamental unit of charge is called the *coulomb* and represents a charge of 6.24×10^{18} electrons or 6.24×10^{18} protons. If, for example, our rod of ebonite were to gain 6.24×10^{18} electrons above its neutral charge, it would have a charge of -1 coulomb (C). In a similar manner, if our rod had 6.24×10^{18} less electrons than it had when it was electrically neutral, it would have a charge of $+1$ C, for this would mean there were 6.24×10^{18} more protons than electrons in the rod. Keep in mind that 1 C merely represents a "bucket of electrons," where the bucket has a capacity to hold 6.24×10^{18} of the critters.

2-2. CURRENT FLOW

Next, let us examine the behavior of electrons in a copper wire. At room temperature, the valence electrons have acquired sufficient energy to break free of their orbit and float randomly throughout the wire, Fig. 2-1. Some will float to the left and some to the right, but the algebraic sum of all motion would be zero. Note also that the total number of electrons equals the total number of protons. Thus, the wire is still electrically neutral, even with all its free electrons.

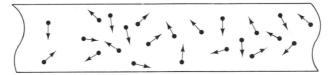

FIGURE 2-1: *Random electron flow within a wire.*

Next, let us, by some magical means, inject one electron into the left end of the wire. We have now disturbed the balance of charges within the wire, and other electrons, since they have a negative charge also, will be repelled by this unwelcome visitor and drift toward the right end until the charge has been evenly distributed along the wire.

Next, assume that a device at the right end captures one electron per second from the wire and a device at the left end injects one electron per second, Fig. 2-2. There would be a drift of electrons toward the right. This drift is called an *electric current*, and the current is said to flow to the right.

One device that would accomplish the electron injection and electron capturing

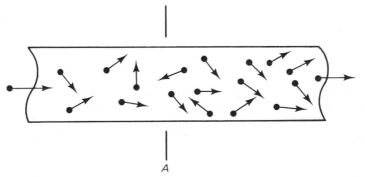

FIGURE 2-2: *Current flow.*

is a battery. The negative end of a battery has an excess of electrons caused by its chemical action; similarly, the positive side of the battery has an excess of protons stripped of their electrons. Thus, when our wire is connected to the battery, current will flow. This current flow must be limited by some means or it would discharge the battery or melt the wire.

Let us now apply a number to this concept of current. One ampere (A) is defined as a flow of 1 C/second(s) past a point on a wire. If I were to count the number of electrons passing point A in Fig. 2-2 and I counted 12.48×10^{18} flowing by that point in 1 s, the current is said to be 2 A. Note the difference between current flow and charge. *Current flow*, measured in amperes, represents the quantity of charge (coulombs) flowing past a particular point per unit time, whereas *charge* is unrelated to time. Current flow is buckets of electrons per second; charge is simply buckets of electrons.

Velocity of Current

Current flowing along a wire can be thought of as one electron entering a wire causing a chain of events along that wire, Fig. 2-2. As the initial electron enters the wire, it repels the one nearest it toward the opposite end of the wire, for its negative charge repels the charge of other electrons. This second electron, in turn, repels the one next to it toward the right end in the figure, and this, in turn, repels others. Thus, our initial electron causes action at the right end even though it travels only a short distance. This action we shall call a *wavefront*. There are now two questions about velocity we must ask ourselves:

1. How quickly does an electron pop off the right end after we inject an electron into the left end? This figure forms the wavefront's velocity.

2. Keeping your eyeball on a single electron, how fast does that electron travel along the wire?

We know now that the velocity of the wavefront is the same as the velocity of light, 300,000,000 meters per second (m/s), or 186,000 miles per second (mi/s), whereas an individual electron travels at inches per second or less. It is, therefore, more convenient to speak in terms of a wavefront traveling along a wire than of individual electrons, for it is the wavefront velocity that effects action within a circuit.

2-3. VOLTAGE

In order to lift a weight, it is necessary to do work. Similarly, to move an electron from one end of a wire to the other, it is necessary to do work, that is, to provide a force, Fig. 2-3. This force is called an *electromotive force* (abbreviated emf) and is measured in *volts*. After all, these electrons are lazy and must be pushed from one end

FIGURE 2-3: *Moving electrons requires work (drawing by Kirishian).*

of the wire to the other. Voltage can be thought of as the push necessary to move electrons. However, it is easier to push electrons along a conductor than an insulator. Therefore, the amount of work necessary to move a charge is dependent upon the material. If, in the mechanical world, we were to lift a 0.7276-lb weight 1 ft, we would have done work defined as 1 *joule* (J). Relating this quantity of work to electricity, a volt (V) is defined as 1 J of work moving 1 C of charge. If, for example, it were necessary to expend 2 J of work to move 1 C, this would represent 2 V.

> **EXAMPLE 2-1:** On the planet K9-4, a material known as dogonite requires 1 J of work be expended to move 3 C of charge. What voltage does this represent?
>
> *SOLUTION:* Since volts is joules per coulomb:

$$\text{volts} = \frac{\text{joules}}{\text{coulombs}} = \frac{1}{3} = 0.3333 \text{ V}$$

Voltage Sources

This voltage we are discussing has to originate somewhere, and many sources are used, Fig. 2-4. One important type is the *battery*, a device that changes chemical action to voltage. It is, literally, a supplier of electrons on its negative terminal and a remover of electrons on its positive terminal. The typical car battery supplies 12 V; thus, it has a "push" of 12 V to the equipment it supplies.

(a)

(b)

(c)

FIGURE 2-4: *(a) Lantern battery (Courtesy, Union Carbide Inc.); (b) A 30 V 300 A dc generator for aircraft (Courtesy, Bendix Electric and Fluid Power Division); (c) brushless ac generator for aircraft use that develops 120/208 V 400 Hz at 60 KVA. The generator is cooled by blast air (Courtesy, Bendix Electric and Fluid Power Division).*

Another type of voltage source is the *generator*. It converts rotational energy to electrical energy, which results in an excess of electrons on its negative terminal and a deficiency of electrons on its positive terminal. The generators of many military aircraft supply 28 V to the various loads on board the vehicle.

Solar cells are also a voltage source, and convert the sun's rays to voltage. They are extensively used in space vehicles to generate electricity for communications and other equipment.

Voltage sources can be divided into two major types: direct-current (dc) and alternating-current (ac) sources. Direct-current sources (such as the battery, solar cell, and dc generator) supply a constant, fixed voltage. Alternating current is used to power our homes and factories and originates at an ac generator. This type of current reverses itself every $\frac{1}{120}$ second (s); that is, the electrons head east for $\frac{1}{120}$ s, then reverse and head west for $\frac{1}{120}$ s, then east, and so forth. The voltage that such generators supply is called a *sine wave* and is shown in Fig. 2-5. We shall discuss alternating current more in detail in Part II.

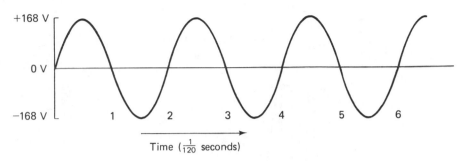

FIGURE 2-5: *Sine wave.*

2-4. RESISTANCE AND CONDUCTANCE

Any time an object must be pushed, it is inferred that there is resistance to movement. In electricity, it is easier to push electrons through a conductor than through an insulator. This characteristic is called *resistance* and is measured in a quantity called an *ohm* (abbreviated Ω using the Greek capital letter omega). We can define 1 Ω as the amount of resistance in a circuit that would limit current flow to 1 A with an applied voltage of 1 V. Note that, by this definition, the greater the resistance, the smaller the current.

On the other hand, the inverse of resistance is conductance. That is, *conductance*, G, equals 1/R, where R is resistance. Thus, the greater the conductance, the greater current flow. Conductance is measured in a unit called the siemen (S); this unit was formerly known as the mho.

2-5. FLUID ANALOGY

The three characteristics of electricity—current, voltage, and resistance—can be more easily visualized by considering an analogy of fluids.[1] Assume that a large tank of water were installed on top of a tower, and that the water flowed through a pipe into a stream below, Fig. 2-6. The height of the tank, *h*, would be analogous to the voltage in a circuit. If *h* were only 1 ft, much less water flow could be generated than if *h* were 100 ft. In a similar manner, a voltage of 1 V will result in much less current that a voltage of 100 V. The height of the tower, like the voltage in a circuit, represents the amount of "push" in the fluidic circuit.

On the other hand, the larger the pipe, the larger the water flow. Thus, the diameter of the pipe is analogous to the resistance in an electrical circuit with a larger pipe representing a smaller resistance value.

[1]Although the fluid analogy is very useful in understanding the concepts of flow, resistance, and potential, it is not infallible. For example, in fluids, current flows from a positive height to a less positive height, whereas in electricity current flow is from negative to positive.

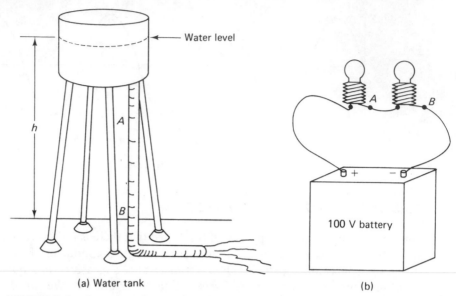

(a) Water tank   (b)

FIGURE 2-6:   *Comparison of fluid and electrical systems: (a) water tank; (b) electrical circuit.*

The amount of water flowing out of the pipe in 1 s represents the current flow through the pipe in the same way that the number of coulombs flowing through a wire in 1 s represents the current flow in a wire. One gallon per second represents less current flow than 10 gal/s, just as 1 A represents less electrical current than 10 A.

Potential Difference

This fluid analogy is a very useful tool in trying to understand the concept of *potential difference* in a circuit. If h were 100 ft, we could say the potential difference (that is, the difference in water head) was 100 ft, as in the electrical circuit shown in Fig. 2-6(b) we could say the potential difference between the terminals of the battery was 100 V. However, we could also say that the potential difference between point A and point B of Fig. 2-6(a) is 50 ft, just as we could say that the potential difference between points A and B in Fig. 2-6(b) is 50 V. Thus, voltage not only represents the amount of "push" in a circuit, it also refers to the potential difference between two points, just as water pressure can be evaluated between any two points of a fluidic circuit.

2-6. RESISTORS

Resistors are devices that display a known value of resistance. They are one of the most widely used components in electronics, appearing in practically every circuit and system. There are two basic types available: fixed and variable. *Fixed* resistors come from the factory with a certain value, measured in ohms, and cannot be changed by the designer or equipment operator. The *variable* types have a movable shaft or other mechanism that allows the value of the device to be changed.

Fixed Resistors

Fixed resistors can be divided into four types: carbon composition, carbon film, metal film, and wirewound. The *carbon-composition* type comes in four physical sizes: $\frac{1}{4}$ watt (W), $\frac{1}{2}$ W, 1 W, and 2 W, the size governing how much heat the resistor is allowed to generate without being destroyed, Fig. 2-7. Each consists of a small rod of powdered carbon that has been mixed with an insulating material, called a binder, in proper proportion to supply the correct value of resistance. Metal leads are then added in order to connect it to its external circuit and the whole assembly encapsulated in an insulating body for mechanical strength and electrical isolation. This type of resistor is by far the most common of all found in electronics.

Carbon-film resistors consist of a film of carbon that has been deposited upon an insulating body; this carbon is then cut in a spiral manner, the angle of the cut determining the resistance value, Fig. 2-8. Finally, the device is sprayed with an insulating material. These resistors have more accurate values in ohms than the carbon composition and are, therefore, a little higher in cost.

Metal-film resistors provide even more accuracy than that of the carbon-film type. They are constructed by depositing a metallic conductor on a glass form.

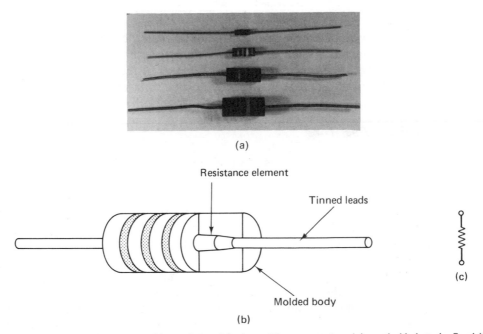

(a)

(b)

(c)

FIGURE 2-7: *Carbon-composition resistors: (a) photo; (b) cutaway view; (c) symbol (photo by Ruple).*

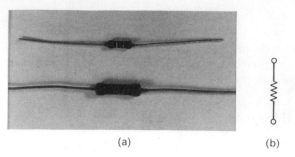

(a)   (b)

FIGURE 2-8: *Metal-film (top) and carbon-film (bottom) resistors (photo by Ruple).*

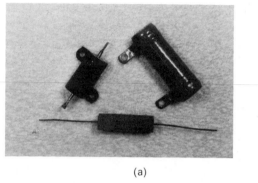

(a)   (b)

FIGURE 2-9: *Power resistors (photo by Ruple).*

Wirewound resistors consist of a length of resistance wire, such as manganin, that has been wrapped around an insulating body, then sprayed or dipped in an insulating coating, Fig. 2-9. These resistors are capable of dissipating much more heat than the carbon types and are usually much larger than their smaller brothers.

Variable Resistors

There are two basic types of variable resistors, also: carbon and wirewound. The *carbon* variable type has a shaft that is used to adjust a wiper arm to select the value of resistance that is to appear between the wiper arm and one of the terminals, Fig. 2-10. As the shaft is rotated, the amount of resistance between the arm and the outside terminals is varied. These devices are commonly used in radio and television receivers as volume controls. They come in varied shapes and sizes to suit the purpose of the design.

There are several ways in which the resistance is allowed to vary as the shaft is turned: this is called the *taper* of the device. In a linear taper, the resistance is changed linearly with shaft rotation. If the shaft were rotated exactly 50% of the total possible rotation, the resistance would be 50% from point B to point A and 50% from point B to point C. However, there are other tapers that change the resistance in a non-

28

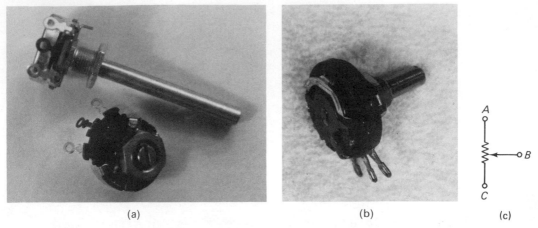

FIGURE 2-10: *Potentiometers: (a) photo; (b) cutaway view; (c) symbol (photos by Ruple).*

linear manner as the shaft is rotated, such as a logarithmic taper or semilogarithmic taper.

These three terminal devices are called *potentiometers*, pots for short. They can be connected in tandem to make one shaft control several controls, or an on–off switch can be provided, Fig. 2-11. The carbon-type variable resistor is capable of dissipating less heat than the wirewound type, typically 2 W for a volume control. They are also less precise, for the resistance of the carbon cannot be controlled as well as the resistance of the wire used in the wirewound.

The *wirewound* variable resistor is available in two configurations, as shown in Fig. 2-12. One is used in a manner similar to that described for the carbon variable resistor. The other is actually a wirewound fixed resistor with an adjustable slider that allows the electronic technician to adjust the value by loosening the adjusting screw and moving the metal slider. Both are used whenever large quantities of heat must be dissipated.

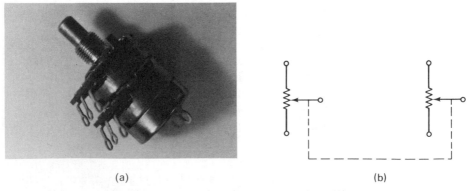

FIGURE 2-11: *Potentiometers connected in tandem: (a) photo; (b) symbol (photo by Ruple).*

(a)   (b)   (c)

FIGURE 2-12: *Wire wound variable devices: (a) potentiometer; (b) variable resistor; (c) symbol (photos by Ruple).*

Resistor Color Code

Carbon composition resistors have a color code on the body of the resistor to indicate the value of the resistor in ohms (Ω), Fig. 2-13. This code consists of five colored bands arranged as shown. The first and second bands indicate a numerical value, the third band a multiplier, the fourth band a tolerance, and the fifth band a reliability factor. Each uses the color code shown in Table 2-1. Assume that a resistor

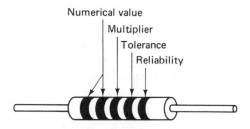

FIGURE 2-13:   *Color bands on a resistor.*

has the following colors, reading left to right: yellow, violet, orange, gold, and red. The yellow and violet bands would yield the number 47 from the table. This value would have the number of zeros placed after it indicated by the third band, orange, in this case three. Thus, the value of the resistor is 47,000 Ω. The fourth band indicates the tolerance, 5%. Therefore, this resistor is 47,000 $\pm$ 5% of 47,000 Ω. Since 5% of 47,000 is 2350, the value of the resistor lies between 47,000 $-$ 2350, or 44,650 Ω, and 47,000 $+$ 2350, or 49,350 Ω. The fifth band indicates that during a test of this type of resistor, 0.1% of the units failed to operate within specifications after a 100-h test period. This fifth band is usually ignored by the technician, being important only in very critical applications.

EXAMPLE 2-2: A resistor has the color bands red, violet, green, silver, and brown. What are its nominal, maximum, and minimum values?

TABLE 2-1: *Resistor color code.*

Color	Significant Figure	Multiplier	Tolerance (%)	100-h Reliability (%)
Black	0	1		
Brown	1	10		1.0
Red	2	10^2		0.1
Orange	3	10^3		0.01
Yellow	4	10^4		0.001
Green	5	10^5		
Blue	6	10^6		
Violet	7	10^7		
Gray	8	10^8		
White	9	10^9		
Gold		0.1	5	
Silver		0.01	10	
No color			20	No test made

SOLUTION: The first two digits yield 27 from the table. There are five zeros to be added as indicated by the green band, making its nominal value 2,700,000 Ω. Its tolerance, from the table, is 10% and 10% of 2,700,000 is 270,000. Therefore, the actual value of the resistor lies between 2,700,000 − 270,000 Ω and 2,700,000 + 270,000 Ω or 2,430,000 and 2,970,000 Ω. It has a 1% reliability factor.

EXAMPLE 2-3: A resistor has the color bands violet, green, and yellow. What are its nominal, maximum, and minimum values?

SOLUTION: Its nominal value is 75 plus four zeros, or 750,000 Ω. Since there is no tolerance band, the tolerance is 20% of 750,000, or 150,000. Thus, its maximum value is 750,000 + 150,000, or 900,000 Ω, and its minimum value is 750,000 − 150,000, or 600,000 Ω. No reliability test was performed.

EXAMPLE 2-4: A resistor has color bands of brown, black, black, silver, and yellow. What are its nominal, maximum, and minimum values?

SOLUTION: Its nominal value is 10 plus zero zeros, or 10 Ω. Its maximum value is 10 + 0.10 × 10, or 11 Ω, and its minimum value is 10 − 0.10 × 10, or 9 Ω. It has 0.001% reliability.

Values Under 10 Ohms

Values of resistors under 10 Ω require moving the decimal point to the left, rather than to the right as was done above. In these resistors, the third stripe is either gold or silver. If it is gold, the decimal point is moved one place to the left, if silver, two places.

EXAMPLE 2-5: What is the nominal value of a resistor with the following color bands: green, blue, gold, and silver?

SOLUTION: The first two bands yield the number 56. This is then multiplied by 0.1, for the multiplier band is gold, giving 5.6 Ω. The tolerance is then figured using this value, giving 5.6 ± 0.56 Ω. No reliability test was made.

EXAMPLE 2-6: What are the nominal, minimum, and maximum values of a resistor with the following color bands: orange, white, silver, gold, and orange?

SOLUTION: The orange and white bands yield 39. This is then multiplied by 0.01, since the third band is silver, giving a nominal value of 0.39 Ω. When the tolerance, 5%, is computed, the result is 0.39 ± 0.0195, giving a minimum of 0.3705 Ω and a maximum of 0.4095 Ω. The reliability factor indicated by the fifth band is 0.01%.

Preferred Values

The resistor manufacturers have standardized the more common values of resistors to lower the cost as much as possible. These values are the ones from which the designer chooses to construct his circuit. Table 2-2 lists the standard values for 5%, 10%, and 20% tolerance resistors. Higher values are merely multiplied by a decimal

TABLE 2-2: *Preferred values.*

20% Tolerance	10% Tolerance	5% Tolerance
10	10	10
		11
	12	12
		13
15	15	15
		16
	18	18
		20
22	22	22
		24
	27	27
		30
33	33	33
		36
	39	39
		43
47	47	47
		51
	56	56
		62
68	68	68
		75
	82	82
		91
100	100	100

factor; there are 0.91-, 9.1-, 91-, 910-, 9100-, 91,000-, 910,000-, and 9,100,000-Ω 5% resistors available, for example. Note that if one chooses a 15-Ω 10% resistor, its actual value lies between 13.5 and 16.5 Ω, because of its tolerance. The next higher value for 10% resistors is nominally 18 Ω but actually lies between 16.2 and 19.8 Ω; the 22-Ω value lies between 19.8 and 24.2 Ω. Thus, each nominal value actually represents a range of values, and the next higher nominal value starts its range at about the same point as the lower range leaves off. This is true for all three tolerance lists; a 20-Ω 5% resistor has the range 19.0 to 21.0 Ω; a 22-Ω 5%, the range 20.9 to 23.1 Ω, a 24-Ω 5%, the range 22.8 to 25.2 Ω, and a 27-Ω 5%, the range 25.65 to 28.35 Ω. When choosing a resistor, one should choose the value closest to the nominal value that was computed after the tolerance has been determined.

> **EXAMPLE 2-7**: Choose the correct nominal value for a resistor that was computed at 2530 Ω.
>
> *SOLUTION:* The closest value for 20% tolerance is 2200 Ω; the closest value for 10% tolerance is 2700 Ω; the closest value for 5% tolerance is 2400 Ω. These were all selected from Table 2-2 using the closest values on the table.
>
> **EXAMPLE 2-8**: Choose the correct nominal value for a resistor that was computed at 779,943 Ω.
>
> *SOLUTION:* The closest values are as follows:
>
> $$5\%—750,000 \ \Omega$$
> $$10\%—820,000 \ \Omega$$
> $$20\%—680,000 \ \Omega$$

2-7. THE ELECTRICAL CIRCUIT

In electronics, a voltage source is used to provide energy to a device for performing some work. That device is said to be a *load*, whether the work is in the form of rotational work, as in the case of a motor, or light, as in the case of a light bulb. The term "load" actually refers to the current through the light bulb or motor; a *heavy* load refers to a large current flow through the load and a *light* load refers to a small current flow.

A *circuit* is defined as a complete path for electrons and consists of a voltage source, conducting wires, and a load, Fig. 2-14, in this case a light bulb. Note that the supplier of electrons in the circuit is the voltage source. Electrons flow out the negative terminal of the battery, through the connecting wire to the load, through the load, through the connecting wire and into the positive terminal of the battery, where chemical action removes them from the wire. Note that the essential parts of the circuit consist of a voltage source to supply the electrons, connecting wires, and a load that restricts current flow—has resistance. Note further that the current is said to

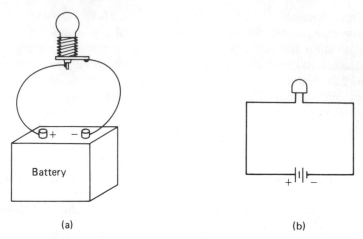

FIGURE 2-14: *Electrical circuit: (a) physical connections; (b) schematic.*

flow through the circuit; the voltage does not flow, but merely exists across the circuit, just as the height of the water tank does not flow in our fluid example, but the water flows.

A circuit in which current is flowing is said to be a *closed circuit*. If a wire in the circuit were to come loose and interrupt a connection, current would cease to flow; such a circuit is said to be an *open circuit*. Note that "closed" and "open" refer to whether current is flowing or not.

Conventional Current Flow

In all the previous examples, we, have stated that electrons flow from negative to positive through the load. In the early development of electricity, man assumed that positive charges did the flowing and that they flowed from positive to negative. All the symbols, analysis, and literature was based upon this assumption. However, later it was discovered that the electron was the actual charge doing the flowing, and it flowed from negative to positive. Thus, two systems of current flow have evolved: *electron flow* and *conventional current flow*, Fig. 2-15. Electron flow is the actual direc-

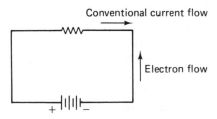

FIGURE 2-15: *Conventional current flow and electron flow.*

tion of the travel of the electrons; conventional current flow can most easily be defined as the opposite of electron flow. It turns out that, except for describing the action of the electron itself, either assumption will yield a correct answer to what the resistance should be or what the values of current and voltage are within a circuit. We shall, however, use electron flow throughout this text. Whenever the term "current flow" is used, it will refer to electron flow. The older system will always be referred to as conventional current flow (the word "conventional" will always precede the word "current").

2-8. MEASURING CURRENT, VOLTAGE, AND RESISTANCE

In order to tell how much voltage there may be across two points within a circuit, it is often necessary to measure it, and this is done using a device called a *voltmeter*. Two types are commonly used: the *digital voltmeter* (DVM) and the *analog meter*, Fig. 2-16. The DVM has the advantage of providing the number of volts that is being read

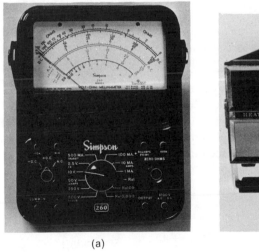
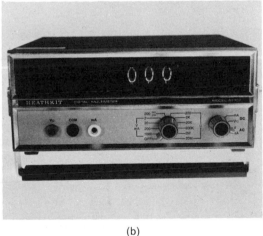

(a)   (b)

FIGURE 2-16: *Measuring devices:* (a) *analog voltmeter;* (b) *digital voltmeter* (photos by Gilchrist).

quite accurately, and is not subject to interpretation on the part of the observer. If the numbers on the DVM read 96.5, there is no doubt whether the number is 96.4 or 96.6. This is not the case in using an analog meter, for the scale must be interpreted by the observer and one person might read a slightly different answer than a second person.

When using a voltmeter, it is always placed across the voltage source or potential difference being determined, Fig. 2-17. Returning to our water analogy, its purpose is to measure potential difference. It must do this by allowing very little current to

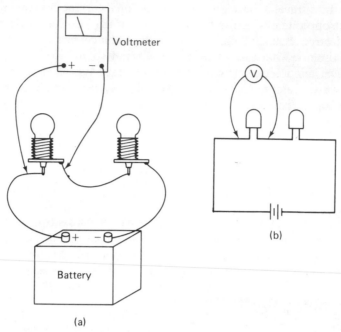

FIGURE 2-17: *Measuring voltage: (a) physical connections; (b) schematic diagram.*

flow through it and therefore has a high resistance. Remember, instruments should disturb the circuit as little as possible.

An ammeter is used to measure current. Again, two types may be used: the analog and the digital meter. The device is placed in the "mainstream" of electron flow to measure the number of coulombs per second flowing, Fig. 2-18. Note that it is equiva-

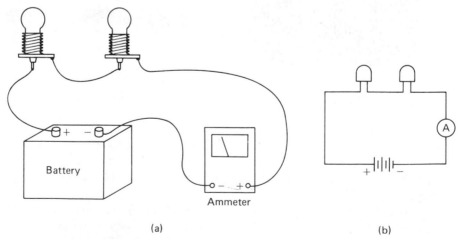

FIGURE 2-18: *Measuring current: (a) physical connections; (b) schematic.*

lent to a flowmeter in our water analogy. Since it must be placed in the main stream of electron flow, it must have a very low resistance; otherwise, it would impede the flow of current and disturb the normal flow.

An ammeter must never be placed across a voltage source, for with its very low resistance, it would conduct a great deal of current, and there would be one less meter in the world.

An ohmmeter is used to measure the resistance of a circuit. The device has within it a small voltage source, usually a dry cell, that forces current into the circuit being tested, Fig. 2-19. It then measures the amount of current that resulted. Note that this device supplies its own voltage. For this reason it must never be used in a circuit that is energized, for it could easily be destroyed by the voltage within the circuit.

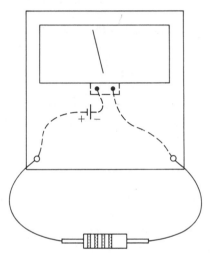

FIGURE 2-19: *Ohmmeter, showing its internal battery.*

Analog meters are subject to an error in reading called *parallax*. This is caused by the observer being off at an angle to the face of the meter rather than looking at it head on, Fig. 2-20. Many analog meters have a mirror behind the needle, permitting the observer to line up the needle with its reflection in the mirror; this greatly reduces parallax.

There are many meters that measure all three electrical characteristics of a circuit: voltage, current, and resistance. An analog meter such as this that does not require power from an outlet is called a *volt-ohm-milliammeter*, usually abbreviated VOM. (A milliamp is 0.001 A.) A digital meter that measures all three characteristics is usually called a *digital multimeter* (DMM). We shall study the construction of these meters in detail in Chapter 9.

Most meters have several ranges of values that can be measured. Whenever a meter is placed in a circuit, the meter should always be placed on its highest range. For example, a common set of voltmeter ranges include 0–1.5 V, 0–5 V, 0–15 V,

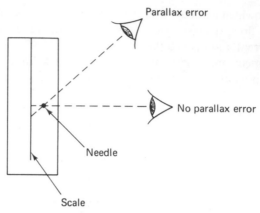

FIGURE 2-20: *Parallax error.*

0–50 V, 0–150 V, 0–500 V, and 0–1500 V. Whenever this meter is placed in a circuit to measure voltage, it should be set on the 1500-V scale. Then, once connected, the meter can be switched downrange to select the proper scale, preventing damage to the meter. If this meter were initially set to 1.5 V, and there were actually 370 V in the circuit, the meter would be destroyed.

In the next section we shall study how to express electrical quantities that are read on such a meter.

2-9. *ENGINEERING NOTATION*

In electronics we will encounter very large numbers and very small numbers. The metric system of designating these numbers by selecting a particular prefix is especially valuable to us, considering how often we must use them. This system is very orderly and fairly simple to learn. It starts by assuming that everything starts with units, such as volts, amperes (amps), ohms, hertz, and others. To express a large number or small number of these, one merely selects the proper prefix from Table 2-3. For example,

TABLE 2-3: *Prefix designations.*

Number of Units	*Prefix*	*Abbreviation*	
1,000,000,000,000	tera	T	
1,000,000,000	giga	G	
1,000,000	mega	M	
1,000	kilo	k	
1	units		
0.001	milli	m	
0.000,001	micro	μ	(Greek lowercase letter mu)
0.000,000,001	nano	n	
0.000,000,000,001	pico	p	

1000 units is 1 kilo-unit; 1000 V is 1 kilovolt (kV); 1000 A is 1 kilampere (kA); 1000 Ω is 1 kilohm (kΩ). In a similar manner, 1 million units is expressed as 1 mega-unit; 1,000,000 V is 1 megavolt (MV); 1,000,000 Ω is 1 megohm (MΩ).

We also deal with very small units. For example, 0.001 A is called 1 milliampere (mA); 0.001 V is 1 millivolt (mV). In a similar manner, 0.000,001 A is 1 microampere (μA); 0.000,000,001 A is 1 nanoampere (nA).

Voltage, current, and resistance are usually expressed using this system of designation such that the numbers 1 through 999 appear to the left of the decimal point. For example, instead of indicating a current of 0.012 A, we would say there are 12 mA through the circuit. This merely requires moving the decimal point in multiples of three places to the left or to the right in order to obtain the numbers 1 through 999 to the left of the point. Moving the point to the left will move up the prefix table. For example, 11,000 A requires moving the point one group of three to the left. Therefore, referring to the table, we must move one prefix up the table from the units designation to the kilo prefix, yielding 11 kA.

> **EXAMPLE 2-9:** Express 12,460 V in engineering notation.
>
> *SOLUTION:* Moving the point three places to the left results in 12.460; this moves us up one prefix on the table, resulting in 12.460 kV.
>
> **EXAMPLE 2-10:** Express 0.000,000,001,33 A in engineering notation.
>
> *SOLUTION:* It is necessary to move the point three groups to the right; consulting the table, we pass through milli, micro, and arrive at nano, three places down from the units place. Therefore, the correct designation is 1.33 nA.
>
> **EXAMPLE 2-11:** Express 0.0009845 A in engineering notation.
>
> *SOLUTION:* We must move two groups to the right and two places down the table from the units designation, resulting in 984.5 μA.

2-10. ELECTRICAL SAFETY

Electronics is fun! It provides enjoyment to those who use electronic devices and to those who build and repair electronic equipment. However, unless one observes some commonsense safety procedures, electronics can be painful, hazardous, and even fatal. This author's 12-year-old uncle was killed because of improper treatment of electricity. It could happen to you or your loved ones unless you are careful.

The human body is filled with an ideally conductive fluid, blood, and its muscles are easily paralyzed by electrical current. The heart muscle is particularly susceptible to this problem, for not only must it beat, but each muscle must beat in proper rhythm. With a certain amount of electrical current, the heart looses its rhythm and becomes a useless, twitching muscle incapable of doing its job. This condition is called *ventricular fibrillation*.

Current flowing through muscle tissue is, then, the danger that must be avoided.

Fortunately, we have been provided with a natural resistive barrier to current flow, our skin. However, it is only effective if it is kept dry. I just measured the electrical resistance between my left hand and my right hand, and it was 300 kΩ. However, when I wet the fingers of my left and right hand, the resistance is reduced to 25 kΩ and by immersing my thumbs in salt water and using spoons for electrodes, it further reduced to 10 kΩ. Under ideal conditions, this can be reduced to much less.

The next question is: Just how much current is dangerous? Safety experts tell us that the average person's arm will convulse at 35 mA, preventing his release from a wire, his heart will go into ventricular fibrillation at 100 mA, and, finally, his heart will be paralyzed at 200 mA. This means that one could die if his body resistance is lowered to a few ohms and he got across a 1.5-V dry cell.

Safety Precautions

The single, most important safety precaution that should be observed when working in electronics is to use common sense. Some detailed precautions include:

1. Do not engage in horseplay around electrical equipment.
2. Remove all rings and watches prior to working on "live" equipment. Even very low voltages can literally fry the skin by passing high currents through jewelry.
3. Make sure that all capacitors are discharged prior to working on equipment that has been turned off.
4. Use a rubber mat on the floor and a nonconductive work surface.
5. Keep one hand in your pocket when working on "live" equipment. This prevents electrical current from passing through the heart.

In summary, enjoy electronics, but enjoy it safely.

2-11. SUMMARY

The unit of charge is the coulomb and represents 6.24×10^{18} electrons or protons. The unit of current is the ampere and represents a charge of 1 C/s flowing past a point on a wire. The current wavefront moves at a velocity of 186,000 mi/s, whereas the individual electrons travel much slower. Voltage is the expending of 1 J of work to move 1 C and batteries, generators, and solar cells are sources of this voltage. Potential difference is the voltage between any two points within a circuit as measured with a voltmeter. Resistance is measured in ohms and represents the amount of opposition to current flow. Resistors come in both fixed and variable types, with the carbon-composition fixed type being the most common. The color code for this type indicates its value in ohms, but only certain preferred values are produced by the manufacturers. Both VOMs and DMMs can be used to measure voltage, current, and resistance.

However, voltmeters must be placed across the circuit to be measured, whereas ammeters must be placed in series with the current flow. Ohmmeters can only be used on a deenergized circuit. Engineering notation is a system of representing very large and very small numbers using a metric prefix. As little as 100 mA of current can be fatal to human beings. We must all observe the proper safety precautions to avoid this danger.

2-12. REVIEW QUESTIONS

2-1. The unit of charge is the _____.

2-2. The unit of electric current is the _____.

2-3. What is the difference between charge and current flow?

2-4. What is a wavefront?

2-5. What is the velocity of a single electron within a wire?

2-6. What is the velocity of a wavefront in both English and metric units?

2-7. The unit of electric potential is the _____.

2-8. Relate joules, coulombs, and voltage to each other.

2-9. Name three sources of voltage.

2-10. The unit of resistance is the _____.

2-11. In a fluid system, the height of the water tower is analogous to what electrical quantity? The diameter of the pipe? The quantity of water in the tank? The amount of water flow per unit time?

2-12. Carbon composition resistors are available in what standard wattage sizes?

2-13. Why would the designer use a 2-W 100-Ω resistor instead of a $\frac{1}{2}$-W 100-Ω resistor?

2-14. When would a wirewound resistor be used?

2-15. What is meant by the taper of a potentiometer?

2-16. List the colors that represent the numbers zero through nine in the color code.

2-17. What is meant by the preferred values of resistors?

2-18. What is meant by the tolerance of a resistor?

2-19. What is an open circuit? A closed circuit?

2-20. Why must a voltage source in a closed circuit have a resistance in series with it?

2-21. What is conventional current flow? Electron flow?

2-22. What advantage does a DMM have over a VOM?

2-23. What precaution must be observed when measuring current in a circuit?

2-24. What precaution must be observed when measuring resistance in a circuit?

2-25. How should a voltmeter be placed within a circuit to measure potential difference?

2-26. What is parallax?

2-27.   Name the prefixes used in engineering notation and how they relate to units. State the abbreviation for each.

2-28.   When the heart muscles fail to operate in rhythm, the heart is said to be in _____ _____ _____.

2-29.   How much current will cause convulsions in the arm?

2-30.   How much current will cause ventricular fibrillation?

2-31.   How much current will cause heart paralysis?

2-32.   Name five safety precautions that we should observe in electronics.

2-13.   PROBLEMS

2-1.   What is the charge on an object if it contains 18.72×10^{18} more electrons than protons?

2-2.   What is the charge on an object if it contains 12.48×10^{18} more protons than electrons?

2-3.   How much current is there in a wire if 40 C flows past a particular point in 10 s?

2-4.   How much current is there in a wire if 55 C flows past a particular point in 11 s?

2-5.   How long would it take a wavefront to travel from Seattle to Chicago, a distance of 1737 mi?

2-6.   How long would it take a wavefront to travel from the earth to the moon, a distance of 216,420 mi?

2-7.   What is the voltage of a circuit if it requires 3 J to move 1 C?

2-8.   What is the voltage in a circuit if it requires 10 J to move 5 C?

2-9.   What are the nominal, maximum, and minimum resistances of resistors with the following color bands:

    (a)   Red, red, red, no color, no color.

    (b)   Brown, black, black, silver, brown.

    (c)   Violet, green, yellow, gold, red.

    (d)   Green, blue, gold, gold, yellow.

2-10.   What are the nominal, maximum, and minimum resistances of resistors with the following color bands:

    (a)   Brown, green, brown, silver, red.

    (b)   Yellow, orange, red, gold, yellow.

    (c)   Brown, black, brown, no color, no color.

    (d)   Blue, gray, silver, gold, brown.

2-11.   What color code would the following resistors have?

    (a)   12 kΩ, 10%.

    (b)   3.6 kΩ, 5%.

    (c) 10 Ω, 20%.

    (d) 0.62 Ω, 5%.

2-12. What color code would the following resistors have?

    (a) 11 kΩ, 5%.

    (b) 470 kΩ, 10%.

    (c) 5.6 MΩ, 5%.

    (d) 6.8 Ω, 20%.

2-13. What preferred value of resistor would be selected for the following calculated values?

    (a) 13,569 Ω, 5%.

    (b) 888,764 Ω, 10%.

    (c) 464 Ω, 20%.

    (d) 2.34 Ω, 5%.

2-14. What preferred value of resistor would be selected for the following calculated values?

    (a) 8,693,000 Ω, 5%.

    (b) 6.3 Ω, 10%.

    (c) 16,531 Ω, 20%.

    (d) 0.252 Ω, 5%.

2-15. Express the following quantities in engineering notation:

    (a) 4,560,000 V.

    (b) 23,643 Ω.

    (c) 0.000,067,450 V.

    (d) 0.000,000,000,067,985 A.

2-16. Express the following quantities in engineering notation:

    (a) 235,400 V.

    (b) 7,315,000,000 Ω.

    (c) 0.0563 A.

    (d) 0.000,000,783,256 A.

2-14. PROJECTS

2-1. Measure the resistance of distilled water. Then add salt and measure again. Explain the results.

2-2. One of the devices used as a lie detector measures the resistance of the skin. Using two electrodes that make contact with the palms of the hands and an ohmmeter, build a lie detector of this type and test it. Can it be beaten?

2-3. Measure the resistance of paper, wire, plastic, wood, cotton cloth, and a signal diode. Which are good insulators? Which are good conductors? Which are semiconductors?

2-4. Measure the resistance of an incandescent light bulb using two different resistance ranges of a VOM. Explain the difference.

2-5. Connect three 1-kΩ resistors in series across a supply voltage of less than 30 V. Next, measure the voltage across each of the three resistors and the voltage across the supply. What conclusions do you reach? Measure the current in the circuit. Remove the power supply from the circuit and measure the resistance of the resistors from one end of the string to the other. Do you see any relationship among the voltage, current, and resistance?

2-6. Measure the resistance of a 1-kΩ resistor. Next, carefully file a groove in the body of the resistor and take resistance readings until the resistor is cut in two. What conclusions do you reach, considering the construction of a carbon-composition resistor?

2-7. Obtain several potentiometers. Determine the total resistance of each and what type taper each has, linear or nonlinear.

2-8. Obtain 10 color-coded resistors and determine the nominal, maximum, minimum, and measured resistances of each.

2-9 If two 10% tolerance resistors are connected in series, what will be the worst-case tolerance of the entire circuit? If they are connected in parallel?

2-10. What is the difference between the volt as described in this chapter, and an electron volt? Are they a measure of the same type of phenomenon? How about an abvolt?

3

OHM'S LAW

In the early 1800s, a Cologne, Germany, high school physics teacher became fascinated with the subject of electricity, spending many hours investigating the relationship between current and voltage. As a result, he published several papers on the subject, the most important of which was "The Galvanic Circuit Investigated Mathematically" (1827). However, the scientific world greeted his work with skepticism, ridicule, and outright hostility, ultimately resulting in his resigning teaching and fleeing to Berlin, where he spent the next six years in obscurity. After several years had passed, the great minds of that time looked closer at this man's work and came to recognize its importance. In 1841, the English Royal Society awarded him the Copley Medal and, in 1849, this once-maligned high school teacher became professor of physics at the University of Munich, where he served until his death in 1854. Finally, in 1881, the International Electric Congress officially declared that the unit of resistance was henceforth to be known as the *ohm*, after the persevering former high school teacher, Georg Simon Ohm. This man dedicated his entire life to developing what is now known as Ohm's law, the subject of this chapter.

The sections include:

3-1. Constant resistance

3-2. Constant voltage

3-3. Constant current

3-4. Summary of Ohm's law

3-5. Power

3-6. Energy

3-1. CONSTANT RESISTANCE

The relationship that Georg Simon Ohm discovered is known as *Ohm's law* and states that the resistance (R) is equal to the voltage (V) divided by the current (I):

$$R = \frac{V}{I}$$

Let us examine Fig. 3-1 and apply Ohm's law to find the resistance. Since we know both V and I:

$$R = \frac{V}{I} = \frac{6}{6} = 1\,\Omega$$

Therefore, the resistance is 1 Ω.

Next, let us assume we have a fixed resistance, 1 Ω, and wish to compare the effect a 6-V battery and a 12-V battery have on the current. We already know from the example above that a 6-V battery will result in 6 A of current. For 12 V:

$$R = \frac{V}{I}$$

$$1 = \frac{12}{I}$$

I must therefore be 12 A. Note that, for this fixed resistance, doubling the voltage doubles the resultant current. Would you not expect this from our water analogy? If we double the height of the tank and keep the pipe diameter the same, would not the water flow rate be doubled? Therefore, Ohm's law seems not only possible, but reasonable. Let us apply it further.

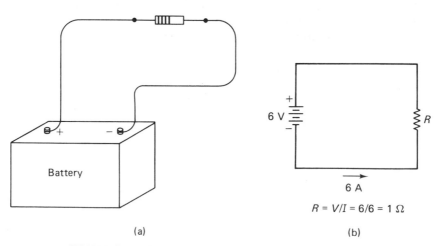

(a)   (b)

FIGURE 3-1: *Series circuit: (a) physical connections; (b) schematic diagram.*

46

EXAMPLE 3-1: A resistor has 10 A passing through it when connected to a voltage source of 25 V. What is its resistance?

SOLUTION: According to Ohm's law,

$$R = \frac{V}{I} = \frac{25}{10} = 2.5\ \Omega$$

EXAMPLE 3-2: A certain electrical heater has a current of 13.5 A when connected to a source voltage of 120 V. What is its resistance?

SOLUTION: According to Ohm's law,

$$R = \frac{V}{I} = \frac{120}{13.5} = 8.889\ \Omega$$

EXAMPLE 3-3: A certain circuit draws 2 mA from its 5-V source, Fig. 3-2. What is the resistance of the circuit?

SOLUTION:

$$R = \frac{V}{I} = \frac{5}{0.002} = 2500\ \Omega = 2.5\ \text{k}\Omega$$

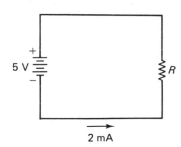

$$R = V/I = 5/2 = 2.5\ \text{k}\Omega$$

FIGURE 3-2: *Example 3-3, unknown resistance.*

Because of the units involved, if resistance is expressed in kΩ, and potential in volts, current will be in mA. As an illustration, let us examine Example 3-3 more closely:

$$R = \frac{V}{I} = \frac{5}{0.002} = \frac{5}{2 \times 10^{-3}}$$

$$= \frac{5}{2} \times 10^3$$

$$= 2.5\ \text{k}\Omega$$

Note that, by bringing the 10^{-3} to the top of the fraction, it becomes 10^{+3}, making the result kΩ.

EXAMPLE 3-4: A certain circuit draws 1.56 mA from its 6-V source. What is the resistance of the circuit?

SOLUTION: Use Ohm's law, with voltage expressed in units, resistance in kΩ, and current in mA:

$$R = \frac{V}{I} = \frac{6}{1.56} = 3.846 \text{ k}\,\Omega$$

This principle also applies to smaller and larger units. If potential is expressed in volts and current in μA, then resistance is in MΩ.

EXAMPLE 3-5: A certain circuit draws 12.3 μA from its 15-V source. What is the resistance of the circuit?

SOLUTION: If unit dimensions are used,

$$R = \frac{V}{I} = \frac{15}{12.3 \times 10^{-6}} = 1.220 \times 10^6 = 1.220 \text{ M}\Omega$$

However, if MΩ and μA are used directly,

$$R = \frac{V}{I} = \frac{15}{12.3} = 1.220 \text{ M}\,\Omega$$

This method saves much fiddling with the decimal point.

Graphing Resistance

Let us now assume the circuit in Fig. 3-3(a) and graph the current as we change the voltage. If voltage is 0 V, the current is 0 mA; if $V_1 = 25$ V, $I = 25$ mA; if

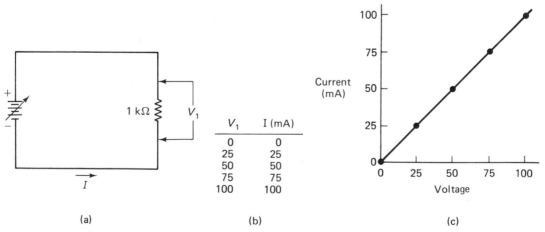

V_1	I (mA)
0	0
25	25
50	50
75	75
100	100

(a)   (b)   (c)

FIGURE 3-3: *Graph of 1-kΩ resistor: (a) schematic diagram: note that the arrow through the battery symbol represents a variable voltage supply; (b) data; (c) graph.*

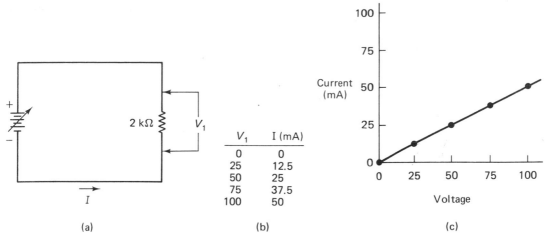

FIGURE 3-4: *Graph of a 2-kΩ resistor: (a) schematic diagram; (b) data; (c) graph.*

$V_1 = 100$ V, $I = 100$ mA. Thus the plot results in a straight line. This will be the result of a plot of all general-purpose resistors in a circuit; all will be straight lines.

Next, let us look at a different-valued resistor and plot this, Fig. 3-4. Note that, again, the plot is that of a straight line. However, this one has a different slope than the one we did previously. Thus, all resistors plot as a straight line; only the slope will vary with the value of the resistor.

3-2. CONSTANT VOLTAGE

Transposing Ohm's law and solving for V, we obtain $V = IR$, the voltage is equal to the current multiplied by the resistance. Thus, given the current and the resistance, we can solve for the voltage. Let us do so in Fig. 3-5; knowing the current and the resistance, $V = IR = 3.0 \times 5 = 15.0$ V. With our 15-V supply, let us now double the resistance and observe what happens to the current. Does it not make sense that

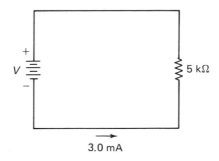

3.0 mA

$V = IR = 3 \times 5 = 15$ V

FIGURE 3-5: *Unknown voltage.*

if we double the resistance in a circuit that the current would be halved? Note that this is also true according to the formula

$$V = IR$$

$$15 = 1.5 \times 10$$

It also seems reasonable that in our fluid example, if we were to cut the diameter of the water pipe and leave the height the same, that the amount of flow would be reduced. Let us now apply the formula to some examples.

EXAMPLE 3-6: A certain circuit draws 5.6 A from its source and has a resistance of 23.1 Ω. What is the source voltage?

SOLUTION: $V = IR = 5.6 \times 23.1 = 129.4$ V

EXAMPLE 3-7: A certain circuit has a resistance of 12.3 kΩ and draws 4.3 mA. What is its voltage?

SOLUTION: $V = IR = 4.3 \times 12.3 = 52.89$ V. Note that the whole problem can be done in kΩ, mA, and V.

3-3. CONSTANT CURRENT

Again, by transposing Ohm's law, we are able to solve for the current, obtaining $I = V/R$. Thus, if we know the voltage and the resistance, we can obtain the current.

EXAMPLE 3-8: A certain clothes iron has a resistance of 9.60 Ω and uses a 120-V source, Fig. 3-6. What is the current drawn by the iron?

SOLUTION: $I = V/R = 120/9.60 = 12.50$ A. Thus, the current is 12.50 A.

EXAMPLE 3-9: A certain 15-kΩ resistor is placed across a 20-V source. How much current will it draw?

SOLUTION: $I = V/R = 20/15 = 1.333$ mA. Note that if voltage is in volts, and the resistance in kΩ, then the current will be in mA.

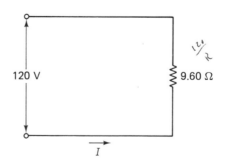

$I = V/R = 120/9.60 = 12.50$ A

FIGURE 3-6: *Unknown current.*

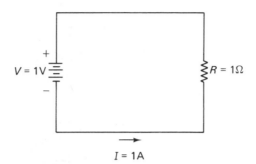

FIGURE 3-7: *Ohm's law, constant current.*

Now let us examine the circuit shown in Fig. 3-7. The current through the resistor is 1 A. According to Ohm's law, if we were to increase the resistance from 1 Ω to 5 Ω, the voltage must go from 1 V to 5 V in order to maintain our 1-A current. Similarly, if we were to increase the resistance to 10 Ω, we would need 10 V to maintain the 1 A of current flow. Thus, a higher voltage is necessary to obtain our 1 A if we increase the resistance.

3-4. SUMMARY OF OHM'S LAW

Georg Simon Ohm left us with one very important law; in fact, it is the foundation of all electrically related sciences. The law has three variables: V, I, and R. Given any two of them, the third can be computed, Table 3-1.

TABLE 3-1: *Forms of Ohm's law.*

$$V = IR \qquad R = \frac{V}{I} \qquad I = \frac{V}{R}$$

EXAMPLE 3-10: A circuit has a resistance of 12.6 kΩ and a current of 7.8 mA. What is the voltage of the circuit?

SOLUTION: We need to solve for V; therefore,

$$V = IR = 7.8 \times 12.6 = 98.28 \text{ V}$$

EXAMPLE 3-11: A circuit has a current of 5.6 mA and a voltage of 2.4 V. What is its resistance?

SOLUTION: We must solve for resistance and

$$R = \frac{V}{I} = \frac{2.4}{5.6} = 0.4286 \text{ k}\Omega = 428.6 \ \Omega$$

EXAMPLE 3-12: A 216-kΩ resistor is connected across a 45-V source. What is its current?

SOLUTION: We must solve for current:

$$I = \frac{V}{R} = \frac{45}{216} = 0.2083 \text{ mA} = 208.3 \ \mu\text{A}$$

3-5. POWER

The term *power* occurs frequently in the mechanical world: it is, for example, applied to a car engine when we speak of horsepower. Perhaps the easiest way to visualize power is to realize that 1 horsepower (hp) equals 550 ft-lb/s. That is, if a motor were to lift 550 lb 1 ft in 1 s and continue to do this indefinitely, it would be expending 1 hp. If the motor were to do this for only 1 s, we could say it expended 1 hp for only 1 s.

In electrical terms, power is expressed in *watts*. In fact, there are 746 W in 1 hp, for watts and horsepower express the same idea. We speak of a 100-W light bulb as using 100 W of power while it is turned on. Power is, therefore, continuously being expended while the circuit is energized.

The formula for relating power to voltage and current is $P = VI$, or power is equal to the voltage times the current. Given a particular resistor, the greater the voltage across it, the more power it is giving off (dissipating). Or, the more current through this resistor, the more heat it is generating and therefore power it is consuming.

As in Ohm's law, this formula has three variables. Given any two of them, we can solve for the third.

EXAMPLE 3-13: A certain battery has a voltage of 12 V and is supplying 5 A to the load, Fig. 3-8. How much power is the battery generating?

SOLUTION: The battery is supplying

$$P = VI = 12 \times 5 = 60 \text{ W}$$

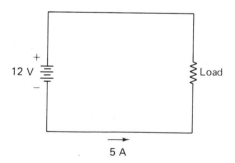

$$P = VI = 12 \times 5 = 60 \text{ W}$$

FIGURE 3-8: *Computing power.*

Note that not only is the battery supplying 60 W, but the load is consuming the 60 W. Thus, power is both supplied and expended.

EXAMPLE 3-14: A light bulb draws 56 mA from its 60-V source. How much power is it drawing from the source?

SOLUTION: P = VI = 60 × 0.056 = 3.36 W.

EXAMPLE 3-15: A resistor draws 5 mA from a 6-V source. What power is it drawing?

SOLUTION: P = VI = 6 × 0.005 = 0.030 = 30 mW.

If we were to combine Ohm's law with this power law, we would have four variables: P, V, I, and R. Using these two formulas, $P = VI$ and $V = IR$, and given any two out of the four, we can solve for the other two. However, it may be necessary to combine and transpose the formulas. Starting with $P = VI$ and substituting IR for V, we obtain $P = I^2R$. Similarly, if $P = VI$ and $I = V/R$, substituting for I, we obtain $P = V^2/R$. These three forms of the power formula ($P = VI$, $P = V^2/R$, and $P = I^2R$) can then be transposed as needed, Table 3-2.

TABLE 3-2: *Forms of Ohm's law and the power law.*

$P = VI$	$P = I^2R$
$V = \dfrac{P}{I}$	$R = \dfrac{P}{I^2}$
$I = \dfrac{P}{V}$	$I = \sqrt{\dfrac{P}{R}}$
$P = \dfrac{V^2}{R}$	$V = IR$
$R = \dfrac{V^2}{P}$	$I = \dfrac{V}{R}$
$V = \sqrt{PR}$	$R = \dfrac{V}{I}$

EXAMPLE 3-16: What is the current and resistance of a 75-W 120-V light bulb?

SOLUTION: The known variables are P and V; therefore, we shall select the power formula that has these two included. There are two: $P = VI$ and $P = V^2/R$:

$$P = VI$$

$$75 = 120 \times I$$

$$I = \frac{75}{120} = 0.6250 \text{ A} - 625 \text{ mA}$$

Since we now know the current, we can either use $V = IR$ or $P = V^2/R$ to solve for the resistance. We shall use the power formula:

$$P = \frac{V^2}{R}$$

$$75 = \frac{120^2}{R}$$

$$R = \frac{120^2}{75} = 192.0\ \Omega$$

EXAMPLE 3-17: A 600-Ω headset receives a 1-mW signal, Fig. 3-9. What is the current through and voltage across the headset?

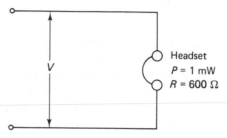

FIGURE 3-9: *Example 3-17, current and voltage in a headset.*

SOLUTION: The two known variables are R and P. The two power formulas that include these variables are $P = V^2/R$ and $P = I^2 R$. Solve for V:

$$P = \frac{V^2}{R}$$

$$0.001 = \frac{V^2}{600}$$

$$V^2 = 600 \times 0.001 = 0.6000$$

$$V = 0.7746\ \text{V} = 774.6\ \text{mV}$$

Let us now solve for I using Ohm's law and save having to take the square root:

$$V = IR$$

$$0.7746 = I \times 600$$

$$I = \frac{0.7746}{600} = 1.291\ \text{mA}$$

Power in Resistors

Power in a resistor is given off as heat that must be transferred to the air around the unit. A large resistor is capable of giving off more heat than a smaller one because it contacts the air over a greater surface area; therefore, a 2-W resistor is physically

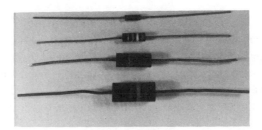

FIGURE 3-10: *Power dissipation determines resistor size (photo by Ruple).*

larger than a 1-W resistor, as can be seen from Fig. 3-10. This power rating has to be taken into account when selecting resistors for a circuit. A 100-Ω $\frac{1}{2}$-W resistor would burn up if it were connected to a 50-V source, for it would dissipate:

$$P = \frac{V^2}{R} = \frac{50^2}{100} = 25 \text{ W}$$

The resistor would be consuming 50 times the heat it is capable of dissipating.

The usual method for selecting carbon resistors is to assume they can dissipate 50% of their rated value; thus, a 2-W device would never be asked to dissipate more than 1 W, a 1-W device no more than $\frac{1}{2}$ W. This *derating*, as it is called, allows the circuit to operate more reliably. It should be noted that the power rating of a resistor is listed by the manufacturer as being for room temperature, not for the high temperatures encountered in actual circuits. This 50% derating rule of thumb allows for this specification.

Wirewound resistors are made to operate at higher temperatures than carbon ones; they can, therefore, operate closer to their nominal rating.

EXAMPLE 3-18: What is the proper size of 4.7-kΩ resistor to use if it is to draw 2.5 mA from the source?

SOLUTION: $P = I^2 R = 0.0025^2 \times 4700 = 29.38$ mW. A $\frac{1}{4}$-W size can be used, since this is well under 125 mW ($\frac{1}{2}$ of $\frac{1}{4}$ W).

EXAMPLE 3-19: What is the proper size of 470-Ω resistor if it is to be connected to a 15-V battery?

SOLUTION: The power to be dissipated by the resistor is

$$P = \frac{V^2}{R} = \frac{15^2}{470} = 478.7 \text{ mW}$$

To select the $\frac{1}{2}$-W size would put it too close to its rated dissipation. Therefore, select the 1-W size.

3-6. ENERGY

Thus far, we have been using two terms rather extensively that we have not yet defined: work and energy. Let us now examine them a little closer.

Energy is the capacity to do work, such as a raised hammer, a tank of gasoline, or a lake of water behind a dam. Each are energy sources, for they can do work, Fig. 3-11. On the other hand, *work* is energy that is being translated into an active force.

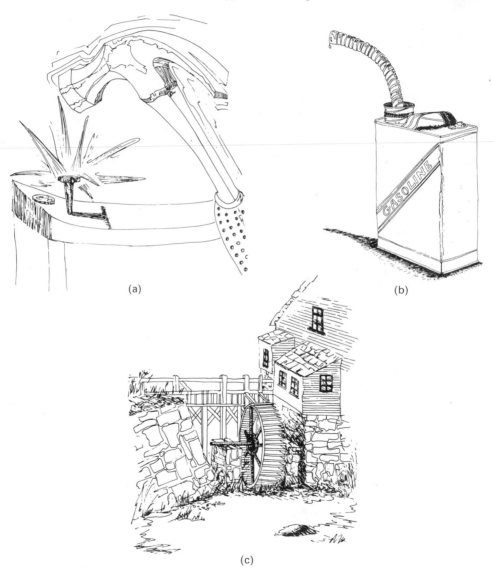

(a)

(b)

(c)

FIGURE 3-11: *Forms of energy: (a) poised hammer; (b) gasoline; (c) water behind a dam (drawings by Kirishian).*

56

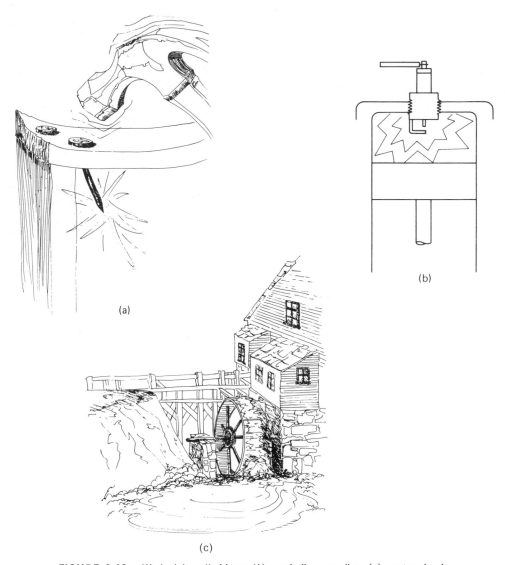

FIGURE 3-12: *Work: (a) nail driven; (b) exploding gasoline; (c) waterwheel turning (drawings by Kirishian).*

In the mechanical world, it is defined as force moving through a distance. Thus, actually moving a nail through a 1-inch plank with a hammer constitutes work, as does exploding gasoline within a cylinder or the water behind a dam actually turning a turbine, Fig. 3-12. Both energy and work are measured in the same units. In the mechanical world, the foot-lb are used; this is the lifting of a 1-lb weight 1 ft in height. Power, however, differs from energy and work, for power is the time rate of work, measured in ft-lb/s in the mechanical realm. Note that, according to this definition, if one were to multiply power by time, one would obtain energy; moving 10 ft-lb/s for 20 s yields 200 ft-lb.

In the electrical world, power is measured in watts (W) and work (or energy) in watt-hours (Wh). Thus, a 100-W lamp that burns for 3 h has consumed 300 Wh of energy. In medical electronics, the amount of energy produced by an X-ray machine is measured in watt-seconds (Ws). Electrical energy consumed by a home is measured in kilowatt-hours (kWh). Note that each of these is represented as power multiplied by time:

$$W = Pt$$

EXAMPLE 3-20: A home dryer uses 10,000 W and $\frac{1}{2}$ h in drying a load of clothes. How much energy does it consume drying one load and what is the cost per load if the electrical rate is $0.05/kWh?

SOLUTION: To calculate energy:

$$W = Pt$$
$$= 10,000 \times 0.5$$
$$= 5000 \text{ Wh}$$
$$= 5.0 \text{ kWh}$$

Since the rate per kWh is $0.05, the total cost is

$$\text{cost} = W \times \text{rate}$$
$$= 5.0 \times 0.05 = \$0.25/\text{load}$$

EXAMPLE 3-21: What is the cost of a 5-W night light left on for 1 yr if the rate is $0.05/kWh?

SOLUTION: The total energy is

$$W = Pt$$
$$= 5 \text{ W} \times \frac{24 \text{ hr}}{\text{day}} \times \frac{365 \text{ days}}{\text{yr}}$$
$$= 43,800 \text{ Wh} = 43.80 \text{ kWh}$$
$$\text{cost} = W \times \text{rate}$$
$$= 43.80 \times 0.05 = \$2.19$$

3-7. SUMMARY

Ohm's law states that $V = IR$. Knowing any two of the variables, we can compute the third. Fixed resistors plot as a straight line on an VI graph. Power is computed by the formula $P = VI$. With these two formulas ($V = IR$ and $P = VI$), and knowing any two of the four variables, we can compute the third and the fourth. Power is the time rate of energy.

3-8. *REVIEW QUESTIONS*

3-1. According to Ohm's law, if resistance were held constant and voltage were to increase, what would happen to current?

3-2. According to Ohm's law, if voltage were held constant and resistance were to decrease, what would happen to current?

3-3. According to Ohm's law, if current were held constant and resistance were to increase, what would happen to voltage?

3-4. In Ohm's law, if current is expressed in mA and potential in V, what would be the unit of resistance?

3-5. In Ohm's law, if resistance is expressed in MΩ and potential in V, what would be the unit of current?

3-6. State three algebraic forms of Ohm's law.

3-7. In the formula $P = VI$, if current is fixed, what would be the effect of doubling the voltage upon the power?

3-8. In the formula $P = I^2 R$, what would be the effect of doubling the current upon the power if the resistance were fixed?

3-9. In the formula $P = V^2/R$, what would be the effect upon the power if the voltage were doubled while the resistance remained fixed?

3-10. How many watts are in 1 hp?

3-11. State 12 forms of the power formula and Ohm's law.

3-12. What is meant by derating a resistor?

3-13. How are energy and power mathematically related?

3-9. *PROBLEMS*

3-1. A type 334 lamp draws 40 mA at 28 V. What is its resistance?

3-2. A type 240 lamp draws 0.36 A at 6.3 V. What is its resistance?

3-3. A 24-V relay has a resistance of 215 Ω. What is its current?

3-4. A 220-V relay has a resistance of 1930 Ω. What is its current?

3-5. A 3.5-mA current passing through a 4.7-kΩ resistor produces what potential across the resistor?

3-6. The 10-kΩ collector resistor of a transistor circuit has 315 μA flowing through it. What is the potential across the resistor?

3-7. What is the maximum current that a 10-kΩ resistor can draw without exceeding 1 W?

4

SERIES CIRCUITS

Imagine that you are in a circus watching the elephants parade around the ring in single file, Fig. 4-1. Note that the trunk of each is connected to the tail of the one ahead. If we were to imagine resistors connected in the same manner, Fig. 4-2, this would be called a *series* circuit. In this chapter we shall analyze series circuits, applying Ohm's law and the power formulas to the solution of unknown quantities within the circuit. The sections include:

4-1. Series versus parallel

4-2. Current in a series circuit

4-3. Resistance in a series circuit

FIGURE 4-1: *Series connection (drawing by Kirishian).*

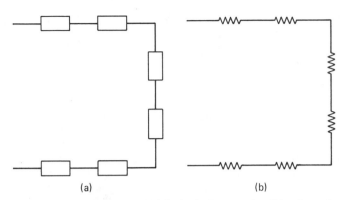

(a)   (b)

FIGURE 4-2: *Series circuit: (a) physical connection; (b) schematic diagram.*

4-1. SERIES vs. PARALLEL

The introductory paragraph above compared the example of elephants parading around a ring with that of a series circuit. However, as any respectable circus performer knows, there is another way in which circuits may be connected: in parallel. If our pachydermal friends were all lined up side by side, heads facing north and tails facing south, they would be standing parallel to one another, Fig. 4-3. Resistors connected so that the heads are all connected together and the tails are connected together are said to be connected in parallel, Fig. 4-4. In this chapter we shall analyze the series circuit and in the next chapter the parallel circuit.

4-2. CURRENT IN A SERIES CIRCUIT

Let us examine the behavior of current in the series circuit shown in Fig. 4-5. In such a circuit, the current is the same at any point along the circuit. For example, if the current is measured as 1 mA at point A, it will also be 1 mA at points B and C. The reason for this behavior is that for each electron injected into the circuit by the power supply's negative terminal, one electron is extracted by its positive terminal. The electron that is injected repels its neighbor toward the positive end and that neighbor repels its neighbor, all toward the end where electrons leave the wire. The action is similar to a freeway, with the number of cars leaving the freeway equal to the number

FIGURE 4-3: *Parallel connection (drawing by Kirishian).*

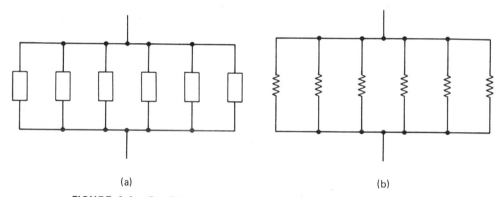

(a)   (b)

FIGURE 4-4: *Parallel circuit: (a) physical connection; (b) schematic diagram.*

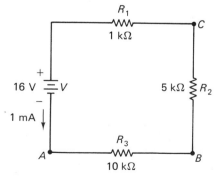

FIGURE 4-5: *Current in a series circuit.*

entering. If this were not so, a traffic jam would result from cars piling up on the road. Thus, our electronic traffic jam is prevented by the positive end extracting the same number of electrons as the negative end supplies.

Calculating Voltage Drop

Since the current is the same at any point along the conducting path, we can calculate the potential difference between one end of each resistor and its other end by Ohm's law. In Fig. 4-5, the current through resistor R_1 is 1 mA and its resistance is 1 kΩ. By Ohm's law, the potential difference across R_1 is

$$V = IR = 1 \times 1 = 1 \text{ V}$$

Therefore, if we were to place a voltmeter across R_1, it would read 1 V of potential difference. The voltage across a resistor is referred to as the resistor's *voltage drop*. This voltage-drop calculation points out one very important factor to consider when using Ohm's law: for $V = IR$ to be true, the V must be across the same R as the I passing through that R. We could not, for example, use the V of Fig. 4-5's power supply and just the resistance of R_1 to calculate the current of the power supply. In this case the potential difference (voltage drop) considered is not across the same circuit as the resistance under consideration.

> **EXAMPLE 4-1:** Calculate the voltage drop across R_2 in Fig. 4-5.
>
> *SOLUTION:* $V = IR = 1 \times 5 = 5$ V.

> **EXAMPLE 4-2:** Calculate the voltage drop across R_3 in Fig. 4-5.
>
> *SOLUTION:* $V = IR = 1 \times 10 = 10$ V.

Let us now examine the polarity of the voltages we have just calculated in Fig. 4-5. First, let us look at just one resistor across a power supply, Fig. 4-6. Note that, if we were to place a voltmeter across the resistor, the positive end of the voltmeter would be connected to the positive end of the power supply and the negative end to

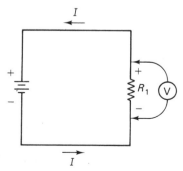

FIGURE 4-6: *Resistor voltage-drop polarity.*

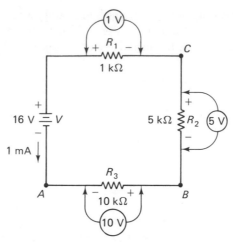

FIGURE 4-7: *Voltage drops and polarities for circuit in Fig. 4-5.*

the negative end of the power supply. Thus, the positive end of the resistor is the end out of which the electrons are flowing and the negative end is the end into which electrons are being injected. Returning to Fig. 4-5, the more positive end of resistor R_1 is the left end, as shown in Fig. 4-7, and the more negative end is the right end of R_1. Considering R_2 and R_3, their values and polarities would be as indicated.

4-3. RESISTANCE IN A SERIES CIRCUIT

If several resistors were connected in a series circuit, the total resistance of the combination would be the sum of the individual resistors:

$$R_t = R_1 + R_2 + R_3 + R_4 + \ldots$$

This can be illustrated by our example of fluids. Using the same pressure, if we were to feed water through a 100-ft length of pipe and a 100-mi length of pipe, more water would flow through the shorter length because of the reduced pipe resistance to flow. Resistors, similarly, reduce current as more are added in series with the supply.

EXAMPLE 4-3: Four resistors, of 3, 4, 5, and 6 kΩ, are connected in series. What is the total resistance?

SOLUTION: Since total resistance is the sum of the individual resistors,

$$R_t = R_1 + R_2 + R_3 + R_4$$
$$= 3 + 4 + 5 + 6 = 18 \text{ k}\Omega \text{ total resistance}$$

EXAMPLE 4-4: What is the total resistance seen by the battery in Fig. 4-7?

SOLUTION: $R_t = R_1 + R_2 + R_3 = 1 + 5 + 10 = 16 \text{ k}\Omega$.

4-4. VOLTAGE IN A SERIES CIRCUIT

Do you remember the definition for voltage? Let me refresh your memory: it is 1 J of work expended in moving 1 C. Let us now examine Fig. 4-7 a little closer by considering what happens to 1 C of charge that is injected into the circuit by the negative terminal of the battery. We must first perform work by pushing it through R_3. Next, we must perform more work by pushing it through R_2, and finally R_1. That little coulomb really gets pushed around. But how much total work have we expended? It is the sum of the work expended in each of the resistors. Thus, we would say that the total voltage across a series circuit is equal to the sum of the voltages across each of the resistors. If we were to assign the voltage drops across the resistors as negative values (actual drops) and the battery a positive sign (for it does the actual generating), the law could be stated as follows:

The sum of all the voltages around a complete circuit is equal to zero volts.

This law, known as *Kirchhoff's voltage law*, is one of the fundamental relationships in electronics; it is an extremely useful tool in calculating unknown values in a series circuit.

Let us now apply the law to Fig. 4-7. We shall travel around the circuit in a counterclockwise (CCW) direction, starting at the power supply. When we encounter a positive sign at the input to a device (whether power supply or resistor), we shall assign a plus sign to it; when we encounter a minus sign at the input to a device, we shall assign it a negative sign. Thus, the total voltage, V_t, is equal to zero:

$$V_t = 0 = +16 - 10 - 5 - 1$$

It works!

This law is very useful in calculating unknown voltages across resistors.

EXAMPLE 4-5: Calculate the voltage drop, V_1, across R_1 in Fig. 4-8.

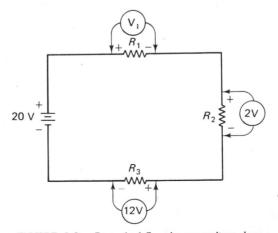

FIGURE 4-8: *Example 4-5, unknown voltage drop.*

SOLUTION: Since the total voltage drop is zero,

$$V_t = 0 = +20 - 12 - 2 - V_1$$
$$V_1 = 20 - 12 - 2 = 6 \text{ V}$$

(4-1)

Note that the battery was assigned a positive voltage and the resistor voltage drops negative signs. V_1 came out positive, indicating that our assumption that it was a voltage drop was correct.

EXAMPLE 4-6: A series circuit has a 120-V battery across resistors that have voltage drops of 20, 16, and 60 V, and one that is unknown, V_u. Calculate V_u.

SOLUTION: Since the sum of all voltages equals zero,

$$120 - 20 - 16 - 60 - V_u = 0$$
$$V_u = 24 \text{ V}$$

Multiple Voltage Supplies

When voltage supplies are connected in series, their effect depends upon the polarities of the connection. In Fig. 4-9, the two batteries are connected so that the negative terminal of one is connected to the positive terminal of the other. Thus, the 10-V battery pushes the electrons east along the wire and the 5-V battery provides some more push eastward, further aiding current flow in that direction. This is said to be a *series-aiding connection*, for the action of one battery aids the action of the other.

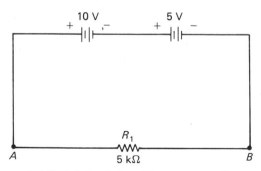

FIGURE 4-9: *Series-aiding power supplies.*

In Fig. 4-10, however, the 10-V supply is forcing electrons east, while the 5-V supply is forcing them west. This is called a *series-opposing connection*, for the action of the one battery opposes the action of the other.

In either case we can use Kirchhoff's voltage law to analyze the current flow through the resistor. Analyzing Fig. 4-9, the total voltage in the circuit is zero. Thus, If V_t represents the total voltage and V_r the drop across the resistor,

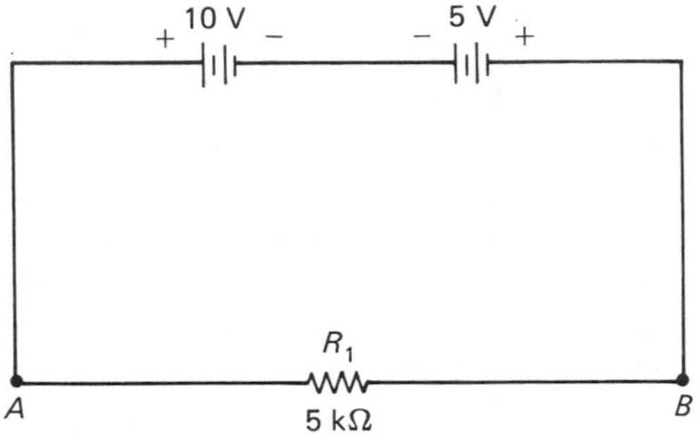

FIGURE 4-10: *Series-opposing power supplies.*

$$V_t = 0 = +10 + 5 - V_r$$
$$V_r = +10 + 5 = 15 \text{ V}$$

Note that we assumed that there would be a voltage drop across the resistor such that the negative end would be to the right and the positive end to the left of the resistor. This assumption is evident, for we used a minus sign before V_r in the equation. If we do not know the polarity across the resistor, we can assume one, and the sign of the answer will tell us whether we assumed right or wrong. Let us purposely assume that electron flow is CCW. Our equation would be

$$0 = -V_r - 5 - 10$$
$$V_r = -5 - 10 = -15 \text{ V}$$

Thus, in the equation we assumed that point A was negative with respect to point B. However, the minus sign of the answer tells us that our assumption was incorrect. Therefore, point A is positive with respect to point B.

> **EXAMPLE 4-7**: Determine the voltage drop and polarity across the resistor in Fig. 4-10.
>
> *SOLUTION:* We shall arbitrarily assume that current flow is CCW. Traveling around the circuit in this direction from point B, we have
>
> $$0 = +5 - 10 - V_r$$
> $$V_r = -5 \text{ V}$$
>
> Thus, there is a 5-V drop across the resistor and, instead of point A being negative with respect to point B as our equation assumed, point A is positive with respect to point B.

4-5. *THE LOOP EQUATION*

The equation that we have been using (Kirchhoff's voltage law) is also called a *loop equation,* for we travel around in a loop. This equation could be generalized as follows:

$$0 = V_1 + V_2 + V_3 + V_4 + V_5 + \cdots$$

The polarity of each of the terms would depend upon the direction current was assumed and whether that V was a resistor or a battery, and the polarity of the battery. In any event, the sign encountered at the input of the device (battery or resistor) would determine the sign of that particular term. However, since $V = IR$, each V across a resistor could be expressed as IR, where I is the current through the resistor and R is its value. Let us apply this to the circuit in Fig. 4-11.

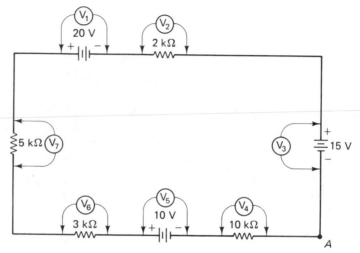

FIGURE 4-11: *Example 4-8, the loop equation.*

EXAMPLE 4-8: Determine the current through and voltage drop across each of the resistors in Fig. 4-11.

SOLUTION: Starting at point A and assuming that current flows in a clockwise (CW) direction:

$$0 = -V_4 - V_5 - V_6 - V_7 + V_1 - V_2 + V_3$$

Note that the polarities of the supplies were assigned by the polarity at the input, where the current enters the device. All voltage drops across the resistors were assumed negative. Substituting IR for each of the resistor voltage drops:

$$0 = -IR_4 - V_5 - IR_6 - IR_7 + V_1 - IR_2 + V_3$$

Note that the current, I, is the same through all resistors. Substituting the values given,

$$0 = -10I - 10 - 3I - 5I + 20 - 2I + 15$$
$$20I = 25$$
$$I = 1.25 \text{ mA}$$

Thus, the current through each resistor is 1.25 mA. Since the sign of the current is positive, we assumed the correct direction of electron flow. If it were negative, the current direction would be the reverse of what we had assumed. Knowing the current, we can, by Ohm's law, calculate the voltage drop across each resistor:

$$V_2 = IR_2 = 1.25 \times 2 = 2.5 \text{ V}$$
$$V_4 = IR_4 = 1.25 \times 10 = 12.5 \text{ V}$$
$$V_6 = IR_6 = 1.25 \times 3 = 3.75 \text{ V}$$
$$V_7 = IR_7 = 1.25 \times 5 = 6.25 \text{ V}$$

The polarity of each was assumed correct; thus, the more negative side of each resistor would be at the CCW end and the more positive side at the CW end.

EXAMPLE 4-9: Compute the current and voltage drops across each of the resistors in Fig. 4-12.

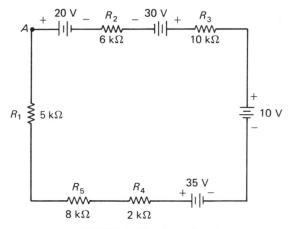

FIGURE 4-12: *Example 4-9.*

SOLUTION: Assuming CW current flow and starting at point A:

$$0 = 20 - V_{R2} - 30 - V_{R3} + 10 - 35 - V_{R4} - V_{R5} - V_{R1}$$
$$0 = 20 - 6I - 30 - 10I + 10 - 35 - 2I - 8I - 5I$$
$$31I = -35$$
$$I = -1.129 \text{ mA}$$

Since the current is negative, we assumed the wrong direction; it really flows in the CCW direction. The voltage drops would be:

$$V_1 = IR_1 = 1.129 \times 5 = 5.645 \text{ V}$$
$$V_2 = IR_2 = 1.129 \times 6 = 6.774 \text{ V}$$
$$V_3 = IR_3 = 1.129 \times 10 = 11.29 \text{ V}$$

$$V_4 = IR_4 = 1.129 \times 2 = 2.258 \text{ V}$$
$$V_5 = IR_5 = 1.129 \times 8 = 9.032 \text{ V}$$

The polarity of the voltage drop across the resistors would be such that the more negative end is the CW end.

4-6. POWER IN A SERIES CIRCUIT

In a circuit with just resistance, the only power that is dissipated is that done by each resistor. Thus, the total power is the sum of the power expended by each component.

EXAMPLE 4-10: Compute the power expended within the circuit shown in Fig. 4-11.

SOLUTION: Since the power is the sum of the individual powers:

$$P_t = P_2 + P_4 + P_6 + P_7$$
$$= V_2 I + V_4 I + V_6 I + V_7 I$$
$$= (2.5 \times 1.25 \times 10^{-3}) + (12.5 \times 1.25 \times 10^{-3})$$
$$\qquad + (3.75 \times 1.25 \times 10^{-3}) + (6.25 \times 1.25 \times 10^{-3})$$
$$= 31.25 \text{ mW}$$

Note that any of the power formulas, I^2R, VI, or V^2/R, could have been substituted for P_2, P_4, P_6, and P_7.

Total power expended within a circuit can also be computed by using total current, total resistance, or total voltage, if they are known. However, be sure to express all quantities in units. Do not try to combine kΩ and mA, MΩ, and μA.

EXAMPLE 4-11: Compute total power dissipated by the circuit shown in Fig. 4-12.

SOLUTION: Since total resistance is the sum of the resistances,

$$R_t = R_1 + R_2 + R_3 + R_4 + R_5$$
$$= 5 + 6 + 10 + 2 + 8 = 31 \text{ k}\Omega = 31.0 \times 10^3 \ \Omega$$
$$P_t = I^2 R_t$$
$$= (1.12903 \times 10^{-3})^2 \times 31.0 \times 10^3$$
$$= 0.03952 \text{ W} = 39.52 \text{ mW}$$

4-7. THE UNLOADED VOLTAGE DIVIDER

A *voltage divider* is a circuit that provides two or more voltages from a single source. Consider the circuit shown in Fig. 4-13. If we were to place a voltmeter with its negative lead on the point designated as ground and measure from ground to point *A*,

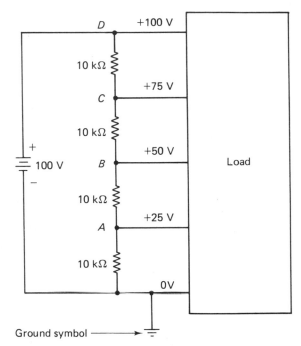

FIGURE 4-13: *Unloaded voltage divider.*

we would read 25 V. In a similar fashion, we could measure from ground to *B*, ground to *C*, and ground to *D*, and would measure 50, 75, and 100 V, respectively. This circuit is called a voltage divider because it divides the 100-V supply into several voltages for a using circuit, which, in this case, would receive four different voltage levels instead of one. If the current supplied the using circuit were less than 10% of the current through the 10-kΩ divider resistors, this load current could be considered insignificant, and we could neglect it. In this case the divider is said to be unloaded, for no current flows to the load (the using circuit).

Knowing the divider string current (called the bleeder current), the supply voltage, and the required load voltages, it is possible to design a divider string to satisfy these specifications.

EXAMPLE 4-12: A 60-V supply is used to supply 5, 20, and 30 V to a load from an unloaded voltage divider with 1-mA bleeder current. Design the circuit.

SOLUTION: The first step is to draw the schematic diagram showing all the resistors and the supply. We shall then fill in the required values, Fig. 4-14. Since the highest voltage required by the load is less than the 60-V supply, we shall need four resistors. Next, examine R_4. We know that it drops 5 V and has 1 mA of current; therefore, its size is

$$R_4 = \frac{V}{I} = \frac{5}{1} = 5 \text{ k}\Omega$$

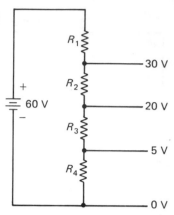

FIGURE 4-14: *Example 4-12, computing resistor string.*

Since the bottom of R_3 is 5 V and the top is 20 V, the potential difference across R_3 is 20 − 5, or 15 V. Its value is, then,

$$R_3 = \frac{V}{I} = \frac{15}{1} = 15 \text{ k}\Omega$$

Similarly,

$$R_2 = \frac{V}{I} = \frac{30 - 20}{1} = 10 \text{ k}\Omega$$

$$R_1 = \frac{V}{I} = \frac{60 - 30}{1} = 30 \text{ k}\Omega$$

However, we are not through, for we must check the power rating required of these resistors:

$$P_4 = VI = 5 \times 1 = 5 \text{ mW}$$
$$P_3 = VI = 15 \times 1 = 15 \text{ mW}$$
$$P_2 = VI = 10 \times 1 = 10 \text{ mW}$$
$$P_1 = VI = 30 \times 1 = 30 \text{ mW}$$

Therefore, $\frac{1}{4}$- or $\frac{1}{2}$-W resistors would suffice.

EXAMPLE 4-13: A 50-V divider must supply 5 V and 40 V to a using circuit. Assuming that it is an unloaded divider and bleeder current is 50 mA, design the circuit.

SOLUTION: The schematic is shown in Fig. 4-15.

$$R_3 = \frac{V}{I} = \frac{5}{50} = 100 \ \Omega$$

$$R_2 = \frac{V}{I} = \frac{40 - 5}{50} = 700 \ \Omega$$

$$R_1 = \frac{V}{I} = \frac{50 - 40}{50} = 200 \ \Omega$$

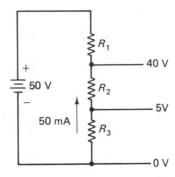

FIGURE 4-15: *Example 4-13.*

The power would be

$$P_3 = VI = 5 \times 50 = 250 \text{ mW, so use a } \tfrac{1}{2}\text{-W resistor}$$
$$P_2 = VI = 35 \times 50 = 1.75 \text{ W, so use at least a 3-W resistor}$$
$$P_1 = VI = 10 \times 50 = 500 \text{ mW, so use a 1-W resistor}$$

Changing the Ground Reference Point

In the examples above, the negative end of the supply voltage was considered 0 V to the using circuit and was labeled ground. However, there are many places where a negative supply is necessary. Figure 4-16 shows such a supply; note that by merely

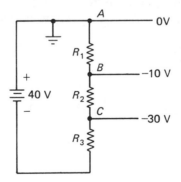

FIGURE 4-16: *Providing nega-tive voltages to the load.*

grounding the positive end of the supply, we can feed a negative voltage to the load. The important thing is what the load sees as zero volts, for it uses this as its reference voltage; it refers all its voltages to that point within the circuit. Thus, point *A* is considered 0 V, point *B* is 10 V below point *A* in potential, and point *C* is 30 V below point *A* in potential.

It is also possible to supply both positive and negative voltages to the load by means of changing the reference (ground) point to a point in the middle of the divider

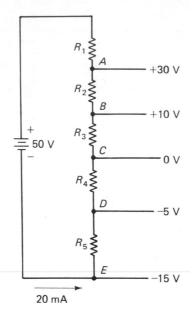

FIGURE 4-17: *Providing posi-tive and negative voltages.*

string, Fig. 4-17. Here, point *A* is 30 V above point *C* in potential, point *B* 10 V above *C*, point *D* 5 V below, and point *E* 15 V below. Note that the potential difference between point *A* and point *E* is 30 − (−15), or 45 V. Thus, with a supply voltage of 50 V, there is 5 V across R_1. We shall now calculate the resistors of the string.

EXAMPLE 4-14: Calculate values of the resistors of the unloaded voltage divider shown in Fig. 4-17.

SOLUTION:

$$R_5 = \frac{V}{I} = \frac{15 - 5}{20} = 500 \ \Omega$$

$$R_4 = \frac{V}{I} = \frac{5}{20} = 250 \ \Omega$$

$$R_3 = \frac{V}{I} = \frac{10}{20} = 500 \ \Omega$$

$$R_2 = \frac{V}{I} = \frac{30 - 10}{20} = 1 \ k\Omega$$

$$R_1 = \frac{V}{I} = \frac{5}{20} = 250 \ \Omega$$

$$P_5 = VI = 10 \times 20 = 200 \ mW, \text{ so use a } \tfrac{1}{2}\text{-W size}$$

$$P_4 = VI = 5 \times 20 = 100 \ mW, \text{ so use at least a } \tfrac{1}{4}\text{-W size}$$

$$P_3 = VI = 10 \times 20 = 200 \ mW, \text{ so use a } \tfrac{1}{2}\text{-W size}$$

$$P_2 = VI = 20 \times 20 = 400 \text{ mW, so use a 1-W size}$$

$$P_1 = VI = 5 \times 20 = 100 \text{ mW, so use at least a } \tfrac{1}{4}\text{-W size}$$

EXAMPLE 4-15: An unloaded voltage divider provides $+20$, $+10$, and -30 V to its load. Assuming 25 mA in the string and a 60-V supply, design the circuit.

SOLUTION: Use the most negative value as the negative side of the 60-V supply and draw the circuit, Fig. 4-18. Since the potential difference between the $+20$-V output and the -30-V output is 50 V, and the supply voltage is 60 V, R_1 must drop 10 V. Thus,

$$R_4 = \frac{V}{I} = \frac{30}{25} = 1.2 \text{ k}\Omega$$

$$R_3 = \frac{V}{I} = \frac{10}{25} = 400 \ \Omega$$

$$R_2 = \frac{V}{I} = \frac{10}{25} = 400 \ \Omega$$

$$R_1 = \frac{V}{I} = \frac{10}{25} = 400 \ \Omega$$

$$P_4 = VI = 30 \times 25 = 750 \text{ mW, so use a 2-W size}$$

$$P_3 = P_2 = P_1 = VI = 10 \times 25 = 250 \text{ mW, so use a } \tfrac{1}{2}\text{-W size}$$

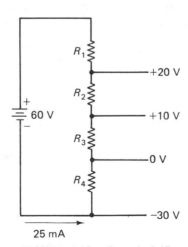

FIGURE 4-18: *Example 4-15.*

4-8. THE OPEN CIRCUIT

Any series path that has been broken is called an *open circuit*. One of the difficultie encountered by many students is that of knowing what the voltage should be withi an open circuit, such as that shown in Fig. 4-19. Here, a voltmeter is in series with voltage supply and a resistor.

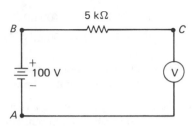

FIGURE 4-19: *Measuring volt-age in an open circuit.*

First, let us remember that a perfect voltmeter must draw no current from the power supply; it must be a very high resistance. Therefore, since we have a very high resistance (the voltmeter) in series with the 100-V supply, we will have practically no current through either the voltmeter or the 5-kΩ resistor. Thus, the voltage from *B* to *A* is 100 V and the voltage drop from *B* to *C* is

$$V = IR = 0 \times 5 = 0 \text{ V}$$

Subtracting this 0-V drop from the 100-V supply voltage, we find that the voltage across the voltmeter will be 100 V. Therefore, unless the resistance of the series resistor is substantial, the voltage across an open circuit will be the same as across the supply. That is not as hard as it looks, is it?

4-9. SUMMARY

A series circuit is one in which each resistor is connected head to tail. A parallel circuit is one in which the head of each resistor is connected to the head of each of the other resistors, and the tails of each resistor are connected together. Current is the same throughout a series circuit. Resistances within a series circuit can be added to obtain the total resistance of the circuit. The sum of all voltages around a circuit is zero volts (Kirchhoff's voltage law). Power supplies are said to be connected in series aiding if they push the electrons in the same direction along the wire; they are said to be connected in series opposing if the push of one supply opposes the push of the second supply. Loop equations formed from applying Kirchhoff's voltage law can be used to compute unknown variables within a circuit. Power dissipated within a series circuit is the sum of the power dissipated by the individual resistances. A voltage divider is a circuit composed of resistors, which supplies several voltages to a load from a single power supply. A voltmeter placed in series with a power supply and a resistance will read the voltage of the supply, assuming that the value of the resistance is insignificant compared with the resistance of the voltmeter.

4-10. REVIEW QUESTIONS

4-1. In Fig. 4-20, which resistors are in parallel?

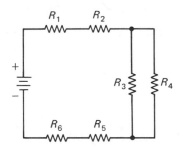

FIGURE 4-20: *Review Questions 4-1 and 4-2.*

4-2. In Fig. 4-20, which resistors are in series?

4-3. If, in a series circuit containing a 5-, 1-, and a 2-kΩ resistor, 1 mA flows through the 1-kΩ resistor, how much current flows through the 5-kΩ resistor?

4-4. Voltage across a resistor in a series circuit is equal to _____ multiplied by _____.

4-5. If lead A of a resistor has current flowing into it, what will the polarity of its second lead, B, be with respect to A?

4-6. What is the formula for finding the total resistance of three resistors, R_1, R_2, and R_3, in series?

4-7. State Kirchhoff's voltage law.

4-8. If the negative terminals of two batteries in series are wired together, the batteries are connected in series _____.

4-9. If the negative terminal of one battery is wired to the positive terminal of a second, series battery, the batteries are connected in series _____.

4-10. What does the sign of the calculated current indicate?

4-11. State the relationship between total power dissipated in a series circuit and the power dissipated by the individual resistors.

4-12. Under what conditions is a voltage divider considered unloaded?

4-13. A voltmeter in series with a power supply and a low resistance measures _____ _____.

4-11. PROBLEMS

4-1. In Fig. 4-21, $I = 1$ mA, $R_1 = 2$ kΩ, $R_2 = 5$ kΩ, $R_3 = 6$ kΩ, and $R_4 = 8$ kΩ. Calculate the voltage drops across each of the resistors and state which end is more positive, CCW or CW.

4-2. In Fig. 4-21, $I = 6.3$ mA, $R_1 = 4.7$ kΩ, $R_2 = 6.8$ kΩ, $R_3 = 1.2$ kΩ, and $R_4 = 150$ kΩ. Calculate the voltage drops across each of the resistors and state which end of each resistor is more positive, CCW or CW.

4-3. What is the total resistance seen by the power supply in Prob. 4-1?

4-4. What is the total resistance seen by the power supply in Prob. 4-2?

4-5. In Fig. 4-21, $V_1 = V_2 = V_3 = 20$ V, and $V_s = 75$ V. Calculate V_4.

4-6. In Fig. 4-21, $V_1 = 36$ V, $V_2 = 28$ V, $V_4 = 65$ V, and $V_s = 150$ V. Calculate V_3.

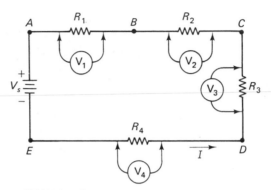

FIGURE 4-21: *Problems 4-1, 4-2, 4-5, and 4-6.*

4-7. Calculate I and V_1 of Fig. 4-22.

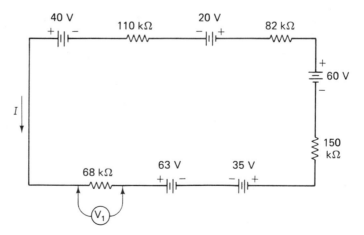

FIGURE 4-22: *Problem 4-7.*

Problems 4-8 through 4-13 refer to Fig. 4-23.

	I	R_1	R_2	R_3	V_A	V_B	V_C	V_1	V_2	V_3
4-8.	1 mA	2 kΩ	4 kΩ	3 kΩ	5 V	20 V	____	____	____	____
4-9.	2 mA	82 kΩ	120 kΩ	68 kΩ	20 V	5 V	____	____	____	____
4-10.	____	27 kΩ	33 kΩ	68 kΩ	40 V	60 V	12 V	____	____	____
4-11.	____	1 MΩ	6.8 MΩ	3.6 MΩ	60 V	20 V	5 V	____	____	____
4-12.	____	5 kΩ	10 kΩ	20 kΩ	____	5 V	60 V	40 V	____	____
4-13.	____	7 MΩ	6 MΩ	2 MΩ	____	20 V	80 V	30 V	____	____

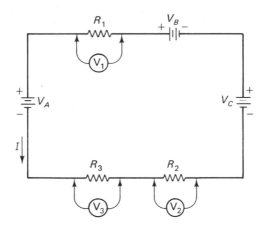

FIGURE 4-23: *Problems 4-8 through 4-13.*

4-14. Calculate P_t in Prob. 4-8.

4-15. Calculate P_t in Prob. 4-9.

4-16. Calculate P_t in Prob. 4-10.

4-17. Calculate P_t in Prob. 4-11.

4-18. An unloaded voltage divider must supply 25, 40, and 60 V from a 90-V source. Design the circuit assuming a 10-mA bleeder current.

4-19. An unloaded voltage divider must supply 20, 45, 60, and 100 V from a 120-V source. Design the circuit assuming a 20-mA bleeder current.

4-20. An unloaded voltage divider must supply −30, +20, and +40 V from a 100-V source. Design the circuit assuming a 15-mA bleeder current.

4-21. An unloaded voltage divider must supply −60, −20, +10, +40, and +60 V from a 150-V source. Design the circuit assuming a 10-mA bleeder current.

4-12. PROJECTS

4-1. Pick any five carbon-composition resistors from your stock and measure each one with an ohmmeter. Also read the rated value. Next, connect all five in series and measure the total resistance. Compare with the sum of the individual measured values and the sum of the individual rated values.

4-2. (a) Wire the circuit shown in Fig. 4-5. Measure current through points A, B, and C. Compare the results.

(b) Measure the voltage drops across R_1, R_2, and R_3. Compare the sum with the power-supply voltage.

(c) Connect a voltmeter in series at point B. Explain the voltage reading.

4-3. (a) Connect two 6-V power supplies in series opposing with a 10-kΩ resistor. Measure the current in the circuit and the voltage across each of the three circuit elements. Explain the results in terms of Kirchhoff's voltage law.

(b) Connect the 6-V power supplies in series aiding and repeat the experiment.

4-4. Design and build an unloaded voltage divider that supplies $+10$, $+16$, and $+20$ V from a 25-V source. Use a 3-mA bleeder current.

4-5. Design and build an unloaded voltage divider that supplies -10, -5, $+5$, and $+12$ V from a 25-V source. Use a 10-mA bleeder current.

4-6. Connect a 1-MΩ potentiometer in series with a VOM used as a voltmeter and a 20-V source. Adjust the potentiometer until the voltmeter reads 10 V. How is the resistance of the potentiometer under this condition related to the resistance of the voltmeter? Next, move the VOM to the next-higher scale and repeat the experiment. What are your conclusions?

4-7. Assume that a black box contained an unknown voltage supply in series with an unknown resistance. Devise a system of finding the value of the voltage and the resistance using external measurements. Verify your conclusions on a test circuit.

5

PARALLEL CIRCUITS

In Chapter 4 we studied the series circuit: the current through it, its resistance, and the voltage relationships within it. In this chapter we shall examine the parallel circuit in a similar manner, discovering that electrons now have a choice of which path to follow— sort of an electronic democracy. No longer are they subject to the dictatorial constraints of a single series path.

The sections include:

5-1. The parallel circuit

5-2. Voltage across a parallel circuit

5-3. Current in a parallel circuit

5-4. Resistance in a parallel circuit

5-5. Power in a parallel circuit

5-6. Power-supply loads

5-1. THE PARALLEL CIRCUIT

A *parallel circuit* is formed by connecting the ends of components together as shown in Fig. 5-1. Note that in this figure, the top ends of the resistors are all connected together and the bottom ends of the resistors are connected together. It is important to recognize that both ends must be so connected to call it a parallel circuit.

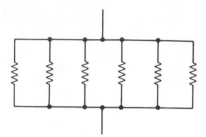

FIGURE 5-1: *Parallel resistors.*

5-2. VOLTAGE ACROSS A PARALLEL CIRCUIT

Consider the voltage across each of the resistors in Fig. 5-2. Note that potential will be the same across each of the resistors; this is always true in such a parallel circuit. Voltage is constant across the elements. This contrasts to the series circuit, in which the voltage across each of the resistors is dependent upon the value of the resistance.

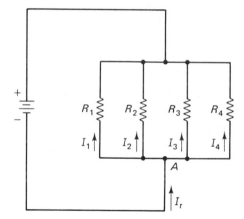

FIGURE 5-2: *Current and voltage of a parallel circuit.*

5-3. CURRENT IN A PARALLEL CIRCUIT

However, consider how the current behaves in the parallel circuit. If an electron approaches point *A* from the power supply, Fig. 5-2, it is afforded an opportunity to go through any one of four resistors to arrive safely at the positive end of the power supply. Thus, current divides in a parallel circuit, some flowing through each of the branches. A branch that has very little resistance will have more current than one that has much resistance. This is analogous to a river encountering an island; the river divides so that current flows on either side of the island. However, the wider the channel, the more current will flow. Note that the sum of the current flowing on either side of the island is equal to the total current upstream of the island and also the total current downstream of the island. So it is in electronics. The sum of all the currents through the resistive branches is equal to the current flowing out of the negative terminal of the power supply and is also equal to the current flowing into the positive side of the power supply. Stated mathematically,

$$I_t = I_1 + I_2 + I_3 + I_4 \tag{5-1}$$

Let us now examine all the wires connected to point A. Since these wires have very little resistance, the conductive path from the negative terminal of the power supply to each of the resistors can all be considered the same electrical point; it has the same potential with reference to any other part of the circuit. This electrical point is called a *node*. Thus, this circuit has two nodes, one at the positive side of the supply and one at the negative side of the supply. If we assign current entering a node as positive and current leaving the node as negative, we can express Eq. (5-1) as follows:

$$I_t + I_1 + I_2 + I_3 + I_4 = 0$$

This is known as *Kirchhoff's current law* and basically states that the algebraic sum of the current within a node is zero.

Knowing this remarkable bit of information, we can compute branch currents and total current if there is only one unknown.

EXAMPLE 5-1: In Fig. 5-2, $I_1 = 20$ mA, $I_2 = 30$ mA, $I_4 = 15$ mA, and $I_t = 100$ mA. Compute I_3.

SOLUTION: According to Kirchhoff's current law,

$$I_t + I_1 + I_2 + I_3 + I_4 = 0$$
$$100 - 20 - 30 - I_3 - 15 = 0$$
$$I_3 = 35 \text{ mA}$$

The positive sign means that we assumed the current direction correctly, that is, flowing out of the node.

EXAMPLE 5-2: In Fig. 5-2, $I_1 = 220$ μA, $I_2 = 145$ μA, $I_3 = 460$ μA, and $I_4 = 390$ μA. Compute I_t.

SOLUTION: According to Kirchhoff's law,

$$I_t - 220 - 145 - 460 - 390 = 0$$
$$I_t = 1215 \text{ } \mu\text{A} = 1.215 \text{ mA}$$

5-4. RESISTANCE IN A PARALLEL CIRCUIT

Let us compare the two circuits shown in Fig. 5-3. In the left circuit, the resistance is 10 kΩ, whereas in the right circuit, there are two 10-kΩ resistors in parallel. Since the power supply is 10 V, there will be 1 mA through the 10-kΩ resistor in the left

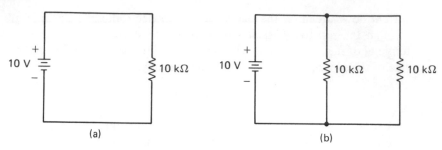

FIGURE 5-3: *Adding resistors in parallel: (a) single resistor; (b) two resistors.*

figure. But there will also be 1 mA through each of the resistors in the right figure; therefore, the right figure will have less resistance. In fact, the resistance of a parallel circuit will always be less than the resistance of the smallest resistor. With the aid of some algebraic gymnastics, we can now develop a general formula for finding the total resistance of a parallel network. The total current would be

$$I_t = \frac{V}{R_1} + \frac{V}{R_2} + \frac{V}{R_3} + \cdots$$

$$\frac{I_t}{V} = \frac{1}{R_1} + \frac{1}{R_2} + \frac{1}{R_3} + \cdots$$

But $V/I_t = R_t$ (total resistance). Therefore,

$$\frac{1}{R_t} = \frac{1}{R_1} + \frac{1}{R_2} + \frac{1}{R_3} + \cdots$$

This, then, is the formula for finding the total resistance of a parallel network.

EXAMPLE 5-3: Find the total resistance of three 10-kΩ resistors connected in parallel.

SOLUTION: According to the general formula,

$$\frac{1}{R_t} = \frac{1}{R_1} + \frac{1}{R_2} + \frac{1}{R_3}$$

$$= \frac{1}{10} + \frac{1}{10} + \frac{1}{10}$$

$$= \frac{3}{10}$$

$$R_t = 3.333 \text{ k}\Omega$$

Note that in Example 5-3 we used kΩ in all our calculations. The parallel formula can be used with any convenient system of units as long as they are consistent. For example, if we were to use tens of ohms for the right side of the equation, the resultant total resistance would be in tens of ohms.

EXAMPLE 5-4: Compute the total resistance seen by the power supply in Fig. 5-4.

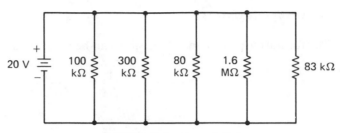

FIGURE 5-4: *Example 5-4.*

SOLUTION: We shall use units of 100s of $k\Omega$:

$$\frac{1}{R_t} = \frac{1}{R_1} + \frac{1}{R_2} + \frac{1}{R_3} + \frac{1}{R_4} + \frac{1}{R_5}$$

$$= \frac{1}{1} + \frac{1}{3} + \frac{1}{0.8} + \frac{1}{16} + \frac{1}{0.83}$$

$$= 1.0 + 0.333 + 1.250 + 0.0625 + 1.205 = 3.851$$

$$R_t = 0.2597 \text{ 100s of } k\Omega = 25.97 \text{ } k\Omega$$

EXAMPLE 5-5: Compute the total resistance seen by the power supply shown in Fig. 5-5.

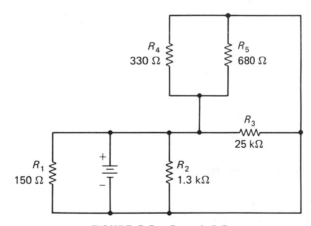

FIGURE 5-5: *Example 5-5.*

SOLUTION: Note that all the resistors are connected in parallel, for the leads of one end of all the resistors are connected and the leads of the other end of all the resistors are connected. Therefore, using $k\Omega$:

$$\frac{1}{R_t} = \frac{1}{R_1} + \frac{1}{R_2} + \frac{1}{R_3} + \frac{1}{R_4} + \frac{1}{R_5}$$

$$= \frac{1}{0.15} + \frac{1}{1.3} + \frac{1}{25} + \frac{1}{0.33} + \frac{1}{0.68} = 11.98$$

$$R_t = 0.08349 \text{ } k\Omega = 83.49 \text{ } \Omega$$

Two Parallel Resistors

The formula for two parallel resistors can be modified as follows:

$$\frac{1}{R_t} = \frac{1}{R_1} + \frac{1}{R_2} = \frac{R_1 + R_2}{R_1 R_2}$$

$$R_t = \frac{R_1 R_2}{R_1 + R_2}$$

This is usually easier to manipulate and can, many times, be done mentally.

EXAMPLE 5-6: Two resistors, 10 kΩ and 4 kΩ, are connected in parallel. What is their total resistance?

SOLUTION: Use kΩ:

$$R_t = \frac{R_1 R_2}{R_1 + R_2} = \frac{10 \times 4}{10 + 4} = \frac{40}{14} = 2.857 \text{ k}\Omega$$

EXAMPLE 5-7: A resistor must be connected in parallel with a 1-kΩ resistor such that the total resistance is 860 Ω. Find the value of this resistor.

SOLUTION:

$$R_t = \frac{R_1 R_2}{R_1 + R_2}$$

$$860 = \frac{1000 \times R_2}{1000 + R_2}$$

$$860{,}000 + 860 R_2 = 1000 R_2$$

$$140 R_2 = 860{,}000$$

$$R_2 = 6143 \ \Omega = 6.143 \text{ k}\Omega$$

Identical Parallel Resistors

Where resistors of identical value are in parallel, the total resistance can be found by dividing the resistance of one of the resistors by the number of resistors. This can be illustrated by use of the formula for four resistors:

$$\frac{1}{R_t} = \frac{1}{R} + \frac{1}{R} + \frac{1}{R} + \frac{1}{R} = \frac{4}{R}$$

$$R_t = \frac{R}{4}$$

EXAMPLE 5-8: Five 10-kΩ resistors are connected in parallel. Compute the total resistance.

SOLUTION: $R_t = R/n = 10/5 = 2$ kΩ.

5-5. POWER IN A PARALLEL CIRCUIT

Note that, in a parallel circuit, work is done within each resistor. Thus, total power dissipated within a parallel circuit is the sum of that dissipated by each resistor. This is, of course, the same rule as that used in a series circuit.

EXAMPLE 5-9: Compute the total power dissipated by the resistors in Fig. 5-4.

SOLUTION: The total power is

$$P_t = P_1 + P_2 + P_3 + P_4 + P_5$$

$$= \frac{V^2}{R_1} + \frac{V^2}{R_2} + \frac{V^2}{R_3} + \frac{V^2}{R_4} + \frac{V^2}{R_5}$$

$$= \frac{20^2}{10^5} + \frac{20^2}{3.0 \times 10^5} + \frac{20^2}{8.0 \times 10^4} + \frac{20^2}{1.6 \times 10^6} + \frac{20^2}{8.3 \times 10^4}$$

$$= 4.00 \times 10^{-3} + 1.333 \times 10^{-3} + 5.000 \times 10^{-3}$$

$$+ 2.500 \times 10^{-4} + 4.819 \times 10^{-3}$$

$$= 1.540 \times 10^{-2} = 15.40 \text{ mW}$$

A much easier method of solving for total power of Fig. 5-4 would be to recognize that for total power we need only know total resistance and total voltage.

EXAMPLE 5-10: Compute total power in Fig. 5-4 from total resistance and total voltage.

SOLUTION: From Example 5-4, $R_t = 25.97 \text{ k}\Omega$. Therefore,

$$P_t = \frac{V^2}{R_t} = \frac{20^2}{25,970} = 15.40 \text{ mW}$$

5-6. POWER-SUPPLY LOADS

One excellent example of a parallel circuit is that of the typical voltage supply feeding several loads, Fig. 5-6. In some of these loads the resistance is known; in others, just the power dissipated or the current through the branch.

EXAMPLE 5-11: Compute total power and total current supplied by the source in Fig. 5-6.

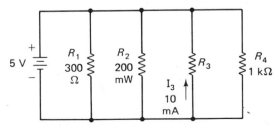

FIGURE 5-6: *Example 5-11, power-supply distribution.*

SOLUTION: The total current through R_1 is

$$I_1 = \frac{V}{R_1} = \frac{5}{0.3} = 16.667 \text{ mA}$$

The current through R_4 is

$$I_4 = \frac{V}{R_4} = \frac{5}{1} = 5.00 \text{ mA}$$

The current through R_2 is

$$I_2 = \frac{P_2}{V} = \frac{0.200}{5} = 40 \text{ mA}$$

Thus, the total current is

$$I_t = I_1 + I_2 + I_3 + I_4$$
$$= 16.67 + 40.00 + 10.00 + 5.00$$
$$= 71.67 \text{ mA}$$

The total power is

$$P = V_t I_t = 5 \times 0.07167 = 358.3 \text{ mW}$$

Incidentally, the other unknown values within the circuit can be computed as follows:

$$R_2 = \frac{V}{I_2} = \frac{5}{40} = 125 \, \Omega$$

$$R_3 = \frac{V}{I_3} = \frac{5}{10} = 500 \, \Omega$$

$$P_1 = VI_1 = 5 \times 16.667 = 83.33 \text{ mW}$$
$$P_3 = VI_3 = 5 \times 10 = 50 \text{ mW}$$
$$P_4 = VI_4 = 5 \times 5 = 25 \text{ mW}$$

Note that the total power is equal to the sum of the individual power-dissipation values.

$$P_t = P_1 + P_2 + P_3 + P_4$$
$$= 83.3 + 200 + 50 + 25 = 358.3 \text{ mW}$$

which is what we calculated using the total power and total current.

EXAMPLE 5-12: Compute the total current and power supplied by the 5-V source in Fig. 5-7.

SOLUTION:

$$I_3 = \frac{P_3}{V} = \frac{0.600}{5} = 120 \text{ mA}$$

$$I_4 = \frac{P_4}{V} = \frac{0.900}{5} = 180 \text{ mA}$$

$$I_5 = \frac{V}{R_5} = \frac{5}{0.5} = 10 \text{ mA}$$

$$I_6 = \frac{V}{R_6} = \frac{5}{60} = 83.33 \text{ mA}$$

The total current is

$$I_t = I_1 + I_2 + I_3 + I_4 + I_5 + I_6$$
$$= 70 + 30 + 120 + 180 + 10 + 83.33 = 493.3 \text{ mA}$$

The total power is

$$P_t = VI_t = 5 \times 0.4933 = 2.467 \text{ W}$$

The remaining unknown values within the circuit can be calculated as follows:

$$R_1 = \frac{V}{I_1} = \frac{5}{0.07} = 71.43 \ \Omega$$

$$R_2 = \frac{V}{I_2} = \frac{5}{0.03} = 166.67 \ \Omega$$

$$R_3 = \frac{V}{I_3} = \frac{5}{0.120} = 41.67 \ \Omega$$

$$R_4 = \frac{V}{I_4} = \frac{5}{0.180} = 27.78 \ \Omega$$

$$P_1 = VI_1 = 5 \times 0.07 = 350 \text{ mW}$$
$$P_2 = VI_2 = 5 \times 0.03 = 150 \text{ mW}$$
$$P_5 = VI_5 = 5 \times 10 = 50 \text{ mW}$$
$$P_6 = VI_6 = 5 \times 83.33 = 416.7 \text{ mW}$$

Computing total power using the individual power dissipations:

$$P_t = P_1 + P_2 + P_3 + P_4 + P_5 + P_6$$
$$= 350 + 150 + 600 + 900 + 50 + 416.7 = 2.467 \text{ W}$$

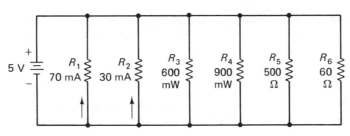

FIGURE 5-7: *Example 5-12.*

5-7. SUMMARY

Elements are considered to be connected in parallel if all the heads of the elements are connected together and all the tails are connected together. Voltage is the same across all parallel elements, whereas current divides in inverse proportion to the resistance of the branch. The sum of branch currents is equal to the total current. The inverse of total resistance of a parallel circuit is equal to the sum of the inverses of the branch resistances, whereas total power is equal to the sum of power dissipated in the individual resistances. Power-supply loads are usually in parallel with the source.

5-8. REVIEW QUESTIONS

5-1.   Define a parallel circuit.

5-2.   What is the rule for voltage across a parallel circuit?

5-3.   How does current react to a parallel circuit?

5-4.   State Kirchhoff's current law.

5-5.   What is a node?

5-6.   What is the formula for computing the total resistance of many resistors in parallel?

5-7.   What is the short formula for finding the total resistance of just two resistors in parallel?

5-8.   What is the formula for computing the total resistance of identical resistors in parallel?

5-9.   Name two methods of computing the total power in a parallel circuit?

5-9. PROBLEMS

5-1.   A parallel circuit has branch currents of 1, 5, and 10 mA. What is the total current of the circuit?

5-2.   A soldering iron that draws 420 mA, a light bulb that draws 85 mA, and an aquarium pump that draws 25 mA are connected in parallel to a 120-V source. What is the total current?

5-3.   A 250-Ω integrated circuit, a 210-Ω integrated circuit, and a 500-Ω transistor circuit are connected in parallel with a 5-V source voltage. What is the total resistance seen by the source?

5-4.   A 1-MΩ integrated circuit, a 5-MΩ integrated circuit, and a 2-MΩ resistor are connected in parallel with a 15-V source. What is the total resistance seen by the source?

5-5.   A transistor sees a load of 10 kΩ and one of 5 kΩ in parallel. What is the total resistance seen by the transistor?

5-6.   The biasing resistors of a transistor stage are, effectively, a 50-kΩ resistor in parallel with a 10-kΩ. What is the resistance of this biasing circuit?

5-7.   To get proper power dissipation, three 47-kΩ resistors are connected in parallel. What is the total resistance of the circuit?

5-8.   Ten 20-kΩ resistors are connected in parallel. What is the total resistance of the circuit?

5-9.   What is the total power drawn from the source in Prob. 5-2?

5-10.   What is the total power drawn from the source in Prob. 5-3?

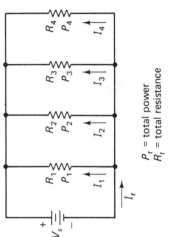

P_t = total power
R_t = total resistance

FIGURE 5-8: *Problems 5-11 through 5-16.*

Compute the missing items, referring to Fig. 5-8.

	V_s	I_t	P_t	R_t	R_1	R_2	R_3	R_4	P_1	P_2	P_3	P_4	I_1	I_2	I_3	I_4
5-11.	10 V	20 mA	___	___	___	___	___	___	___	___	___	___	10 mA	1 mA	2 mA	___
5-12.	10 V	___	___	___	5 kΩ	5 kΩ	2 kΩ	___	___	___	___	___	___	___	___	3 mA
5-13.	20 V	___	___	___	100 Ω	300 Ω	___	___	___	___	1.5 W	___	___	___	___	90 mA
5-14.	___	___	___	___	12 kΩ	5 kΩ	4 kΩ	___	___	___	___	1 W	5 mA	___	___	___
5-15.	___	___	___	5 kΩ	100 kΩ	150 kΩ	20 kΩ	___	___	___	___	___	1 mA	___	___	___
5-16.	___	___	5 W	___	___	___	___	___	1 W	1.5 W	1 W	___	10 mA	___	___	___

5-10. PROJECTS

5-1. Connect one 10-kΩ resistor to an ohmmeter and take the reading. Next, connect two 10-kΩ resistors in parallel and repeat, then three, then four, then five. Do your results verify the parallel resistance formula?

5-2. A 4.7-kΩ resistor must be connected in parallel with another resistor such that the total resistance is 2.1 kΩ. Compute the unknown value, then connect the nearest standard value resistor in parallel with the 4.7 kΩ. Do your results agree with your computed value?

5-3. Three resistors—$\frac{1}{2}$, 1, and 2 W—must be connected in parallel with a 10-V supply such that each resistor dissipates its rated power. Compute the necessary resistors, then connect them to the supply and observe the results.

5-4. Connect a 4.7-, 5.6-, and 10-kΩ resistor in parallel with a 10-V supply. Measure the branch currents and the total current. Also measure the voltage across each resistor. Do your results agree with those of Kirchhoff?

6

SERIES-PARALLEL CIRCUITS

In Chapters 4 and 5 we discussed first the series, then the parallel circuit. We are now going to combine them and analyze the resultant circuit, for most circuits are a combination of these two. We will then cross over from theory to practice and use our series–parallel analysis techniques in solving bridge and loaded-voltage-divider circuit problems. The sections include:

6-1. The series–parallel circuit

6-2. Resistance of a series–parallel circuit

6-3. Current and voltage in a series–parallel circuit

6-4. Power in a series–parallel circuit

6-5. The balanced bridge

6-6. The loaded voltage divider

6-1. THE SERIES–PARALLEL CIRCUIT

Circuits that contain both series and parallel elements are called *series–parallel circuits*. Figure 6-1(a) illustrates such a circuit, for R_1, R_2, and R_3 are in parallel, as are R_4 and R_5. These two parallel networks are in series with each other, with R_6, and with the power supply. Analysis of a series–parallel circuit requires recognition of which parts of the circuit are in parallel with each other and which are in series with each other. The basic ground rules are those set forth in the discussions of series and parallel circuits in the previous chapters, namely that a series circuit is connected head to tail, whereas in a parallel circuit, all heads are tied together and all tails together.

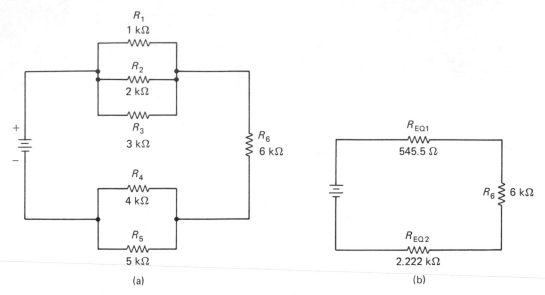

FIGURE 6-1: *Example 6-1, resistance of a series–parallel circuit:* (a) *original circuit;* (b) *equivalent circuit.*

6-2. *RESISTANCE OF A SERIES–PARALLEL CIRCUIT*

In this section we shall discuss how to find the total resistance of a series–parallel network. However, before examining any circuits in detail, let us discuss the concept of equivalent resistance. When we found the total resistance of a parallel network, we could have replaced the entire network with a resistor of equivalent value. For example, if we had four 10-kΩ resistors in parallel, they could be replaced by one 2.5-kΩ resistor, for this is the resistance of these four when they are so connected. In a similar manner, if we had four 10-kΩ resistors in series, we could replace them with one 40-kΩ resistor, for this is the resistance of these four when so connected. This "lumped resistance" we shall call an *equivalent resistance*. Analysis of series–parallel resistors then becomes merely finding equivalent resistances by parallel or series combination of resistors until the equivalent resistance is only one resistor. Sounds easy, doesn't it? Let us now do some examples to illustrate the procedure.

EXAMPLE 6-1: Find the total resistance seen by the power supply in Fig. 6-1(a).

SOLUTION: We must first find the equivalent resistance of R_1, R_2, and R_3, a parallel network:

$$\frac{1}{R_{EQ1}} = \frac{1}{R_1} + \frac{1}{R_2} + \frac{1}{R_3}$$

$$= \frac{1}{1} + \frac{1}{2} + \frac{1}{3} = 1.8333$$

$$R_{EQ1} = 545.5\ \Omega$$

Next, we shall find the equivalent resistance of R_4 and R_5, another parallel network:

$$R_{EQ2} = \frac{R_4 R_5}{R_4 + R_5} = \frac{4 \times 5}{4 + 5} = 2.222 \text{ k}\Omega$$

We can now redraw the circuit, replacing the two parallel networks with R_{EQ1} and R_{EQ2}, Fig. 6-1(b). We are left with a circuit with three resistors in series, R_{EQ1}, R_{EQ2}, and R_6. Adding these, we obtain

$$R_t = R_{EQ1} + R_6 + R_{EQ2} = 545 + 6000 + 2222 = 8.767 \text{ k}\Omega$$

This is the total resistance of the circuit.

EXAMPLE 6-2: Compute the total resistance of the series–parallel circuit shown in Fig. 6-2(a).

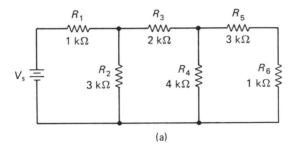

(a)

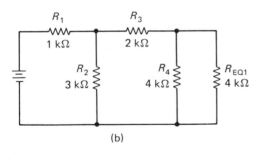

(b)

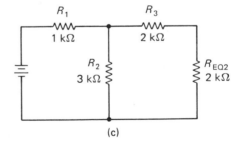

(c)

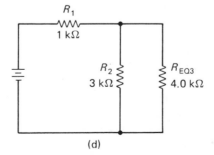

(d)

FIGURE 6-2: *Example 6-2: (a) original circuit; (b) combining R_5 and R_6; (c) combining R_4 and R_{EQ1}; (d) combining R_3 and R_{EQ2}; (e) combining R_2 and R_{EQ3}; (f) combining R_1 and R_{EQ4}.*

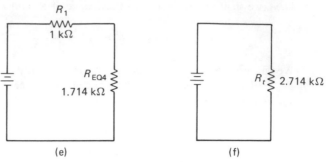

FIGURE 6-2: (*Cont.*)

SOLUTION: Resistors R_5 and R_6 are in series, and this series string in parallel with R_4, Fig. 6-2(b):

$$R_{EQ1} = R_5 + R_6 = 3 + 1 = 4 \text{ k}\Omega$$

Next, R_{EQ1} is in parallel with R_4, Fig. 6-2(c):

$$R_{EQ2} = \frac{R_4 R_{EQ1}}{R_4 + R_{EQ1}} = \frac{4 \times 4}{4 + 4} = 2.000 \text{ k}\Omega$$

But this is in series with R_3, Fig. 6-2(d):

$$R_{EQ3} = R_{EQ2} + R_3 = 2.000 + 2.000 = 4.000 \text{ k}\Omega$$

This is now in parallel with R_2, Fig. 6-2(e):

$$R_{EQ4} = \frac{R_2 R_{EQ3}}{R_2 + R_{EQ3}} = \frac{3 \times 4}{3 + 4} = 1.714 \text{ k}\Omega$$

Finally, R_{EQ4} is in series with R_1, Fig. 6-2(f):

$$R_t = R_1 + R_{EQ4} = 1 + 1.714 = 2.714 \text{ k}\Omega$$

This, then, is the total resistance.

6-3. CURRENT AND VOLTAGE IN A SERIES–PARALLEL CIRCUIT

Current and voltage in a series–parallel circuit obey the laws of parallel and series circuits as discussed in previous chapters. However, each part of the circuit must now be analyzed to see if it is a series network or a parallel network. Using Ohm's law and our knowledge of circuits, we can calculate values that need to be found. The basic approach that is followed is to:

1. Calculate total resistance.

2. Knowing this, calculate total current.

3. Knowing this, compute voltage drops across the series elements or equivalent series resistance formed by combining parallel elements.

4. Knowing the voltage drop across a parallel element, compute the current within the individual branches.

EXAMPLE 6-3: Calculate V_1, V_2, I_1, I_2, and I_t for the circuit shown in Fig. 6-3(a).

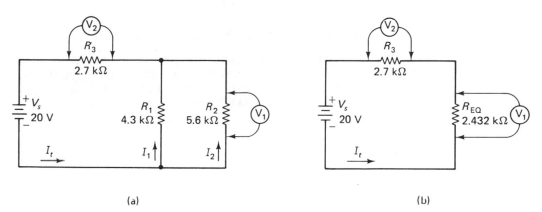

(a)   (b)

FIGURE 6-3: *Example 6-3, computing currents and voltages: (a) original circuit; (b) combining R_1 and R_2.*

SOLUTION: We must first calculate R_t, but this requires finding the R_{EQ} of resistors R_1 and R_2:

$$R_{EQ} = \frac{R_1 R_2}{R_1 + R_2} = \frac{4.3 \times 5.6}{4.3 + 5.6} = 2.432 \text{ k}\Omega$$

We can now redraw the circuit, Fig. 6-3(b). Next we can calculate R_t:

$$R_t = R_{EQ} + R_3 = 2.432 + 2.7 = 5.132 \text{ k}\Omega$$

Knowing this, we can calculate total current, I_t:

$$I_t = \frac{V}{R_t} = \frac{20}{5.132} = 3.897 \text{ mA}$$

We can now calculate V_2:

$$V_2 = I_t R_2 = 3.897 \times 2.7 = 10.522 \text{ V}$$

Knowing V_2, by Kirchhoff's voltage law we can calculate V_1:

$$V_s + V_2 + V_1 = 0$$
$$20 - 10.522 - V_1 = 0$$
$$V_1 = 9.478 \text{ V}$$

Knowing V_1, R_1, and R_2, we can calculate I_1 and I_2:

$$I_1 = \frac{V_1}{R_1} = \frac{9.478}{4.3} = 2.204 \text{ mA}$$

$$I_2 = \frac{V_1}{R_2} = \frac{9.478}{5.6} = 1.693 \text{ mA}$$

EXAMPLE 6-4: Find the unknown values shown in Fig. 6-4(a).

SOLUTION: First, the circuit is quite difficult to understand because of the manner in which it is drawn. Therefore, we shall redraw it, Fig. 6-4(b), where the series and parallel relationships can be clearly seen. Next, we shall find an equivalent of R_3, R_4, and R_6, Fig. 6-4(c). We are using 100s of kΩs for units.

$$\frac{1}{R_{EQ1}} = \frac{1}{R_3} + \frac{1}{R_4} + \frac{1}{R_6} = \frac{1}{10} + \frac{1}{8.9} + \frac{1}{7.5} = 0.3457$$

$$R_{EQ1} = 2.893 \text{ 100s of k}\Omega = 289.3 \text{ k}\Omega$$

Add the 220-kΩ in series, Fig. 6-4(d):

$$R_{EQ2} = R_5 + R_{EQ1} = 220 + 289.3 = 509.3 \text{ k}\Omega$$

The current through R_1, R_2, and R_{EQ2} is

$$I_1 = \frac{V}{R_1} = \frac{40}{0.36} = 111.1 \text{ } \mu\text{A}$$

$$I_2 = \frac{V}{R_2} = \frac{40}{0.43} = 93.02 \text{ } \mu\text{A}$$

$$I_{EQ2} = I_5 = \frac{V}{R_{EQ2}} = \frac{40}{0.5093} = 78.54 \text{ } \mu\text{A}$$

Knowing I_5, we can compute V_5:

$$V_5 = I_5 R_5 = 78.54 \times 0.22 = 17.28 \text{ V}$$

Knowing R_{EQ1} and that $I_5 = I_{EQ1} = I_{EQ2}$, we can calculate V_3:

$$V_3 = I_{EQ1} R_{EQ1} = 78.54 \times 0.2893 = 22.72 \text{ V}$$

Knowing V_3, we can calculate I_3, I_4, and I_6:

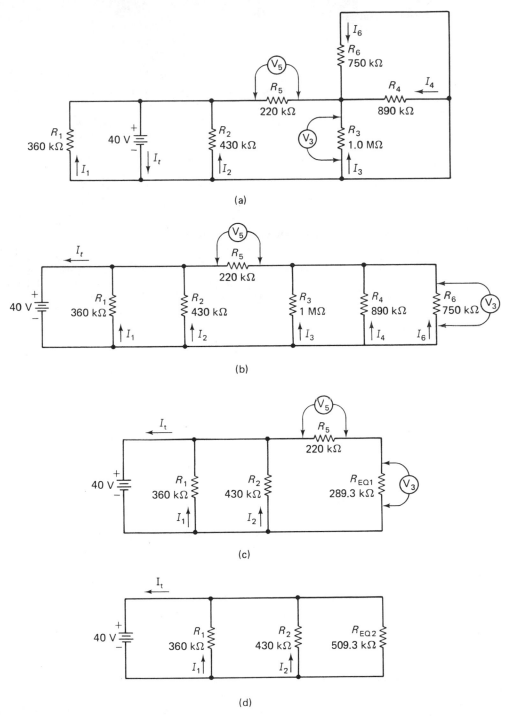

FIGURE 6-4: *Example 6-4: (a) original circuit; (b) redrawn circuit; (c) combining R_3, R_4, and R_6; (d) combining R_5 and R_{EQ1}.*

$$I_3 = \frac{V_3}{R_3} = \frac{22.72}{1.0} = 22.72 \ \mu A$$

$$I_4 = \frac{V_3}{R_4} = \frac{22.72}{0.89} = 25.53 \ \mu A$$

$$I_6 = \frac{V_3}{R_6} = \frac{22.72}{0.75} = 30.29 \ \mu A$$

We can calculate I_t by summing the branch currents:

$$I_t = I_1 + I_2 + I_5 = 111.1 + 93.02 + 78.54 = 282.7 \ \mu A$$

6-4. POWER IN A SERIES–PARALLEL CIRCUIT

Since the power laws for series and parallel circuits are identical, power in a series–parallel circuit is found by summing the power dissipated by each resistor. It can also be found by using total voltage, total resistance, and total current.

EXAMPLE 6-5: Find the total power dissipated by the resistors in Fig. 6-3.

SOLUTION: From Example 6-3, I_t is known. Therefore,

$$P_t = V_t I_t = 20 \times 3.897 = 77.94 \ \text{mW}$$

EXAMPLE 6-6: Find the total power furnished by the 40-V supply in Fig. 6-4(a).

SOLUTION: I_t is known from Example 6-4. Therefore,

$$P_t = V_t I_t = 40 \times 282.7 = 11.31 \ \text{mW}$$

6-5. THE BALANCED BRIDGE

The *Wheatstone bridge*, Fig. 6-5, is a device that contains a series of dials with numerical readouts and a sensitive, zero-center ammeter and is used to measure unknown resistances. To operate it, the unknown resistance is connected to the terminals, the dials manipulated until the meter reads zero, and the value of the unknown resistance read directly off the dials. Variations of the Wheatstone bridge are used extensively in electronic measurements. The basic device consists of the circuit shown in Fig. 6-5(b), where R_D is a calibrated potentiometer that reads out in ohms and R_x is the unknown resistance value. When R_D equal R_x, the current through the $R_1 R_D$ path will equal the current through the $R_2 R_x$ path; therefore, the voltage drop across R_D will equal the voltage drop across R_x. Consequently, the voltage from point A to point B will be zero volts and the meter will read 0 mA. Thus, the operator has merely to manipulate R_D until the meter reads zero, and read the resistance value of R_x directly off the R_D dial.

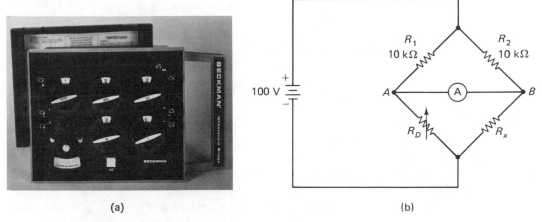

FIGURE 6-5: *Wheatstone bridge: (a) photograph (Courtesy, Beckman Instruments, Inc., Cedar Grove Operations); (b) schematic.*

Assume now that R_1 and R_D are 10 kΩ and R_2 and R_x are each 100 kΩ. Note that there would still be no voltage drop from A to B and the meter would read 0 V, for the voltage drop across both R_D and R_x would be 50 V. Thus, R_2 can be used to select the range of resistance to be measured: 10 times R_D, 100 times R_D, and so forth. Therefore, when the bridge is balanced (that is, the meter reads zero), we have the formula

$$\frac{R_1}{R_D} = \frac{R_2}{R_x}$$

Knowing R_1, R_2, and R_D, we can solve for R_x.

> **EXAMPLE 6-7:** A Wheatstone bridge has the following values when balanced: $R_1 = 10$ kΩ, $R_D = 7.63$ kΩ, and $R_2 = 50$ kΩ. Calculate R_x.
>
> *SOLUTION:* Since
>
> $$\frac{R_1}{R_D} = \frac{R_2}{R_x}$$
>
> then
>
> $$R_x = \frac{R_2 R_D}{R_1} = \frac{50 \times 7.63}{10} = 38.15 \text{ kΩ}$$

Fault Location

The bridge can be used to locate a short in a cable pair by connecting the cable pair to the R_x terminals. Knowing the round-trip resistance of the pair (called the *loop resistance*) and the resistance of that particular type of wire pair per 1000 ft, the fault can be located.

EXAMPLE 6-8: A loop measurement on a size 18 cable pair reads 250 Ω. How far is the short from the bridge?

SOLUTION: A size 18 wire has a loop resistance (R_{ES}) of 13.02 Ω/1000 ft. (This represents the resistance of 2000 ft of wire.) The distance to the fault is

$$D = \frac{R_x}{R_{ES}} \times 1000 = \frac{250}{13.02} \times 1000 = 19{,}201 \text{ ft or } 3.637 \text{ mi}$$

The Wheatstone bridge is used extensively to locate grounded cable faults. The first step in locating a fault is to have a person in the distant office short the cable pair together. The loop resistance (R_L) between the offices is then measured using the procedure discussed in the previous paragraph and in Example 6-8. Once this has been done, the bridge is connected in one of two configurations, Murray or Varley. The *Murray method* uses an adjustable slider, identified as $R_1 R_2$ in Fig. 6-6(a), which

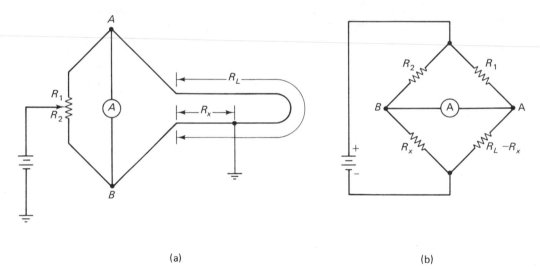

(a)   (b)

FIGURE 6-6: *Murray loop: (a) actual circuit; (b) redrawn.*

is moved until balance is achieved. With R_L representing the loop resistance of the entire pair, the formula becomes

$$\frac{R_1}{R_2} = \frac{R_L - R_x}{R_x}$$

Solve for R_x:

$$R_x = \frac{R_2}{R_1 + R_2} R_L$$

Thus, by taking a simple loop measurement, R_L is obtained, and by taking a Murray loop, R_x can be found and the distance determined.

EXAMPLE 6-9: A grounded wire exists on a pair with a loop measured at 250 Ω and a Murray measured at an R_1 of 650 Ω and an R_2 of 350 Ω. What is the distance to the fault if number 18 wire is used?

SOLUTION: Solve for R_x:

$$R_x = \frac{R_2}{R_1 + R_2} R_L = \frac{350}{1000} \times 250 = 87.5\ \Omega$$

Since 1000 ft of size 18 wire has a resistance of 6.51 Ω,

$$D = \frac{R_x}{R_{ES}} \times 1000 = \frac{87.5}{6.51} \times 1000 = 13{,}440\ \text{ft or } 2.55\ \text{mi}$$

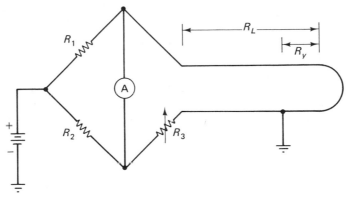

FIGURE 6-7: *Varley measurement.*

The *Varley method* uses the circuit shown in Fig. 6-7. Using loop resistances, the equation becomes

$$\frac{R_1}{R_2} = \frac{\dfrac{R_L}{2} + \dfrac{R_y}{2}}{R_3 + \dfrac{R_L - R_y}{2}}$$

Set $R_1 = R_2$ and solve for R_y:

$$1 = \frac{R_L + R_y}{R_L - R_y + 2R_3}$$

$$R_L - R_y + 2R_3 = R_L + R_y$$

$$2R_y = R_L - R_L + 2R_3$$

$$R_y = R_3$$

Thus, the reading, R_3, will tell the operator the distance the fault is from the distant end of the cable pair.

EXAMPLE 6-10: The Varley reads 426 Ω and the loop 2360 Ω for an 18 gauge wire. What is the distance of the fault from the test end?

SOLUTION: Since 18-gauge wire has a loop resistance of 13.02 Ω/1000 ft, the distant office is

$$D = \frac{R_x}{R_{ES}} \times 1000 = \frac{2360}{13.02} \times 1000 = 181{,}260 \text{ ft or } 34.33 \text{ mi}$$

from the test end. The fault is

$$D = \frac{R_3}{R_{ES}} \times 1000 = \frac{426}{13.02} \times 1000 = 32{,}719 \text{ ft or } 6.197 \text{ mi}$$

from the distant end or $34.33 - 6.20 = 28.13$ mi from the test end.

6-6. THE LOADED VOLTAGE DIVIDER

In Chapter 4 we studied unloaded voltage dividers. We shall now expand the subject to include those dividers whose loads are significant. Figure 6-8 shows a divider that supplies 20, 30, and 45 V to its three separate loads. According to Kirchhoff's current law, the following relationships would be true:

$$I_t = I_1 = I_4 + I_{20} + I_{30} + I_{45}$$
$$I_3 = I_4 + I_{20}$$
$$I_2 = I_3 + I_{30} = I_4 + I_{20} + I_{30}$$

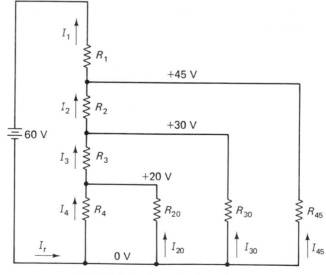

FIGURE 6-8: *Loaded voltage divider.*

I_4 is referred to as the *bleeder current*; it could be defined as the current through the string resistor that has the least amount of current flowing through it.

EXAMPLE 6-11: In Fig. 6-8, $I_4 = 5$ mA, $I_{20} = 20$ mA, $I_{30} = 10$ mA, and $I_{45} = 15$ mA. Calculate the string resistors.

SOLUTION: Since $I_4 = 5$ mA,

$$R_4 = \frac{V_4}{I_4} = \frac{20}{5} = 4 \text{ k}\Omega$$

$$R_3 = \frac{V_3}{I_3} = \frac{V_3}{I_4 + I_{20}} = \frac{30 - 20}{5 + 20} = 400 \ \Omega$$

$$R_2 = \frac{V_2}{I_2} = \frac{V_2}{I_3 + I_{30}} = \frac{45 - 30}{25 + 10} = 429 \ \Omega$$

$$R_1 = \frac{V_1}{I_1} = \frac{V_1}{I_2 + I_{45}} = \frac{60 - 45}{35 + 15} = 300 \ \Omega$$

Calculate the power:

$$P_1 = V_1 I_1 = 15 \times 50 = 750 \text{ mW}$$

$$P_2 = V_2 I_2 = 15 \times 35 = 525 \text{ mW}$$

$$P_3 = V_3 I_3 = 10 \times 25 - 250 \text{ mW}$$

$$P_4 = V_4 I_4 = 20 \times 5 = 100 \text{ mW}$$

EXAMPLE 6-12: A voltage divider must be designed for the following loads: 20 V at 20 mA, 25 V at 5 mA, and 45 V at 30 mA. Bleeder current should be 10 mA and the supply voltage 75 V. Calculate the divider string.

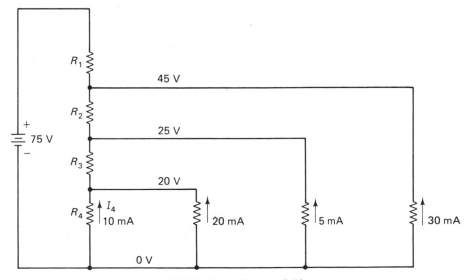

FIGURE 6-9: *Example 6-12.*

SOLUTION: First the schematic is drawn, Fig. 6-9. Then, we calculate the resistances starting at R_4:

$$R_4 = \frac{V_4}{I_4} = \frac{20}{10} = 2 \text{ k}\Omega$$

$$P_4 = V_4 I_4 = 20 \times 10 = 200 \text{ mW}$$

$$R_3 = \frac{V_3}{I_3} = \frac{25 - 20}{10 + 20} = \frac{5}{30} = 166.7 \, \Omega$$

$$P_3 = V_3 I_3 = 5 \times 30 = 150 \text{ mW}$$

$$R_2 = \frac{V_2}{I_2} = \frac{45 - 25}{5 + 30} = \frac{20}{35} = 571.4 \, \Omega$$

$$P_2 = V_2 I_2 = 20 \times 35 = 700 \text{ mW}$$

$$R_1 = \frac{V_1}{I_1} = \frac{75 - 45}{30 + 35} = \frac{30}{65} = 461.5 \, \Omega$$

$$P_1 = V_1 I_1 = 30 \times 65 = 1.95 \text{ W}$$

Changing the Reference Point

As in the unloaded divider, both positive and negative voltages can be supplied if the reference is changed. The only difficulty we will encounter is determining which string resistor has the least amount of current, for this resistor will only have bleeder current flowing through it.

EXAMPLE 6-13: A voltage divider must supply the following: -20 V at 16 mA, -10 V at 3 mA, $+5$ V at 60 mA, and $+15$ V at 20 mA. The bleeder current is 10 mA and the supply voltage is 50 V. Design the divider string.

SOLUTION: First, the schematic is drawn, Fig. 6-10. Note that either R_3 or R_4 must contain the bleeder current only. By applying a dose of Kirchhoff's current law to the 0-V node, we can reveal which resistor has the larger current; then the one with the smaller current must have only bleeder current:

$$I_3 + 60 + 20 = I_4 + 3 + 16$$

$$I_4 - I_3 = 80 - 19 = 61 \text{ mA}$$

Thus, I_4 is greater than I_3 and I_3 must have only the bleeder current, 10 mA. Therefore, I_4 must be

$$I_4 - I_3 = 61 \text{ mA}$$

$$I_4 = 61 + I_3 = 61 + 10 = 71 \text{ mA}$$

Knowing this, we can solve for the resistors:

$$R_3 = \frac{V_3}{I_3} = \frac{5}{10} = 500 \, \Omega$$

$$P_3 = V_3 I_3 = 5 \times 10 = 50 \text{ mW}$$

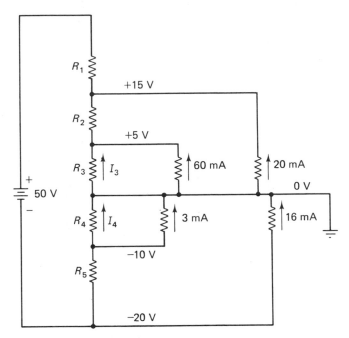

FIGURE 6-10: *Example 6-13, changing the reference point.*

$$R_2 = \frac{V_2}{I_2} = \frac{15 - 5}{10 + 60} = \frac{10}{70} = 142.9 \ \Omega$$

$$P_2 = V_2 I_2 = 10 \times 70 = 700 \ \text{mW}$$

$$R_1 = \frac{V_1}{I_1} = \frac{50 - [15 - (-20)]}{20 + 70} = \frac{15}{90} = 166.7 \ \Omega$$

$$P_1 = V_1 I_1 = 15 \times 90 = 1.350 \ \text{W}$$

$$R_4 = \frac{V_4}{I_4} = \frac{10}{71} = 140.8 \ \Omega$$

$$P_4 = V_4 I_4 = 10 \times 71 = 710.0 \ \text{mW}$$

$$R_5 = \frac{V_5}{I_5} = \frac{20 - 10}{71 + 3} = \frac{10}{74} = 135.1 \ \Omega$$

$$P_5 = V_5 I_5 = 10 \times 74 = 740.0 \ \text{mW}$$

EXAMPLE 6-14: A voltage divider supplies $+9$ V at 100 mA, $+5$ V at 200 mA, -6 V at 120 mA, and -9 V at 250 mA, with a bleeder current of 50 mA and a source of 25 V. Design the divider.

SOLUTION: The schematic is first drawn, Fig. 6-11. Then we determine whether I_3 is greater than I_4:

$$I_3 + 200 + 100 = I_4 + 120 + 250$$

$$I_3 - I_4 = 370 - 300 = 70 \ \text{mA}$$

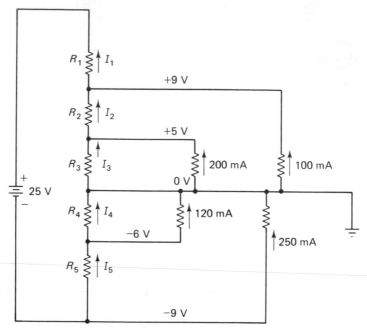

FIGURE 6-11:   *Example 6-14.*

Thus, $I_4 = 50$ mA and

$$I_3 - I_4 = 70 \text{ mA}$$
$$I_3 = 70 + I_4 = 70 + 50 = 120 \text{ mA}$$

Therefore,

$$R_3 = \frac{V_3}{I_3} = \frac{5}{120} = 41.67 \ \Omega$$

$$P_3 = V_3 I_3 = 5 \times 120 = 600 \text{ mW}$$

$$R_2 = \frac{V_2}{I_2} = \frac{9-5}{120+200} = \frac{4}{320} = 12.50 \ \Omega$$

$$P_2 = V_2 I_2 = 4 \times 320 = 1.28 \text{ W}$$

$$R_1 = \frac{V_1}{I_1} = \frac{25 - [9 - (-9)]}{320 + 100} = \frac{7}{420} = 16.67 \ \Omega$$

$$P_1 = V_1 I_1 = 7 \times 0.42 = 2.94 \text{ W}$$

$$R_4 = \frac{V_4}{I_4} = \frac{6}{50} = 120 \ \Omega$$

$$P_4 = V_4 I_4 = 6 \times 50 = 300 \text{ mW}$$

$$R_5 = \frac{V_5}{I_5} = \frac{9-6}{120+50} = \frac{3}{170} = 17.65 \ \Omega$$

$$P_5 = V_5 I_5 = 3 \times 170 = 510 \text{ mW}$$

6-7. SUMMARY

A series–parallel circuit has both series portions and parallel portions. The equivalent resistance of a circuit is that resistance which represents the total resistance of the circuit. Therefore, finding the total resistance of a series–parallel circuit requires finding equivalent resistance for portions of the circuit, then combining these portions until there is only one equivalent resistance, and this represents the total resistance of the circuit. Current and voltage are analyzed in accordance with the series rules and the parallel rules. Power in a series–parallel circuit is equal to the sum of the power dissipated by the individual resistances.

The Wheatstone bridge uses a system of balanced voltages to measure an unknown resistance. Three configurations are used: the loop, the Murray, and the Varley. A loaded voltage divider supplies various voltages to loads from a single voltage source.

6-8. REVIEW QUESTIONS

6-1. What is a series–parallel circuit?

6-2. What is an equivalent resistance?

6-3. Section 6-3 listed a four-step strategy for finding unknown values on a series–parallel network. State the four steps.

6-4. Knowing the power dissipated by each resistor, how is total power found within a series–parallel circuit?

6-5. What is the difference between a balanced bridge and an unbalanced bridge?

6-6. What types of cable faults do the Murray and Varley analyze?

6-7. What is the difference between a loaded voltage divider and an unloaded voltage divider?

6-8. What is a bleeder current?

6-9. PROBLEMS

6-1. In Fig. 6-1, $R_1 = 20$ kΩ, $R_2 = 10$ kΩ, $R_3 = 14$ kΩ, $R_4 = 10$ kΩ, $R_5 = 12$ kΩ, and $R_6 = 5$ kΩ. Find the total resistance.

6-2. In Fig. 6-2(a), $R_1 = 5$ kΩ, $R_2 = 4$ kΩ, $R_3 = 5$ kΩ, $R_4 = 6$ kΩ, $R_5 = 3$ kΩ, and $R_6 = 4$ kΩ. Find the total resistance.

Problems 6-3 through 6-6 refer to Fig. 6-12 using the values specified:

	V_s	I_t	R_1	R_2	R_3	V_1	V_2	I_1	I_2
6-3.	10 V	_____	5.6 kΩ	4.7 kΩ	6.8 kΩ	_____	_____	_____	_____
6-4.	_____	_____	10 kΩ	6 kΩ	4 kΩ	10 V	_____	_____	_____

111

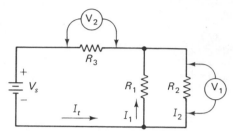

FIGURE 6-12: *Problems 6-3 through 6-6.*

	V_s	I_t	R_1	R_2	R_3	V_1	V_2	I_1	I_2
6-5.	___	___	8 kΩ	___	6.8 kΩ	___	___	3 mA	4 mA
6-6.	50 V	5 mA	___	___	___	35 V	___	3 mA	___

6-7. Compute I_t in Fig. 6-13.

6-8. Compute V_1 and V_2 in Fig. 6-14.

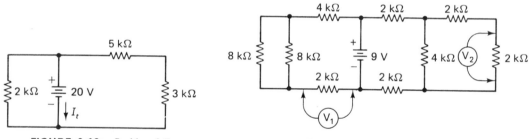

FIGURE 6-13: *Problem 6-7.*   **FIGURE 6-14:** *Problem 6-8.*

6-9. Figure 6-15 shows the signal circuit of a transistor amplifier. Compute V_1.

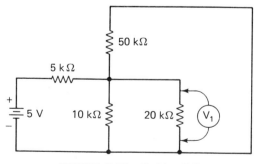

FIGURE 6-15: *Problem 6-9.*

6-10. Compute total power dissipated in Prob. 6-3.

6-11. Compute total power dissipated in Prob. 6-4.

6-12. In the circuit of Fig. 6-5(b), R_1 is 10 kΩ, R_D is 7.63 kΩ, and R_2 is 100 kΩ. Calculate R_x.

6-13. A grounded wire exists on a pair with a loop measured at 365 Ω and a Murray measured at R_1 of 789 Ω and an R_2 of 211 Ω. What is the distance to the fault if number 18 wire is used?

6-14. A Varley measurement reads 972 Ω and the loop 1468 Ω for size 18 wire. What is the distance of the fault from the test end?

6-15. Calculate the string resistors for a loaded voltage divider that supplies 7.5 V at 30 mA, 10 V at 5 mA, and 15 V at 15 mA from an 18-V source. Use a 10-mA bleeder current.

6-16. A loaded voltage divider supplies 80 V at 10 mA, 130 V at 3 mA, and 200 V at 1.5 mA from a 250-V source. Using a bleeder current of 2 mA, calculate the resistor string.

6-17. Calculate the divider string shown in Fig. 6-16.

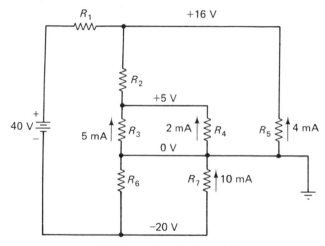

FIGURE 6-16: *Problem 6-17.*

6-18. A divider string supplies −15 V at 10 mA, +5 V at 60 mA, and +15 V at 12 mA. Using a source of 30 V and a bleeder current of 2 mA, design the string.

6-10. *PROJECTS*

6-1. Connect the circuit discussed in Prob. 6-1 and verify the total resistance.

6-2. Connect the circuit discussed in Prob. 6-2 and verify the total resistance.

6-3. Connect the circuit shown in Fig. 6-2(a), using a supply of 10 V. Compute the voltage drops and currents through each resistor, then determine these values experimentally. Do your measured values agree with your computed values?

6-4. Build the Wheatstone bridge shown in Fig. 6-5(b) and use it to measure several resistors. Use a decade box for R_D.

6-5. Build the Varley bridge shown in Fig. 6-7 using $R_1 = R_2 = 10\text{-k}\Omega$ resistors. Simulate a ground fault on a cable and ascertain the location.

6-6. Build a loaded voltage divider that supplies $+5$ V at 20 mA, $+3$ V at 25 mA, $+20$ V at 10 mA, -8 V at 30 mA, and -15 V at 25 mA. Use a bleeder current of 10 mA and whatever source voltage is convenient.

7

SYSTEM HARDWARE

Every electronic system consists of both electronic parts (resistors, capacitors, inductors, and transistors) and those parts that we all have a tendency to ignore—the fuse that stands silent vigil waiting to melt with heat and the fastener that keeps it all together. In this chapter we shall study these silent heros, including:

7-1. Batteries

7-2. Conductors

7-3. Wiring methods

7-4. Switches

7-5. Protective devices

7-6. Indicating devices

7-7. Fasteners

7-8. Miscellaneous hardware

7-1. BATTERIES

Technically, a *battery* is a group of cells, each cell converting chemical energy to electrical energy. One of the most common and most interesting is the automobile battery, also known as the lead–acid storage battery, Fig. 7-1. The positive electrode is formed from a grid of a lead–antimony alloy packed with a substance called lead peroxide (PbO_2). The negative electrode is a similar grid of lead–antimony packed with a spongy type of lead (Pb). The liquid electrolyte is sulfuric acid (H_2SO_4) diluted

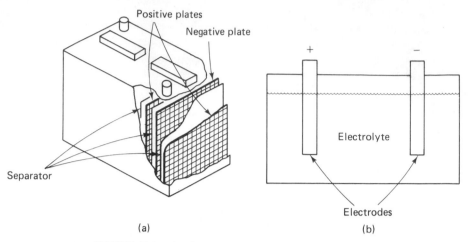

FIGURE 7-1:   *Lead–acid battery: (a) cross section; (b) cell.*

with water (H_2O). The action of discharging is best described using the following chemical equations:

At the negative electrode:

$$Pb \longrightarrow Pb^{2+} + 2e$$
$$Pb^{2+} + SO_4^{2-} \longrightarrow PbSO_4$$

At the positive electrode:

$$PbO_2 \longrightarrow Pb^{2+} + O_2^{2-}$$
$$Pb^{2+} + O_2^{2-} + 2e + 4H^+ + SO_4^{2-} \longrightarrow PbSO_4 + 2H_2O$$

At the negative electrode, lead (Pb) releases its two electrons to the electrode forming a lead ion (Pb^{2+}). This lead ion then combines with a sulfate ion (SO_4^{2-}), which is part of the sulfuric acid in the electrolyte, to form a lead sulfate molecule ($PbSO_4$), a solid, white, crystalline substance that is deposited upon the negative electrode. At the positive electrode, the lead peroxide (PbO_2) separates into a lead ion (Pb^{2+}) and an oxygen ion (O_2^{2-}), which combine with two electrons extracted from the positive electrode, four hydrogen ions (H^+), and one sulfate ion (SO_4^{2-}) to form lead sulfate ($PbSO_4$) and two molecules of water (H_2O). Thus, the negative terminal has two more electrons than it started with, and the positive terminal two less, and a voltage has been generated.

This battery is charged by shoving electrons into its negative terminal and extracting them from the positive terminal, reversing the process.

The lead–acid battery consists of wet cells; that is, the electrolyte is a liquid. A dry cell, however, has a pastelike electrolyte, Fig. 7-2. There are several types available,

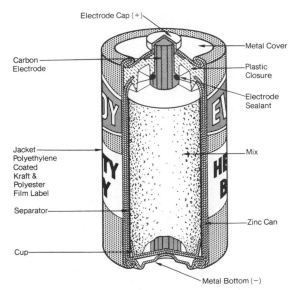

FIGURE 7-2: *Carbon–zinc dry cell (Courtesy, Union Carbide, Inc.).*

using different electrolytes and electrodes, Table 7-1. As a dry cell discharges, gas may form around the electrodes, separating them from the electrolyte and impairing conduction. Most dry cells contain a substance to remove this gas.

Internal Resistance

Every voltage generator, including batteries, has resistance internal to the device. In batteries it is the resistance of the electrolyte and the resistance between the electrolyte and electrodes. Since the battery or cell has this internal resistance, it will have a voltage drop across it, reducing the terminal voltage as shown in Fig. 7-3.

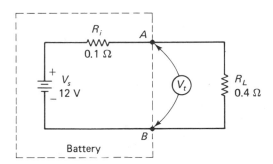

FIGURE 7-3: *Internal resistance of a battery.*

TABLE 7-1: *Wet and dry cells.*

Type	Positive Plate	Negative Plate	Electrolyte	Recharge-able?	Volts/Cell	Dry or Wet	Use	Advantages
Nickel–cadmium	Nickel hydroxide and graphite	Cadmium and iron oxides	Potassium hydroxide	Yes	1.2	Dry	Pocket calculators, electronic flash	Dry, rechargeable
Lead–acid	Lead peroxide	Lead	Sulfuric acid	Yes	2.2	Wet	Automobiles	Large current
Zinc–carbon	Carbon	Zinc	Ammonium and zinc chloride	No	1.5	Dry	Flashlights, toys	Inexpensive
Alkaline	Manganese dioxide	Zinc	Potassium hydroxide and zinc oxide	No	1.5	Dry	Flashlights, toys	Much longer life
Mercury	Mercuric oxide	Zinc	Potassium hydroxide	No	1.35	Dry	Watches, hearing aids	Constant voltage over life of cell
Silver oxide	Silver oxide	Zinc	Potassium hydroxide and zinc hydroxide	No	1.5	Dry	Watches, hearing aids	Higher voltage than mercury type

EXAMPLE 7-1: What is the terminal voltage of the battery shown in Fig. 7-3?

SOLUTION: The total resistance would be

$$R_t = R_i + R_L = 0.1 + 0.4 = 0.5 \, \Omega$$

Therefore, the total current is

$$I_t = \frac{V_s}{R_t} = \frac{12}{0.5} = 24 \text{ A}$$

The voltage drop across R_i is

$$V_i = I_t R_i = 24 \times 0.1 = 2.4 \text{ V}$$

and the terminal voltage is

$$V_t = V_s - V_i = 12 - 2.4 = 9.6 \text{ V}$$

The internal resistance of the battery also limits the amount of current the battery is capable of producing.

EXAMPLE 7-2: Compute the maximum current output by the battery in Fig. 7-3 if its terminals are short-circuited.

SOLUTION: If terminals A and B were shorted,

$$I_t = \frac{V_s}{R_i} = \frac{12}{0.1} = 120 \text{ A}$$

Therefore, the short-circuit current would be 120 A.

7-2. CONDUCTORS

The wires used in electronics come in varied sizes and two common materials, copper and aluminum, with copper being by far the most prevalent. Wire size is specified by an *American wire gauge (AWG) number*, Table 7-2; the higher the number, the smaller the wire. The table refers to several characteristics of wires that we shall discuss: gauge, diameter, area, resistance, weight, and ampacity.

Wire Gauge

The AWG number represents the size of the wire, the higher numbers representing smaller sizes. Note that the cross-sectional area doubles for roughly every three AWG numbers. Although the table lists many wire sizes, as a practical matter only the even-numbered sizes are readily available. When wires larger than zero are needed, 00, 000, and 0000 sizes are available; wires larger than these are measured in millions of circular mils (Mcmil).

TABLE 7-2:   *Copper wire at 20°C.*

Gauge	Diameter (mils)	Cross Section (cmils)	Resistance/ 1000 ft (Ω)	Ampacity[a] (A)
0000	460.2	211,600	0.04901	225
000	409.6	167,800	0.06180	175
00	364.8	133,100	0.07793	150
0	324.9	105,500	0.09827	125
1	289.3	83,690	0.1239	100
2	257.6	66,370	0.1563	90
3	229.4	52,640	0.1970	80
4	204.3	41,740	0.2485	70
5	181.9	33,100	0.3133	55
6	162.0	26,250	0.3951	50
7	144.3	20,820	0.4982	
8	128.5	16,510	0.6282	35
9	114.4	13,090	0.7921	
10	101.9	10,380	0.9989	25
11	90.74	8234	1.260	
12	80.81	6530	1.588	20
13	71.96	5178	2.003	
14	64.08	4107	2.525	15
15	57.07	3257	3.184	
16	50.82	2583	4.016	6
17	45.26	2048	5.064	
18	40.30	1624	6.385	3
19	35.89	1288	8.051	
20	31.96	1022	10.15	
21	28.45	810.1	12.80	
22	25.35	642.4	16.14	
23	22.57	509.5	20.36	
24	20.10	404.0	25.67	
25	17.90	320.4	32.37	
26	15.94	254.1	40.81	
27	14.20	201.5	51.47	
28	12.64	159.8	64.90	
29	11.26	126.7	81.83	
30	10.03	100.5	103.2	

[a] Ampacity varies considerably with insulation used.

Diameter

The mil is 0.001 in.; thus, a $\frac{1}{4}$-in.-diameter wire is 250 mil in diameter (about AWG 2).

Area

Cross-sectional area of wires is measured in circular mils (cmil), with 1 cmil being equal to the area of a circle 1 mil in diameter, or 0.7854 mil². Using this definition, the area of a wire in cmils is

$$A = d^2$$

where A is in cmils and d is the diameter of the wire in mils.

> **EXAMPLE 7-3:** What is the cross-sectional area of a wire 0.5 in. in diameter?
>
> *SOLUTION:* Since 0.5 in. = 500 mils,
>
> $$A = d^2 = 500^2 = 250,000 \text{ cmils}$$

Resistance

The resistance of wire is measured in ohms/1000 ft. Thus, to compute total resistance, we must first find out how many thousands of feet of wire are under consideration.

> **EXAMPLE 7-4:** Compute the resistance of 1 mi of AWG 14 copper wire.
>
> *SOLUTION:* One mile equals
>
> $$1 \text{ mi} = 5280 \text{ ft} = 5.28 \text{ thousands of ft}$$
>
> However, 1000 ft of AWG 14 wire has a resistance (R_{ES}) of 2.525 Ω from the table. Thus,
>
> $$R_t = R_{ES} \times \text{length} = 2.525 \times 5.28 = 13.33 \ \Omega$$

> **EXAMPLE 7-5:** A certain data terminal is installed 200 ft from its signal source, from which it receives signals over AWG 22 wire. What is the resistance of the wire?
>
> *SOLUTION:* Note that the signal must travel to the terminal and back, a distance of 400 ft. Therefore,
>
> $$R_t = R_{ES} \times \text{length} = 16.14 \times 0.40 = 6.456 \ \Omega$$

Ampacity

The larger the diameter of the wire, the more current the wire is capable of handling. It is dangerous to exceed the safe current-carrying capacity (ampacity) of a wire, for many fires have been started by allowing an extension cord or other conductor to supply more current than it can handle. Because of their knowledge of electricity, electronic technicians should be especially alert to this hazard.

The safe current-carrying capacities of wires are dependent upon adequate ventilation or heat conduction away from the wire. If many wires were tied into a bundle, each carrying a large current, heat could build up in the center wires, causing disastrous results. Thus, the environment of the wire must be considered.

Wire size is selected by determining actual current, then selecting the minimum even size capable of carrying this current.

EXAMPLE 7-6: A cord must be selected for an electric iron drawing 1100 W at 120 V. What is the minimum safe size of wire that could be used?

SOLUTION: We must first know the current:

$$I = \frac{P}{V} = \frac{1100}{120} = 9.167 \text{ A}$$

From the wire tables, a size 16 is too small, so we should use a size 14 wire.

EXAMPLE 7-7: A 5-V power supply must provide 17 A to its load. What wire size should be selected?

SOLUTION: From the table, an AWG 12 will supply at least 17 A.

Voltage Drop

Since wire has resistance and it has current flowing through it, there is a voltage drop across the length of any wire. This voltage drop could be substantial or negligible, depending upon its size relative to the drop across the load.

EXAMPLE 7-8: A transmitter receives data from a source 200 ft away over AWG 20 wire. The resistance at the transmitter is 50 Ω and the source provides a 20-mA signal. What is the voltage drop across the wires and across the load, Fig. 7-4?

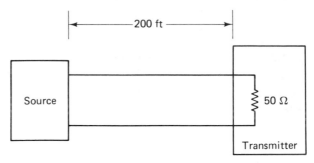

FIGURE 7-4: *Example 7-8.*

SOLUTION: The total length of the wire is 400 ft. Therefore,

$$R = \text{R}_{\text{ES}} \times \text{length} = 10.15 \times 0.4 = 4.060 \text{ Ω}$$

The voltage drop across the wire is

$$V = IR = 0.02 \times 4.060 = 81.20 \text{ mV}$$

Half of this would be across the top wire and half across the bottom wire in the figure. The voltage drop across the load is

$$V = IR = 0.02 \times 50 = 1.000 \text{ V}$$

Thus, the wire voltage drop appears to be insignificant.

EXAMPLE 7-9: An electric saw rated at 9.5 A, 120 V is to be operated using a 300-ft extension cord of AWG 16 wire. What is the actual voltage across the saw if the extension cord is plugged into a 120-V outlet?

SOLUTION: Although the saw is rated at 120 V, 9.5 A, it will actually receive neither because of the voltage drop across the extension cord. The saw's resistance is

$$R = \frac{V}{I} = \frac{120}{9.5} = 12.63 \ \Omega$$

The resistance of the wire is

$$R = R_{ES} \times \text{length} = 4.016 \times 0.6 = 2.410 \ \Omega$$

One-half of this resistance would be in each wire; a schematic is shown in Fig. 7-5.

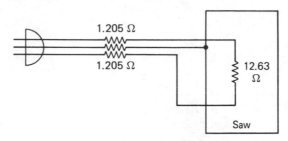

FIGURE 7-5: *Example 7-9.*

The total current would be

$$I = \frac{V}{R} = \frac{120}{2.410 + 12.63} = 7.979 \ \text{A}$$

Note that this is considerably less than the 9.5 A rated. The voltage across the saw would be

$$V = IR = 7.979 \times 12.63 = 100.8 \ \text{V}$$

This is also much less than rated. Just for curiosity, let us compare the actual power with the rated power:

$$P_{\text{rated}} = VI = 120 \times 9.5 = 1140 \ \text{W}$$
$$P_{\text{actual}} = VI = 100.8 \times 7.979 = 804.3 \ \text{W}$$

Thus, the saw is actually operating at 804/1140, or about 70% of its rated power because of the voltage drop across the extension cord.

Types of Wire

Wire is available in several configurations, Fig. 7-6. *Solid conductor wire* is used extensively in consumer electronics, whereas *stranded wire* is used whenever flexibility is desired or the wire is subject to vibration, such as in a car or aircraft. The stranded

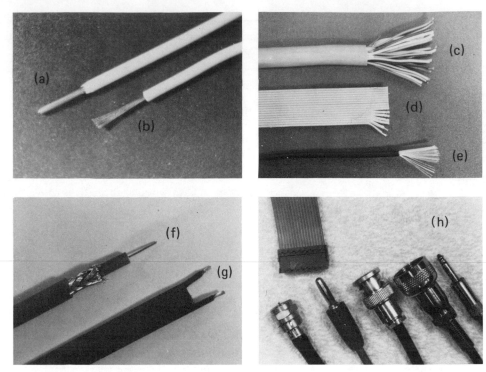

FIGURE 7-6: *Wire types: (a) solid conductor; (b) stranded; (c) cable; (d) ribbon cable; (e) fiber optic cable; (f) coaxial cable; (g) twin lead; (h) cable connectors (photos by Ruple).*

type consists of several very small wires woven together to form a larger wire, with the final size determined by the sum of the cross-sectional areas of the individual strands. For example, a certain brand of size 22 wire consists of 19 strands of size 34 wire; a size 12 of 19 strands of size 25 wire.

Where a number of wires are required, they can be bound into a bundle, called a *cable*, and enclosed in a nonconductive cover, Fig. 7-6(c). Telephone companies use cables extensively throughout their systems. The *ribbon* type of cable is commonly used in computer equipment for connecting one part of the system to another, Fig. 7-6(d).

More recently, *fiber optic* cables have been introduced that carry information in the form of light, rather than electrical current, Fig. 7-6(e).

Coaxial cable is used extensively for radio and television signals. It consists of a center conductor surrounded by an insulator, then a conductor, and, finally, an outside insulator, Fig. 7-6(f).

Twin-lead cable consists of two conductors separated by an insulator, Fig. 7-6(g), and is used for carrying television signals from the antenna to the receiving set.

Cables are connected to equipment and other cables by means of cable connectors, Fig. 7-6(h). These devices allow quick and easy connection and disconnection.

7-3. WIRING METHODS

Wire is fantastic stuff, but it must be connected to form a system. As a result of systems becoming more and more miniaturized, wiring methods have changed in the last few years. In this section we shall discuss the ways in which systems are formed and interconnected.

Soldering

A great many of today's electronic systems use the method of soldering to connect either components or subsystems together, Fig. 7-7. This is a technique that every technician is required to use, so it is essential that he become proficient at it.

Solder is an alloy of tin and lead, 60% tin and 40% lead being common. It melts at a relatively low temperature and makes an excellent, permanent, corrosion-resistant bond between copper conductors, if done properly. Most solder used in electronics comes with a rosin flux core, that is, rosin is embedded in the solder. This flux chemically cleans the conductor leads, allowing the solder to form a good bond. Acid is used as flux in soldering copper pipe together for plumbing. It should *never* be used for electronic work, for it is more corrosive than rosin.

FIGURE 7-7: *Solder connections on a printed-circuit board (photo by Ruple).*

Several types of solder heating devices are available, Fig. 7-8(a)–(c). The soldering iron (50 to 300 W) is used on large power-type connections. The soldering gun (50 to 150 W) has a trigger on the device, heats quite rapidly, and is useful for soldering wires and components to terminals. The pencil iron (25 to 40 W) is used in delicate printed-circuit work, where too much heat will destroy the circuit.

The trick to good soldering is to heat the joint, not the solder. Then, when the joint is hot enough, touch the solder to the component lead (not the iron) and allow it to flow onto the joint. The joint should be immobilized until the solder freezes, forming a shiny, smooth bond. A joint has the correct amount of solder if the component leads are barely visible beneath the solder.

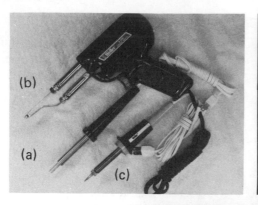

FIGURE 7-8: *Soldering tools: (a) iron; (b) gun; (c) pencil; (d) solder sucker; (e) solder wick (photos by Ruple).*

Two methods are used for removing a soldered component. One uses a "solder sucker," a device that sucks heated solder from the joint leaving it "dry," Fig. 7-8(d).

A fairly recent method has become quite popular using a special wire braid called a *solder wick*, Fig. 7-8(e). The wick is applied to the connection, and both the wick and the connection heated. The solder will then flow from the connection to the wick; by repeating this process, the joint can be dried up of solder and the component lifted out.

Point-to-Point Wiring

In the 1930s, 40s, and 50s, *point-to-point wiring* was the only method used for forming a system, and it is still used in many systems, Fig. 7-9. In such a system, vacuum-tube sockets, transformers, and other components are mounted on a metal box, the chassis, and wires soldered from one part of the system to the other. Resistors and other small components are soldered to either the sockets themselves or to specially mounted terminal strips. When several wires must be routed from one part of the system to another, a special waxed string can be used to tie with wires into

FIGURE 7-9: *Point-to-point wiring (photo by Ruple).*

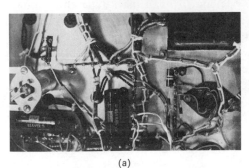

(a)

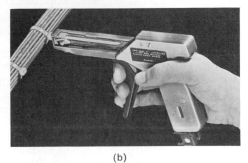

(b)

FIGURE 7-10: *Forming cables: (a) using string (photo by Ruple); (b) using cable wrapping ties (Courtesy, Panduit Corp.).*

cables, Fig. 7-10(a). However, today, plastic tie-wraps that are installed using a special "gun" have replaced hand tying, Fig. 7-10(b).

When many wires need to be plugged into or unplugged from a system, connectors are used, Fig. 7-11. Wires used to all be soldered to the individual pins of these con-

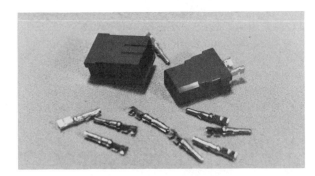

FIGURE 7-11: *Solder- or crimp-type connectors (photo by Ruple).*

nectors, but now a special tool can be used to clamp the pin to the wire using mechanical pressure; the pin is inserted into the connector, where it is locked into place.

The amount of point-to-point wiring has been drastically reduced by the introduction of the printed-circuit card, which is discussed next.

Point-Circuit-Card Fabrication

Printed-circuit (pc) cards are now used almost universally to mount small components. The pc card itself is formed of phenolic or glass laminate, usually $\frac{1}{8}$ or $\frac{1}{16}$ in. thick, with one or both sides coated with copper material, 0.00135 in. being common. The glass laminate material is becoming more common because of its superior mechanical quality and declining relative cost.

The manufacturing of a pc card for mounting components starts with a draftsman, who draws a schematic from the engineer's sketches, Fig. 7-12(a). Next, he designs the physical layout of the board by laying strips of black tape and component patterns representing the conductive paths on a Mylar base, Fig. 7-12(b). This *tape up*, as it is called, is then photographed and reduced.

Next, the copper side of the card is coated with a hard, paintlike substance called *resist*. There are two ways by which the resist can be applied: as photoresist and by silk screen. In the photoresist method, the resist is photosensitive; therefore, the photographic negative is placed on top of the coated copper and the assembly exposed to light, where the exposed resist hardens. Finally, the resultant board is washed in a solvent such as trichlorethane, which dissolves the resist in those areas unexposed to the light, leaving only those parts of the board which are to be conductors.

The second method requires making a silk screen, a device very similar to a screen door except that it has very fine holes. Formed using a photographic process, the screen is coated with a film of plastic where the card is to be nonconductive, and uncoated where a conductive path is to be formed on the pc card. This screen is then placed on the copper surface of the pc card and a special type of ink is squeezed onto it. Where there is to be a conductive path, the ink passes through the tiny holes in the screen and flows onto the card, filling in the gaps formed by the threads of the screen, resulting in solid, painted lines of resist on the copper surface.

Once a card has been coated with resist, it is placed in an etchant bath, such as ferric chloride, where all the copper is dissolved except for those paths formed by the resist. The card is then cleaned using chemicals and/or abrasives, a very important step if the solder is to properly bond to the card.

After a card has been made, in many companies it is coated with a protective paint, leaving only the soldering connections exposed, Fig. 7-12(c). Finally, holes are drilled to accommodate the components and *assemblers* then place the components on the card, Fig. 7-12(d). The parts are then soldered to the card by hand or by wave soldering machine, Fig. 7-12(e). This machine produces a continuous wave of solder and the card is merely passed over it. The underside of the finished card is shown in Fig. 7-12(f).

Where higher component densities are required, printed circuit cards can be

(a)

(b)

(c)

(d)

(e)

(f)

FIGURE 7-12: *Printed-circuit-card fabrication: (a) logic diagram; (b) tape up;*
(c) drilled and painted card; (d) component insertion; (e) flow solder machine,
(f) finished card (solder side) (photos by Vento).

printed on two surfaces. Where even higher densities are needed, special very thin pc cards can be bonded together to form a multilayer card, Fig. 7-13.

Printed-Circuit-Card Installation

Printed circuit cards can be installed into a system in two ways: point-to-point wiring and plug-in. Where point-to-point wiring is used, terminals are installed on the card and wires soldered to the terminals.

129

FIGURE 7-13: *Six-layered printed-circuit card (photo by Ruple).*

(a)   (b)

FIGURE 7-14: *System assembly: (a) card cage (photo by Vento); (b) wire wrap (Courtesy, Gardner-Denver Co.).*

Cards can also be plugged into connectors, and wires run from connector to connector to form the system, Fig. 7-14. These wires may be either soldered to the connectors or wire-wrapped. Wire wrapping is the usual method used in computer-type systems. A special gun is used to wrap an AWG 28 or 30 wire around a 0.025-in.² terminal post, resulting in a very tight, reliable connection. Since wire wrapping is much faster than conventional soldering, it is being used for an increasing number of systems.

Integrated Circuits

This section would be incomplete unless we mentioned the importance of *integrated circuits* (ICs) in reducing the size of systems. Instead of large, expensive, hand-wired circuits, many have been converted to large-scale integrated (LSI) circuits containing thousands of transistors in one package. Figure 7-15 illustrates three stages in the reduction of a circuit, this one a microprocessor. The first stage was that

FIGURE 7-15: *Evolution of the Motorola 6800 Micro-processor (Courtesy, Motorola Semiconductor Products, Inc.).*

of small-scale integrated (SSI), circuit, with fewer than 12 transistors per IC package. Next, medium-scale integration was used, with between 12 and 100 transistors per package; and finally, LSI was used—all for the same circuit performing the same function. We should expect to see many more LSI circuits in the future, further reducing the component count and wiring of these electronic systems.

7-4. SWITCHES

Switches are used to provide an open or a closed path by the equipment operator. Several types of switches in common use are shown in Fig. 7-16. The *slide switch* is the least expensive and the least reliable. The *toggle switch* and the *pushbutton switch*

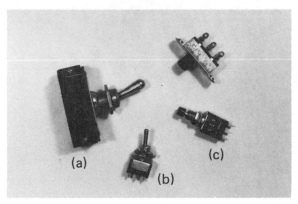

FIGURE 7-16: *Switches: (a) slide; (b) toggle; (c) push-button (photos by Ruple).*

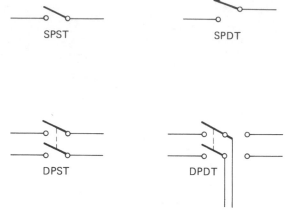

FIGURE 7-17: *Switch schematics.*

are both popular, reliable devices. The pushbutton switch is available in two configurations. The momentary type remains in its switched position until released, when it reverts back to its original position, similar to the starter switch on a car. The alternate action type switches positions every time the button is depressed and released. Press it once and it will go to position A; press it again and it will go to B; another time and it will go to A; and so forth. A power switch on an electronic system would be of this type.

The number of positions that a wire fed the switch can be switched to is called the *number of throws*, Fig. 7-17. Thus, a single-pole double-throw (SPDT) switch can be thrown to one of two positions. The number of independent conducting paths controlled by the switch is called the *number of poles*. A double-pole single-throw (DPST)

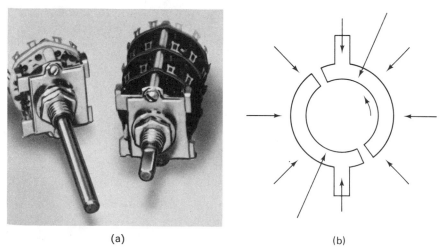

(a)   (b)

FIGURE 7-18: *Rotary switch: (a) photograph (Courtesy, Centralab, Electronics Division, Globe-Union, Inc.); (b) schematic.*

switch controls two independent conducting paths. The dashed line in the schematic represents a mechanical (but not an electrical) connection.

A *rotary switch* is used whenever more than two throws are required, Fig. 7-18. With this type one can select many different positions, each making a separate contact. Additional *wafers*, as each section is called, can be added so that many poles can be switched at once. The variety of contact arrangements on a rotary switch is virtually endless.

7-5. PROTECTIVE DEVICES

Two basic types of *protective devices* are used to prevent components from being destroyed or wires from getting too warm and causing a fire—fuses and circuit breakers. *Fuses*, Fig. 7-19, are used extensively throughout electronic equipment to prevent damage to a power supply by a component that may short the output. They operate by melting a wire inside the device and opening the circuit if current exceeds the rated value of the fuse. Rated in amperes, fuses are available as slow-blow or fast-blow. The slow-blow type will take a surge of current without blowing out, whereas fast-blow fuses react more quickly to an overload.

Circuit breakers are also used, Fig. 7-20, and open the circuit when the current exceeds the rated value. They may then be reset to the closed position if the fault has

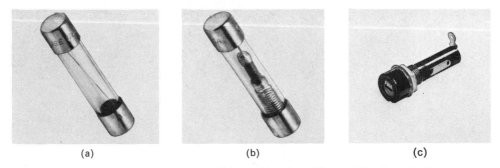

| (a) | (b) | (c) |

FIGURE 7-19: *Fusing devices: (a) fast-blow fuse; (b) slow-blow fuse; (c) fuse holder (Courtesy, Littelfuse, Inc.).*

| (a) | (b) |

FIGURE 7-20: *Heavy-duty protection equipment: (a) circuit breaker; (b) ground-fault interrupter (photos by Ruple).*

been cleared. A recent addition to the protection field is the ground-fault-interrupter (GFI) type of circuit breaker. Not only does it act as an ordinary circuit breaker, but it is a valuable device in preventing electrical shock. The GFI, in essence, measures the current in a grounded circuit leaving the source over the hot lead and the current returning to the source over the return lead. If these two are not the same, it will open the circuit. This would occur if someone were to touch the hot lead and a ground point and the current returned to the source through the ground.

7-6. INDICATING DEVICES

Indicating devices are used to supply information about the system to the operator. Small, incandescent light bulbs called *pilot lamps* are used for lighting dials at night or providing off–on status, Fig. 7-21(a). Neon lamps are also used, for they require very little current and have no filament to deteriorate with time, Fig. 7-21(b).

Light-emitting diodes (LEDs) are very popular indicators, Fig. 7-21(c), and are semiconductor devices (see Chapter 20) that emit light when electrical current is passed through them. Several LEDs can be formed into a matrix of dots that displays

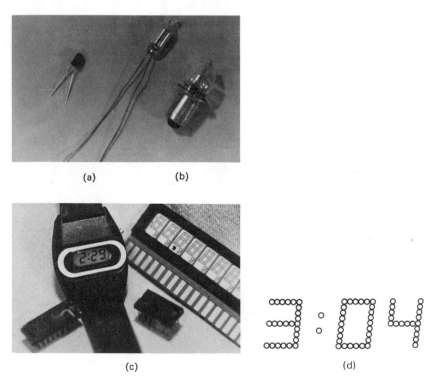

(a)   (b)

(c)   (d)

FIGURE 7-21: *Display devices: (a) incandescent lamp; (b) neon lamp; (c) light-emitting diode; (d) LED numeric display; (e) liquid crystal; (f) plasma display (photos by Ruple).*

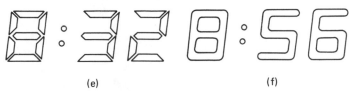

(e)   (f)

FIGURE 7-21:   *(Cont.)*

letters, numbers, or other characters; hand-held calculators use this type extensively, Fig. 7-21(d).

Liquid crystal displays (LCDs) are now used extensively for electronic watches. Often requiring less power than do watch electronics, they merely reflect incident light on a dark background, much as the moon on a clear night.

A third type of numeric indicator is the *plasma display.* Very similar to a neon tube, it uses very little power and provides a crisp image.

7-7. FASTENERS

A *fastener* is a device that holds two materials together. Screws and nuts are used as fasteners on electronic systems and we shall discuss four specifications of them: diameter, threads per inch, head type, and length. The diameter is expressed in a number from zero to 12, Table 7-3, with the larger sizes having the larger number. The number of threads per linear inch of shaft, Fig. 7-22(a), is also shown. Thus, a 6-32 screw is a number 6 diameter screw with a shaft consisting of 32 threads per inch.

Screws are available in flat, oval, and round heads accommodating Phillips, slotted, or hex screwdrivers. The technician must always be careful to use the proper screwdriver size; otherwise, the head of the screw may be damaged. Screw lengths are available in many sizes, ranging from $\frac{1}{4}$ in. to 3 in. or greater.

TABLE 7-3:   *Commonly used screws.*

Size	Diameter (in.)	Threads/Inch
0	0.060	80
1	0.073	64
2	0.086	56
3	0.099	48
4	0.112	40
5	0.125	40
6	0.138	32
8	0.164	32
10	0.190	24
12	0.216	24

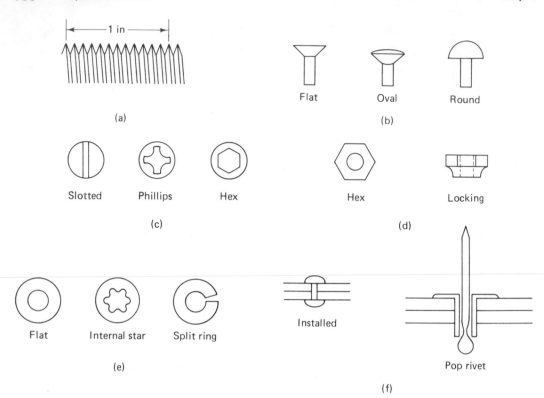

FIGURE 7-22:   *Fasteners: (a) 10 threads per inch; (b) head types; (c) head cuts; (d) nuts; (e) washers; (f) rivets.*

Nuts are available in the same sizes as the screws, such as 4-40 or 6-32. Locking nuts have mechanisms preventing the nut from unscrewing by itself. Three types of washers are popularly used as shown in Fig. 7-22(e). The flat washer is used to provide strength to the connection; the locking washers prevent the nut from unscrewing by itself.

Rivets are another very common type of fastener, Fig. 7-22(f). When installed, they form an excellent bond. The pop rivet is a type that is installed as shown, then the center, nail-like shaft pulled upward, forming a flange at the bottom of the device, until the weak point in the shaft breaks.

7-8.   MISCELLANEOUS HARDWARE

There are many other devices used in the assembly of electronic equipment, some of which are shown in Fig. 7-23. *Grommets* are used to prevent wires chafing when they pass through a hole in a chassis. Numerous types of terminals and terminal strips furnish points to which components can be soldered. *Sockets* are used to permit easy removal of components. However, many transistors and ICs are soldered directly

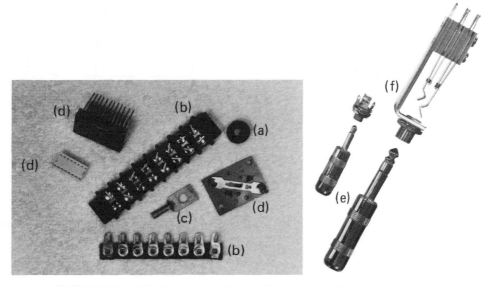

FIGURE 7-23: *Miscellaneous hardware:* (a) *grommet;* (b) *terminal strip;* (c) *solder lugs;* (d) *sockets (photo by Ruple);* (e) *plugs;* (f) *jacks [photo for parts (e) and (f), Courtesy, Switchcraft®, Inc., Chicago, Illinois].*

to the pc board, for not only does the cost of the socket exceed the cost of most transistors and ICs, but the socket has a higher failure rate than the component plugged into it.

Plugs and *jacks* are available in several sizes and shapes, allowing quick disconnect of portions of the system. A plug is the male component; the jack is the female counterpart.

7-9. SUMMARY

Batteries convert chemical energy into electricity and contain a positive electrode, a negative electrode, and an electrolyte, the liquid or paste substance in which the electrodes are placed. Batteries have internal resistance, causing the terminal voltage to be less than the open-circuit voltage.

Conductors are sized according to AWG number, with the larger numbers representing the smaller wires. Wire specifications include size, resistance per 1000 ft, ampacity, and type of material. Many types are available, including solid conductor, stranded, cables, coaxial cables, twin-lead, and ribbon. These may be soldered using point-to-point wiring or wire-wrapped using printed-circuit cards. Pc cards greatly reduce the component-to-component wiring by developing conducting paths using a photographic process. ICs reduce the amount of wiring to an even greater extent, cutting the cost of the system.

Slide, toggle, pushbutton, and rotary switches provide a variety of available throws and poles. Fuses, circuit breakers, and the GFI provide current protection. Pilot lamps and LEDs supply indications of system status to the operator.

Screws are specified by length, diameter, and threads per inch, and locking or nonlocking washers are available. Grommets, sockets, plugs, and jacks are also used extensively.

7-10. REVIEW QUESTIONS

7-1. What are the three parts of a cell?

7-2. What is the difference between a cell and a battery?

7-3. Why can some batteries be recharged and others cannot?

7-4. What effect does internal resistance of a battery have upon terminal voltage?

7-5. Is a size 16 wire larger or smaller than a size 18 wire?

7-6. Define a mil; a circular mil.

7-7. What is the ampacity of a wire?

7-8. Why does a wire have a voltage drop? What is its effect upon load voltage?

7-9. What is stranded wire? Solid conductor wire?

7-10. What is ribbon cable? Coaxial cable? Twin lead?

7-11. What type of flux is used for soldering electronic components?

7-12. What is a soldering pencil used for?

7-13. Where should heat be directed to obtain a good solder joint?

7-14. Name two methods of unsoldering a joint.

7-15. What is tie wrap and what two types are available?

7-16. What is a tape-up of a pc card?

7-17. What is a silk screen used for in pc card fabrication?

7-18. What is a wave solder machine used for?

7-19. What is wire wrap used for?

7-20. What are the differences among SSI, MSI, and LSI?

7-21. Name four types of switches.

7-22. What is a double-throw switch?

7-23. What is a double-pole switch?

7-24. Name two basic types of fuses.

7-25. What is a GFI?

7-26. Name three types of indicating devices.

7-27. What is a 4-40 screw?

7-28. Name three shapes of screw heads.

7-29. Name three types of slotting in screw heads.

7-30. What is a grommet used for?

7-31. Why don't all transistors have sockets?

7-32. What is the difference between a plug and a jack?

7-11. PROBLEMS

7-1. A 16-V battery has an internal resistance of 0.2 Ω and 8 A flowing through it. What is its terminal voltage?

7-2. Compute the short-circuit current of the battery in Prob. 7-1.

7-3. A 12-V battery with an internal resistance of 0.15 Ω is in series with a 1-Ω load. Compute the total current.

7-4. A 1.5-V battery has an internal resistance of 0.25 Ω and is in series with a 1.5-Ω load. Compute the total current.

7-5. What is the cross-sectional area in cmil of a wire 0.100 in. in diameter?

7-6. What is the cross-sectional area in cmil of a wire 0.150 in. in diameter?

7-7. Compute the resistance at 20°C of a 20 gauge wire 3 mi long.

7-8. Compute the resistance at 20°C of an 0000 wire 3 mi long.

7-9. A certain terminal is 200 ft from its signal source. What is the resistance at 20°C of the size 24 wire used for the signal path and return?

7-10. A 5-V power supply is 40 ft from its load. What is the resistance at 20°C of the size 18 wires used for the pair of wires feeding the load?

7-11. Wire must be selected to feed a 10-kW electric dryer operating on 220 V. What size should be used?

7-12. A size 18 extension cord is used to supply a 1400-W 120-V frypan. It is safe?

7-13. A video terminal receives data from a source 450 ft away over AWG 18 wire. The source provides a 60-mA current. What is the voltage drop across the wire?

7-14. A source supplies 5 V to a 40-ft pair of size 22 wires for a load rated at 5 V, 1 A. What is the actual voltage across and current through the load?

7-15. A 120-V generator supplies current to a 250-ft 14 gauge extension cord for use on an electric heater rated at 1500 W, 120 V. What is the actual wattage dissipated by the heater when so connected? How much power is dissipated in the wires?

7-12. PROJECTS

7-1. It has been said that recharging a mercury cell can cause an explosion. Can it and, if so, why?

7-2. Measure the internal resistance of a dry cell, and compute the short-circuit current.

7-3. Determine the wire size of five wires given to you by your instructor.

7-4. What types of insulations are used for wires and what is the characteristic of each?

7-5. How much current is needed to melt a size 14 wire according to the wire tables? Why is rated current so much lower than melting current?

7-6. Measure the voltage drop across 100 ft of size 30 wire carrying 100 mA.

7-7. Compare cmil per rated ampere of size 000, 0, 4, 8, 12, and 16 wires. Why is it not a constant?

7-8. Determine the wire color code used in telephone company cables.

7-9. Wire wrap a circuit given to you by your instructor.

7-10. Solder a circuit given to you by your instructor.

7-11. Solder the printed circuit card given to you by your instructor.

7-12. Investigate the chemical process used in etching pc cards.

7-13. What is a snap-action switch and what are its advantages?

7-14. What materials are used for switch contacts and why?

7-15. What are the voltage, current, and power requirements of incandescent, neon, and LED pilot lamps?

8

NETWORK ANALYSIS

Thus far, we have been using only Ohm's law for solving for unknown values within circuits. However, there are many cases where this method cannot be used and we must rely upon those presented in this chapter. In addition, many of the problems that we have been solving using Ohm's law could more easily have been solved using these methods. Thus, these techniques provide more tools that we can use in analysis of circuits, and, like the mechanic, the more tools we have, the better we can suit the tool to the job.

In this chapter we shall examine six methods of circuit analysis. The first two, loops and nodes, are based upon Kirchhoff's laws, and are used when many unknown values must be determined. The third and fourth, Thévenin's and Norton's theorems, allow an entire circuit—any circuit—to be considered as a single source and a single resistance. These are extremely important in analysis of transistor circuits. The fifth method, superposition, is a shortcut method for finding current in a circuit branch when multiple supplies are present, and the sixth method, wye–delta, is a cookbook method for converting one three-resistor circuit into another, equivalent three-resistor circuit. The sections within this chapter include:

8-1. The loop method

8-2. The nodal method

8-3. Thévenin's theorem

8-4. Norton's theorem

8-5. The superposition method

8-6. Wye–delta transformations

8-1. THE LOOP METHOD

This method of circuit analysis is based upon Kirchhoff's voltage law and is especially useful when:

1. There are few parallel branches; or

2. There are multiple supplies.

Figure 8-1(a) illustrates such a circuit.

The first step in solving a loop problem is to assign algebraic current designators to all branches, such as I_1, I_2, and I_3. Remember, it is unimportant which direction of current flow we assume, for if the sign of the answer is positive, our assumed direction will have been correct; if negative, our assumed direction will have been incorrect. In assigning currents we should assign a minimum number of unknowns by combining them whenever possible. Note that, instead of assigning the bottom branch I_3, a combination of I_1 and I_2 was used, Fig. 8-1(b). Note further that it is easier to assign currents by looking at the nodes rather than at the resistances and power supplies.

Next, we shall follow Kirchhoff's voltage law around a complete path, Fig. 8-1(c). There are three that we might choose: the top closed path, the bottom closed path, or all around the outside of the circuit; each path is referred to as a *mesh*. We must choose the same number of meshes as there are unknowns within the circuit; since there are two unknowns, I_1 and I_2, we must choose two meshes. We shall arbitrarily choose M_1 and M_2.

We have used Kirchhoff's voltage law before around a closed path. However, we now have an additional dimension, for there may be times that the direction of the mesh opposes the direction of the current. When this happens, we shall call a resistor voltage drop positive instead of negative. One way to remember this is that if we were to affix polarity signs to the resistors in accordance with the assumed current flows, Fig. 8-1(d), we would encounter a plus sign whenever we travel through a resistor in a direction that is opposite to current flow. Thus, our signs will be totally determined by the sign we first encounter when we travel the meshy route.

Let us now develop the equation for mesh M_1 starting at point A:

$$\text{①} \quad -16 - 5I_1 + 4I_2 - 12 = 0$$

Note that resistor voltage drops are equal to the resistance multiplied by the current. Thus, the voltage across R_1 is

$$V_{R1} = IR_1 = I_1R_1 = 5I_1$$

Note further that the voltage drop across R_2 is taken as positive because our mesh direction was opposite the assumed current direction, I_2. To prevent mixing up simultaneous equations, they will be labeled with a circle, as ① for mesh M_1.

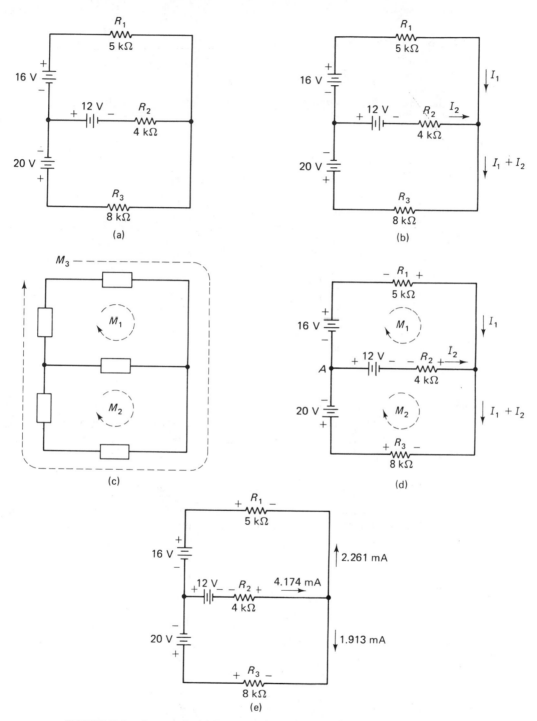

FIGURE 8-1: *Loop method: (a) original circuit; (b) assigning currents; (c) possible meshes; (d) ready for computation; (e) actual currents and voltage polarities.*

Follow M_2 from point A:

$$\textcircled{2} \quad +12 - 4I_2 - 8(I_1 + I_2) + 20 = 0$$

Note that the current through R_3 is $I_1 + I_2$. Rearrange these equations and reduce $\textcircled{2}$:

$$\textcircled{1} \quad 5I_1 - 4I_2 = -28$$
$$\textcircled{2} \quad 2I_1 + 3I_2 = 8$$

Solve for I_1 and I_2

$$I_1 = \frac{\begin{vmatrix} -28 & -4 \\ 8 & 3 \end{vmatrix}}{\begin{vmatrix} 5 & -4 \\ 2 & 3 \end{vmatrix}} = \frac{-52}{23} = -2.261 \text{ mA}$$

$$I_2 = \frac{\begin{vmatrix} 5 & -28 \\ 2 & 8 \end{vmatrix}}{23} = \frac{96}{23} = 4.174 \text{ mA}$$

The current through R_3 is

$$I_{R3} = I_1 + I_2 = -2.261 + 4.174 = 1.913 \text{ mA}$$

Since the sign of I_1 was negative, actual current flow is opposite to that we assumed, Fig. 8-1(e); the others were assumed correctly. Note that with the currents computed, all the voltage drops can be obtained, the actual polarities being determined by the actual current directions.

EXAMPLE 8-1: Compute the voltage across R_1, R_2, and R_3 of Fig. 8-2(a).

SOLUTION: The circuit is shown ready for mathematical analysis in Fig. 8-2(b). Since there are three unknowns, three meshes must be used. First we analyze M_1,

$$\textcircled{1} \quad 4.7I_2 - 2.7I_3 - 5.1I_3 = 0$$
$$\textcircled{1} \quad 0I_1 + 4.7I_2 - 7.8I_3 = 0$$

Then, analyze M_2,

$$\textcircled{2} \quad -4.7I_2 - 2.4(I_2 + I_3) + 3.3I_1 = 0$$
$$\textcircled{2} \quad 3.3I_1 - 7.1I_2 - 2.4I_3 = 0$$

Next, analyze M_3,

$$\textcircled{3} \quad 30 - 3.3I_1 - 1(I_1 + I_2 + I_3) = 0$$
$$\textcircled{3} \quad 4.3I_1 + 1I_2 + 1I_3 = 30$$

Solve for the currents:

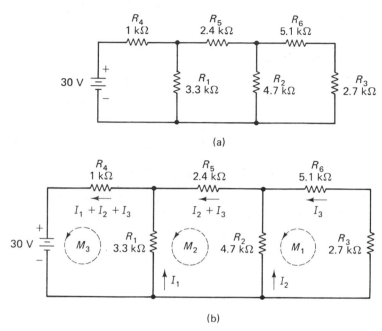

FIGURE 8-2: *Example 8-1: (a) original circuit; (b) ready for computation.*

$$I_1 = \frac{\begin{vmatrix} 0 & 4.7 & -7.8 \\ 0 & -7.1 & -2.4 \\ 30 & 1 & 1 \end{vmatrix}}{\begin{vmatrix} 0 & 4.7 & -7.8 \\ 3.3 & -7.1 & -2.4 \\ 4.3 & 1 & 1 \end{vmatrix}} = \frac{-1999.8}{-327.89} = 6.099 \text{ mA}$$

$$I_2 = \frac{\begin{vmatrix} 0 & 0 & -7.8 \\ 3.3 & 0 & -2.4 \\ 4.3 & 30 & 1 \end{vmatrix}}{-327.89} = \frac{-772.2}{-327.89} = 2.355 \text{ mA}$$

$$I_3 = \frac{\begin{vmatrix} 0 & 4.7 & 0 \\ 3.3 & -7.1 & 0 \\ 4.3 & 1 & 30 \end{vmatrix}}{-327.89} = \frac{-465.3}{-327.89} = 1.419 \text{ mA}$$

Given this, the voltages would be

$$V_{R1} = I_1 R_1 = 6.099 \times 3.3 = 20.13 \text{ V}$$
$$V_{R2} = I_2 R_2 = 2.355 \times 4.7 = 11.07 \text{ V}$$
$$V_{R3} = I_3 R_3 = 1.419 \times 2.7 = 3.831 \text{ V}$$

The Distribution Problem

One problem that is often encountered is that of distribution of both positive and negative voltages over three wires to a load, Fig. 8-3(a). In this figure, R_1 and R_2 represent the load and R_3, R_4, and R_5 the resistance of the wire connecting the load to the supply.

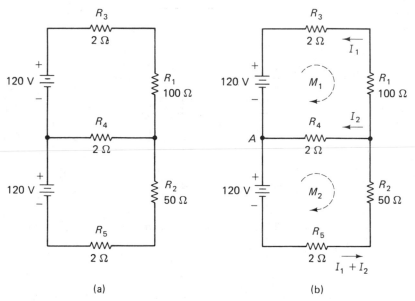

(a)   (b)

FIGURE 8-3: *Example 8-2, voltage distribution: (a) original circuit; (b) ready for computation.*

EXAMPLE 8-2: Compute the voltages across the loads R_1 and R_2 of Fig. 8-3(a).

SOLUTION: Figure 8-3(b) shows the assumed currents, polarities, and mesh directions. Start at point A; from mesh M_1,

ⓐ $\quad -120 + 2I_1 + 100I_1 - 2I_2 = 0$

ⓐ $\quad 102I_1 - 2I_2 = 120$

From mesh M_2,

ⓑ $\quad 2I_2 + 50(I_1 + I_2) + 2(I_1 + I_2) - 120 = 0$

ⓑ $\quad 52I_1 + 54I_2 = 120$

Solve for the branch currents:

$$I_1 = \frac{\begin{vmatrix} 120 & -2 \\ 120 & 54 \end{vmatrix}}{\begin{vmatrix} 102 & -2 \\ 52 & 54 \end{vmatrix}} = \frac{6720}{5612} = 1.197 \text{ A}$$

146

$$I_2 = \frac{\begin{vmatrix} 102 & 120 \\ 52 & 120 \end{vmatrix}}{5612} = \frac{6000}{5612} = 1.069 \text{ A}$$

$$I_1 + I_2 = 1.197 + 1.069 = 2.266 \text{ A}$$

Thus, the load voltage drops are

$$V_{R1} = I_1 R_1 = 1.197 \times 100 = 119.7 \text{ V}$$
$$V_{R2} = (I_1 + I_2)(R_2) = 2.266 \times 50 = 113.3 \text{ V}$$

Note that all currents and voltages were assumed correctly.

Let us now assume that R_1 is much lower and observe V_{R1} and V_{R2}.

EXAMPLE 8-3: Assume that R_1 is 4 Ω and repeat Example 8-2.

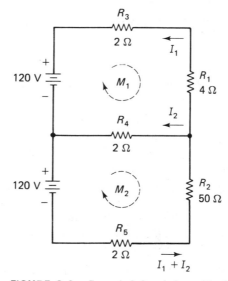

FIGURE 8-4: *Example 8-3, unbalanced loads.*

SOLUTION: The redrawn circuit is shown in Fig. 8-4.
Analyze mesh M_1:

   ① $-120 + 2I_1 + 4I_1 - 2I_2 = 0$
   ① $6I_1 - 2I_2 = 120$

Analyze mesh M_2:

   ② $2I_2 + 50(I_1 + I_2) + 2(I_1 + I_2) - 120 = 0$
   ② $52I_1 + 54I_2 = 120$

Solve for currents:

$$I_1 = \frac{\begin{vmatrix} 120 & -2 \\ 120 & 54 \end{vmatrix}}{\begin{vmatrix} 6 & -2 \\ 52 & 54 \end{vmatrix}} = \frac{6720}{428} = 15.701 \text{ A}$$

$$I_2 = \frac{\begin{vmatrix} 6 & 120 \\ 52 & 120 \end{vmatrix}}{428} = \frac{-5520}{428} = -12.897 \text{ A}$$

$$I_1 + I_2 = 15.701 - 12.897 = 2.804 \text{ A}$$

Solve for V_{R1} and V_{R2}:

$$V_{R1} = I_1 R_1 = 15.701 \times 4 = 62.80 \text{ V}$$

$$V_{R2} = (I_1 + I_2)(R_2) = 2.804 \times 50 = 140.2 \text{ V}$$

Note that, because of the excessive current drawn by R_1, the voltage across R_2 far exceeds 120 V.

8-2. THE NODAL METHOD

Based upon Kirchhoff's current law, the method of nodes is especially useful in analyzing circuits with many parallel branches. Although it can be used with circuits having multiple voltage supplies, the loop method is usually easier to use for this type of problem.

Let us apply the nodal method to Fig. 8-5(a). The first step is to assign algebraic labels to all voltages across all resistors using a minimum number of labels, Fig. 8-5(b). Note that each resistor has a label, and that, instead of labeling the voltage across R_4 as V_2, we can label it in terms of V_1 and the supply voltage. This keeps the number of variables to one instead of two.

Next, we shall need to select as many nodes as we have unknown voltages, in this case, one. The nodes should be selected so that each is completely surrounded by resistors; thus, node A is a good choice, but node B is not, for it is directly connected to the voltage supply.

The third step is to recognize Kirchhoff's current law within each node: the currents flowing into the node must equal the currents flowing out of the node. Finally, each current is set equal to a voltage divided by a resistance:

$$I_1 + I_2 + I_3 = I_4$$

$$\frac{V_1}{R_1} + \frac{V_1}{R_2} + \frac{V_1}{R_3} = \frac{5 - V_1}{R_4}$$

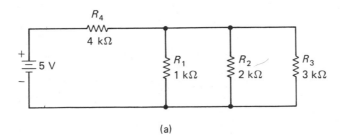

(a)

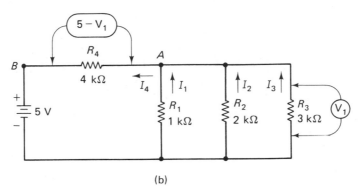

(b)

FIGURE 8-5: *Nodal analysis: (a) original circuit; (b) ready for computation.*

Substitute the known resistance values:

$$\frac{V_1}{1} + \frac{V_1}{2} + \frac{V_1}{3} = \frac{5 - V_1}{4}$$

$$12V_1 + 6V_1 + 4V_1 = 15 - 3V_1$$

$$25V_1 = 15$$

$$V_1 = 0.600 \text{ V}$$

Thus,

$$V_{R4} = 5 - V_1 = 5 - 0.6 = 4.40 \text{ V}$$

Knowing the voltages, each of the branch currents can be found.

> **EXAMPLE 8-4:** Compute the current through the 4-kΩ and 8-kΩ resistors of Fig. 8-6(a).
>
> *SOLUTION:* Current directions and voltages are first assigned, Fig. 8-6(b). Since two variables are required, V_1 and V_2, two nodes must be chosen. Nodes A and B are good choices for they are surrounded by resistors. Analyze node A:

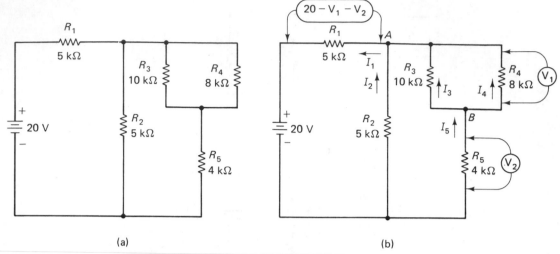

FIGURE 8-6:   *Example 8-4: (a) original circuit; (b) ready for computation.*

① $I_1 = I_2 + I_3 + I_4$

① $\dfrac{20 - V_1 - V_2}{5} = \dfrac{V_1 + V_2}{5} + \dfrac{V_1}{10} + \dfrac{V_1}{8}$

① $160 - 8V_1 - 8V_2 = 8V_1 + 8V_2 + 4V_1 + 5V_1$

① $25V_1 + 16V_2 = 160$

Analyze node B:

② $I_3 + I_4 = I_5$

② $\dfrac{V_1}{10} + \dfrac{V_1}{8} = \dfrac{V_2}{4}$

② $4V_1 + 5V_1 = 10V_2$

② $9V_1 - 10V_2 = 0$

Solve for V_1:

$$V_1 = \frac{\begin{vmatrix} 160 & 16 \\ 0 & -10 \end{vmatrix}}{\begin{vmatrix} 25 & 16 \\ 9 & -10 \end{vmatrix}} = \frac{-1600}{-394} = 4.061 \text{ V}$$

$$V_2 = \frac{\begin{vmatrix} 25 & 160 \\ 9 & 0 \end{vmatrix}}{-394} = \frac{-1440}{-394} = 3.655 \text{ V}$$

Solve for I_4 and I_5:

$$I_4 = \frac{V_1}{R} = \frac{4.061}{8} = 507.6 \ \mu A$$

$$I_5 = \frac{V_2}{R} = \frac{3.655}{4} = 913.8 \ \mu A$$

The Unbalanced Bridge

In Chapter 6 we studied the balanced bridge, where no current flowed through the center branch of the circuit. We shall now consider an unbalanced bridge, Fig. 8-7(a). Note that

$$\frac{R_1}{R_3} \neq \frac{R_2}{R_4}$$

Therefore, current will flow through R_5. Using the nodal method, let us analyze such a circuit.

EXAMPLE 8-5: Compute currents through all the resistors in Fig. 8-7(a).

SOLUTION: First, current directions must be assigned, then voltage designations, Fig. 8-7(b). Note that, by assigning current through R_5 as shown, node A is at a higher voltage than node B. Thus, the voltage across R_3 is $V_4 + V_5$. Note further that the problem is easier to work if the V's are made additive (that is, if V_{R3} were assigned V_3 and V_{R5} were assigned V_5, then V_{R4} would be $V_3 - V_5$). The nodes are then selected for analysis; with only two variables, V_4 and V_5, only two are needed. We shall select nodes A and B, since they are both surrounded by resistors. Analyze node A:

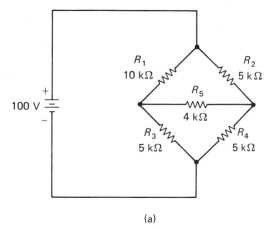

(a)

FIGURE 8-7: *Example 8-5, the unbalanced bridge: (a) original circuit; (b) ready for computation.*

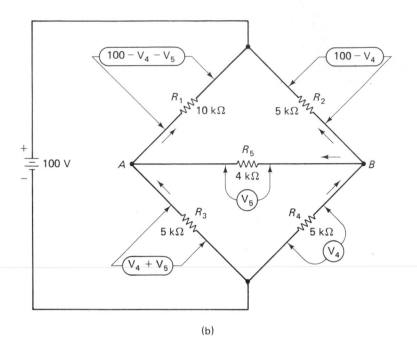

(b)

FIGURE 8-7: (*Cont.*)

① $I_{R3} + I_{R5} = I_{R1}$

① $\dfrac{V_4 + V_5}{5} + \dfrac{V_5}{4} = \dfrac{100 - V_4 - V_5}{10}$

① $4V_4 + 4V_5 + 5V_5 = 200 - 2V_4 - 2V_5$

① $6V_4 + 11V_5 = 200$

Analyze node B:

② $I_{R5} + I_{R2} = I_{R4}$

② $\dfrac{V_5}{4} + \dfrac{100 - V_4}{5} = \dfrac{V_4}{5}$

② $5V_5 + 400 - 4V_4 = 4V_4$

② $8V_4 - 5V_5 = 400$

Solve for V_4:

$$V_4 = \frac{\begin{vmatrix} 200 & 11 \\ 400 & -5 \end{vmatrix}}{\begin{vmatrix} 6 & 11 \\ 8 & -5 \end{vmatrix}} = \frac{-5400}{-118} = 45.763 \text{ V}$$

$$V_5 = \frac{\begin{vmatrix} 6 & 200 \\ 8 & 400 \end{vmatrix}}{-118} = \frac{+800}{-118} = -6.780 \text{ V}$$

Note that the direction of current flow through R_5 was assumed incorrectly, as indicated by the minus sign of V_5. Solve for the resistor currents:

$$I_1 = \frac{V}{R_1} = \frac{100 - V_4 - V_5}{R_1} = \frac{100 - 45.763 + 6.780}{10} = 6.102 \text{ mA}$$

$$I_2 = \frac{V}{R_2} = \frac{100 - V_4}{R_2} = \frac{100 - 45.763}{5} = 10.85 \text{ mA}$$

$$I_3 = \frac{V}{R_3} = \frac{V_4 + V_5}{R_3} = \frac{45.763 - 6.780}{5} = 7.797 \text{ mA}$$

$$I_4 = \frac{V_4}{R_4} = \frac{45.763}{5} = 9.153 \text{ mA}$$

$$I_5 = \frac{V_5}{R_5} = \frac{-6.780}{4} = -1.695 \text{ mA}$$

Thus, I_5 flows from point A to point B.

EXAMPLE 8-6: Compute the value of R_4 in Fig. 8-8(a).

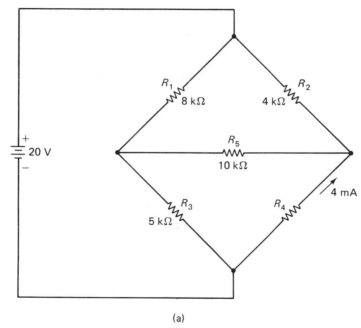

(a)

FIGURE 8-8: *Example 8-8: (a) original circuit; (b) ready for computation.*

SOLUTION: The current directions and voltage designators are first assigned, Fig. 8-8(b). Analyze node A:

    ① $I_{R3} + I_{R5} = I_{R1}$

    ① $\dfrac{V_4 + V_5}{5} + \dfrac{V_5}{10} = \dfrac{20 - V_4 - V_5}{8}$

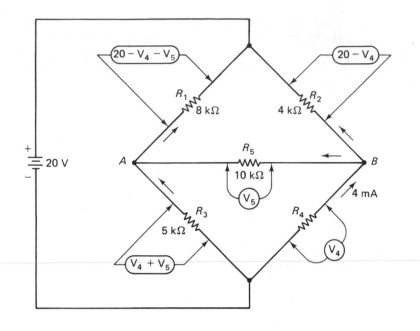

(b)

FIGURE 8-8: *(Cont.)*

$\textcircled{1}$ $8V_4 + 8V_5 + 4V_5 = 100 - 5V_4 - 5V_5$

$\textcircled{1}$ $13V_4 + 17V_5 = 100$

Analyze node B (note that I_{R4} was given):

$\textcircled{2}$ $I_{R4} = I_{R5} + I_{R2}$

$\textcircled{2}$ $4 = \dfrac{V_5}{10} + \dfrac{20 - V_4}{4}$

$\textcircled{2}$ $80 = 2V_5 + 100 - 5V_4$

$\textcircled{2}$ $5V_4 - 2V_5 = 20$

Solve for V_4:

$$V_4 = \frac{\begin{vmatrix} 100 & 17 \\ 20 & -2 \end{vmatrix}}{\begin{vmatrix} 13 & 17 \\ 5 & -2 \end{vmatrix}} = \frac{-540}{-111} = 4.865 \text{ V}$$

Therefore, $R_4 = V_4/I_4 = 4.865/4 = 1.216 \text{ k}\Omega$.

8-3. THÉVENIN'S THEOREM

Thévenin's theorem is both simple to state and simple to implement. It states:

Any two terminal network can be replaced with a Thévenized voltage supply in series with a Thévenized resistance.

Thus, pick any two terminals of any network—it can have as many resistors and as many supplies as you want—and the network can be replaced with a simple voltage supply in series with a single resistor. For example, the very complex network shown in Fig. 8-9(a) as seen by terminals *A* and *B* can be replaced by a V_{th} and an R_{th}, as shown in Fig. 8-9(b). Equipment connected to terminals *A* and *B* of the original circuit could not tell the difference between it and the equivalent circuit.

Not only is the theorem simple to state, it is simple to implement. However, it is easier to picture this implementation as happening on the test bench than on paper. To use the theorem:

1. Remove the load.

2. To find V_{th}, measure the voltage between terminals *A* and *B*.

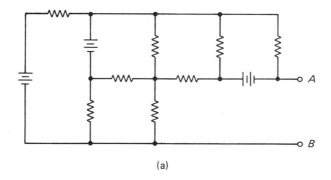

(a)

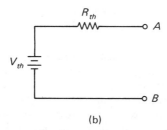

(b)

FIGURE 8-9: *Thévenin's theorem: (a) original network; (b) Thévenized equivalent.*

155

3. To find R_{th}:
   a. Remove all voltage and current supplies, leaving only their internal resistance (if known).
   b. Short the former terminals of each voltage supply and open the terminals of each current supply.
   c. Measure the resistance between A and B; this is R_{th}.

EXAMPLE 8-7: Find the current through R_L of Fig. 8-10(a) using Thévenin's theorem.

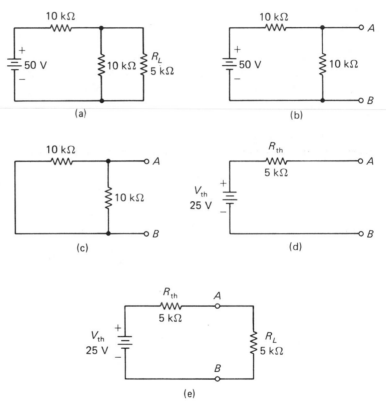

FIGURE 8-10: *Example 8-7, Thévenin's theorem: (a) original circuit; (b) finding V_{th}; (c) finding R_{th}; (d) Thévenized equivalent; (e) loaded equivalent.*

SOLUTION: First, we shall select terminals A and B, then discard the load, R_L, Fig. 8-10(b). To find V_{th}, we must now measure between A and B, obviously 25 V in this circuit. To find R_{th}, remove the 50-V supply and short its terminals, Fig. 8-10(c). Measure R_{th} between A and B, finding 5 kΩ; thus, we have the equivalent circuit shown in Fig. 8-10(d). Since this circuit will react the same as the original circuit to a load, we can now reconnect the load and find the current through it, Fig. 8-10(e), resulting in

$$I = \frac{V}{R} = \frac{25}{5+5} = 2.5 \text{ mA}$$

Thus, the load has 2.5 mA when connected to the original circuit or when connected to the Thévenized circuit.

Thévenin's theorem is very useful for finding the current through the center branch of an unbalanced bridge.

EXAMPLE 8-8: Find the current through R_5 of Fig. 8-11(a).

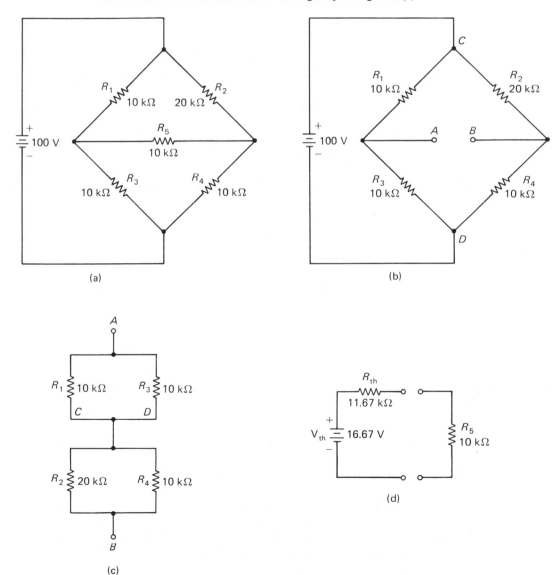

FIGURE 8-11: *Example 8-8, Thévenizing a bridge: (a) original circuit; (b) finding V_{th}; (c) finding R_{th}; (d) Thévenized circuit.*

SOLUTION: We must first select terminals, then discard the load, R_5, Fig. 8-11(b). To find V_{th}, find the difference in voltage between point A and point B; point A is 50 V, point B is

$$I = \frac{V}{R} = \frac{100}{20 + 10} = \frac{100}{30} = 3.333 \text{ mA}$$

$$V_{R4} = IR_4 = 3.333 \times 10 = 33.33 \text{ V}$$

Thus,

$$V_{th} = V_A - V_B = 50 - 33.33 = 16.67 \text{ V}$$

To find R_{th}, remove V_s and short its former terminals, Fig. 8-11(c), resulting in C and D being the same electrical point. We can now compute the resistance, R_{th}:

$$R_{th} = [R_1 \| R_3] + [R_2 \| R_4]$$
$$= 5 + 6.667 = 11.67 \text{ k}\Omega$$

The Thévenized circuit is shown in Fig. 8-11(d). By connecting the load back to terminals A and B, we can compute current through R_5:

$$I = \frac{V}{R} = \frac{16.67}{11.67 + 10} = 769.2 \text{ } \mu\text{A}$$

The single, most important point that should be realized from this discussion is that all circuits can be thought of as a Thévenized voltage source in series with a Thévenized resistance. A 5-V bench power supply (its R_{th} is near zero), a VOM (its V_{th} is zero), the speaker terminals on an amplifier output—all two-terminal devices are merely a V_{th} in series with an R_{th}.

Maximum Power Transfer

Examine Fig. 8-11(d) again. Keeping V_{th} and R_{th} as shown, if R_5 were zero ohms, its power would be

$$P = I^2 R = I^2 \times 0 = 0 \text{ W}$$

Next, assume that R_5 is infinitely large. Current would be zero and the power would be

$$P = VI = V \times 0 = 0 \text{ W}$$

Now we know that when R_5 is nonzero and finite (such as 10 kΩ), it dissipates power. Therefore, there must be a resistance of R_5 at which it will dissipate more power than at any other value of resistance. This turns out to be when the load resistance is set equal to the Thévenized source resistance, R_{th}. This can be restated as:

Maximum power is transmitted from a source to a load when the load resistance is set equal to the Thévenized source resistance.

This principle is very important in all phases of electronics, particularly in amplifiers, where a very small signal must be made into a large signal. At each stage of amplification, the source and load resistances must be matched to achieve the largest possible signal.

8-4. NORTON'S THEOREM

Norton's theorem is closely related to Thévenin's theorem and states:

Any two terminal network can be replaced with a Nortonized current supply in parallel with a Nortonized resistance.

Note that the primary differences between Norton and Thévenin's theorems are that:

1. Norton uses a constant current supply, whereas Thévenin uses a constant voltage supply.
2. Norton has a resistor in parallel with his supply, whereas Thévenin has a resistor in series with his.

The Constant Current Supply

First, let us ask the question: What is a constant current supply? In terms of power supplies, a constant voltage supply is one that provides a constant voltage for a range of loads. A constant current supply is one that supplies a constant current for a wide variety of load resistances. In practice, the constant voltage supply works over a defined range of loads; it would not supply a constant voltage to a short circuit, for example. In a similar manner, the constant current supply is of practical value only over a defined range. We could not expect it to supply a constant current to an open circuit, for example.

Let us now examine the constant current source shown in Fig. 8-12. This particular source will supply 1 mA to its load, R. If R were 1 kΩ, the voltage between A and B would be 1 V; Similarly, Fig. 8-12(b) lists the voltage between A and B for various values of R. Note that in each case, however, the current is 1 mA.

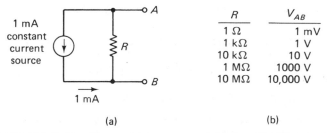

R	V_{AB}
1 Ω	1 mV
1 kΩ	1 V
10 kΩ	10 V
1 MΩ	1000 V
10 MΩ	10,000 V

(a)   (b)

FIGURE 8-12: *Constant current source: (a) schematic; (b) output voltage versus resistance.*

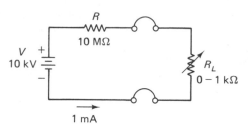

FIGURE 8-13: *Constant current supply from a constant voltage supply.*

The constant current source can be thought of as an infinitely high voltage source in series with an infinitely high resistance, Fig. 8-13. Note that in the figure, the current is exactly 1 mA for an R_L of 0 Ω, and 0.999900 mA for an R_L of 1 kΩ; thus, it is 0.01 % accurate over this range. If E were 10 MV and R 10 GΩ, the accuracy would improve to 0.00001 %.

So, the constant current supply is neat; what good is it? We will find in our study of transistors that the most convenient representation of a transistor is that of a constant current supply in parallel with a resistance. Thus, it is best thought of as a Nortonized equivalent circuit.

Finding the Norton Equivalent

As in Thévenin's theorem, Norton's theorem is both simple to state and simple to implement. The theorem itself states that any two-terminal network can be replaced with a constant current supply, I_n, in parallel with a resistance, R_n. To find I_n and R_n (Fig. 8-14):

1. Remove the load.

2. To find I_n, place an ammeter between terminals A and B and read the meter. Note that the theoretical resistance of an ammeter is zero ohms.

3. To find R_n:
   a. Remove all voltage and current supplies, leaving only their internal resistance (if known).
   b. Short the former terminals of each voltage supply and open the terminals of each current supply.
   c. Measure the resistance, R_n, between terminals A and B.

Note that R_n and R_{th} are identical and are found in the same manner.

    EXAMPLE 8-9: Find the Norton equivalent circuit for Fig. 8-15(a).

    SOLUTION: After discarding the load resistor, terminals A and B are shorted together and the current measured through the short. Measuring would be simpler, but, since

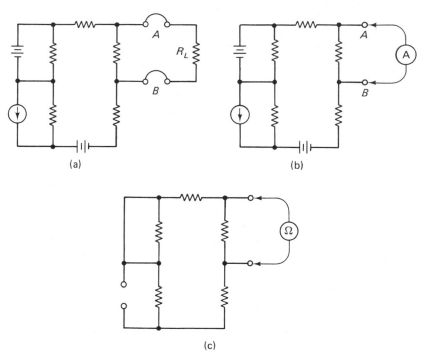

FIGURE 8-14: *Implementing Norton's theorem: (a) original circuit; (b) finding I_n; (c) finding R_n.*

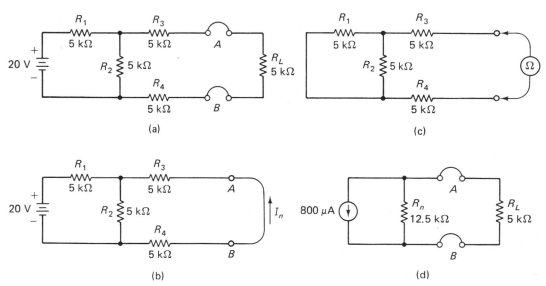

FIGURE 8-15: *Example 8-9, Nortonized equivalent: (a) original circuit; (b) finding I_n; (c) finding R_n; (d) Nortonized circuit.*

we cannot do that here, we shall calculate the voltage across R_2 using nodes, assigning it V_2. Then

$$\frac{20 - V_2}{R_1} = \frac{V_2}{R_2} + \frac{V_2}{R_3 + R_4}$$

$$\frac{20 - V_2}{5} = \frac{V_2}{5} + \frac{V_2}{10}$$

$$40 - 2V_2 = 2V_2 + V_2$$

$$V_2 = 8 \text{ V}$$

Since V_2 is 8 V, the current through R_3 (and thus the short circuit) would be

$$I_n = \frac{V_2}{R_3 + R_4} = \frac{8}{5 + 5} = 800 \ \mu\text{A}$$

To find R_n, Fig. 8-15(c),

$$R_n = 5 + (5 \| 5) + 5 = 12.5 \text{ k}\Omega$$

The final Norton equivalent is shown in Fig. 8-15(d).

A Norton equivalent can be converted to a Thévenin equivalent and a Thévenin equivalent can be converted to a Norton equivalent. Note that R_{th} equal R_n.

EXAMPLE 8-10: Find the Thévenin equivalent for Fig. 8-15(d) looking into the *AB* terminals.

SOLUTION: Since $R_{th} = R_n$, we need only find V_{th}. Discarding the load and measuring the voltage between *A* and *B*:

$$V_{th} = IR = 0.8 \times 12.5 = 10.00 \text{ V}$$

Figure 8-16 shows the resultant circuit.

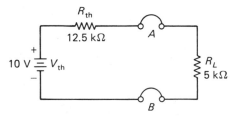

FIGURE 8-16: *Example 8-10, Thévenin equivalent of Fig. 8-15(d).*

EXAMPLE 8-11: Find the Norton equivalent of Fig. 8-16.

SOLUTION: Since $R_n = R_{th}$, we need only discard the 5-kΩ load, short *A* and *B*, and find the current through the short:

$$I_n = \frac{V}{R} = \frac{10}{12.5} = 800 \ \mu A$$

Thus, the result is as shown in Fig. 8-15(d).

8-5. THE SUPERPOSITION METHOD

Where current is to be computed in a circuit having multiple supplies, the *superposition method* can be used. The principle of the method is:

> The current along a particular path is the algebraic sum of the currents produced by the individual supplies along the path.

Thus, connecting only one supply at a time, current is found along a path. The total current is then the sum of all these computed currents.

EXAMPLE 8-12: Find I_1, I_2, and I_3 of Fig. 8-17(a).

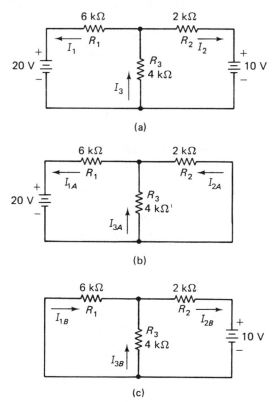

FIGURE 8-17: *Example 8-12, superposition method: (a) original circuit, (b) effect of 20-V supply; (c) effect of 10-V supply.*

SOLUTION: In order to use the superposition method, all supplies except one must be removed from the circuit, Fig. 8-17(b).[1] The currents are then computed in this figure:

$$R_t = R_1 + R_2 \| R_3$$

$$= 6 + \frac{4 \times 2}{4 + 2} = 7.333 \text{ k}\Omega$$

$$I_{1A} = \frac{V}{R} = \frac{20}{7.333} = 2.727 \text{ mA}$$

Because of circuit values, two-thirds of the current will flow through R_2 and one-third through R_3.

$$I_{2A} = \tfrac{2}{3} \times 2.727 = 1.818 \text{ mA}$$

$$I_{3A} = \tfrac{1}{3} \times 2.727 = 0.909 \text{ mA}$$

Next, the 20-V supply is removed and the effect of the 10-V supply determined Fig. 8-17(c):

$$R_t = R_2 + R_1 \| R_3 = 2 + 4 \| 6 = 4.400 \text{ k}\Omega$$

$$I_{2B} = \frac{V}{R_t} = \frac{10}{4.400} = 2.273 \text{ mA}$$

$$V_{R3} = 10 - V_{R2} = 10 - IR_2 = 10 - (2.273 \times 2) = 5.454 \text{ V}$$

$$I_{1B} = \frac{V}{R_3} = \frac{5.454}{6} = 0.909 \text{ mA}$$

$$I_{3B} = \frac{V}{R_3} = \frac{5.454}{4} = 1.364 \text{ mA}$$

The total current in each branch is the algebraic sum of each of the individual currents computed above. Note that current direction is important.

$$I_1 = I_{1A} - I_{1B} = 2.727 - 0.909 = 1.818 \text{ mA}$$

$$I_2 = I_{2B} - I_{2A} = 2.273 - 1.818 = 0.455 \text{ mA}$$

$$I_3 = I_{3B} + I_{3A} = 1.364 + 0.909 = 2.273 \text{ mA}$$

EXAMPLE 8-13: A distribution system feeds loads as shown in Fig. 8-18(a). Compute the voltage across the loads if the wires used are size 14 with a R_{ES} of 2.58 Ω at 25°C and the distance to the load is 1000 ft.

SOLUTION: A 1000-ft length of size 14 wire has a resistance of 2.58 Ω. Compute the effect of V_A, Fig. 8-18(b), using node A:

$$I_{2A} + I_{3A} = I_{1A}$$

$$\frac{V}{2.58} + \frac{V}{100 + 2.58} = \frac{120 - V}{10 + 2.58}$$

$$V = 20.00 \text{ V}$$

[1]Leave the internal resistances of the supplies in the circuit if they are known.

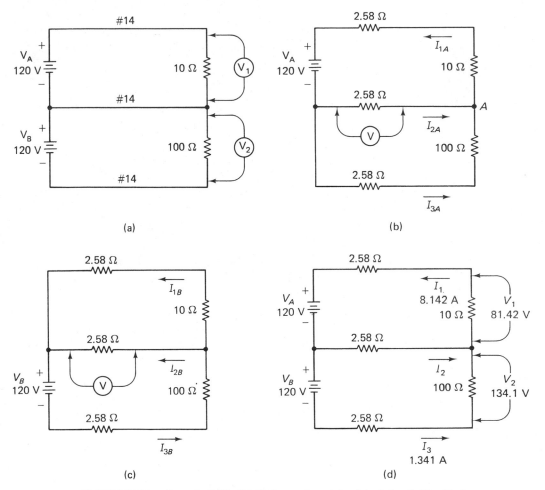

FIGURE 8-18: *Example 8-13, distribution system: (a) original circuit; (b) effect of V_A; (c) effect of V_B; (d) final computed results.*

Thus,

$$I_{2A} = \frac{V}{2.58} = \frac{20}{2.58} = 7.752 \text{ A}$$

$$I_{3A} = \frac{V}{2.58 + 100} = \frac{20}{102.58} = 0.195 \text{ A}$$

$$I_{1A} = I_{2A} + I_{3A} = 7.752 + 0.195 = 7.947 \text{ A}$$

Repeat for the effect of V_B, Fig. 8-18(c):

$$I_{3B} = I_{2B} + I_{1B}$$

$$\frac{120 - V}{100 + 2.58} = \frac{V}{2.58} + \frac{V}{10 + 2.58}$$

$$V = 2.453 \text{ V}$$

$$I_{2B} = \frac{V}{2.58} = \frac{2.453}{2.58} = 0.951 \text{ A}$$

$$I_{1B} = \frac{V}{2.58 + 10} = \frac{2.453}{12.58} = 0.195 \text{ A}$$

$$I_{3B} = I_{1B} + I_{2B} = 0.195 + 0.951 = 1.146 \text{ A}$$

Compute the total I_1 and I_3:

$$I_1 = I_{1A} + I_{1B} = 7.947 + 0.195 = 8.142 \text{ A}$$

$$I_3 = I_{3A} + I_{3B} = 0.195 + 1.146 = 1.341 \text{ A}$$

Thus, the load voltages are

$$V_1 = I_1 R = 8.142 \times 10 = 81.42 \text{ V}$$

$$V_2 = I_3 R = 1.341 \times 100 = 134.1 \text{ V (note this exceeds 120 V)}$$

8-6. WYE–DELTA TRANSFORMATIONS

The *wye–delta transformation* is a set of cookbook formulas for converting a wye network, Fig. 8-19(a), to a delta, Fig. 8-19(b), or vice versa. The formulas for these conversions are shown in Fig. 8-19(c).

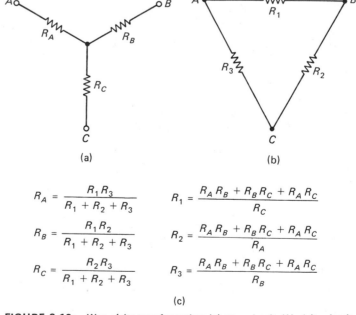

$$R_A = \frac{R_1 R_3}{R_1 + R_2 + R_3} \qquad R_1 = \frac{R_A R_B + R_B R_C + R_A R_C}{R_C}$$

$$R_B = \frac{R_1 R_2}{R_1 + R_2 + R_3} \qquad R_2 = \frac{R_A R_B + R_B R_C + R_A R_C}{R_A}$$

$$R_C = \frac{R_2 R_3}{R_1 + R_2 + R_3} \qquad R_3 = \frac{R_A R_B + R_B R_C + R_A R_C}{R_B}$$

(c)

FIGURE 8-19: *Wye–delta transformation: (a) wye circuit; (b) delta circuit; (c) transformation formulas.*

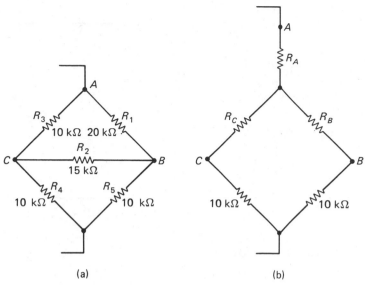

(a)   (b)

FIGURE 8-20: *Example 8-14, wye-delta transformation:* (a) *original circuit;* (b) *transforming the delta to a wye.*

EXAMPLE 8-14: Find the total resistance of the unbalanced bridge shown in Fig. 8-20.

SOLUTION: The top three resistors can be visualized as a delta circuit, labeled as shown, and converted to the wye shown in Fig. 8-20(b). Be very careful to define the nodes and the resistors precisely as shown in Fig. 8-19(a) and (b):

$$R_A = \frac{R_1 R_3}{R_1 + R_2 + R_3} = \frac{20 \times 10}{20 + 15 + 10} = \frac{200}{45} = 4.444 \text{ k}\Omega$$

$$R_B = \frac{R_1 R_2}{R_1 + R_2 + R_3} = \frac{20 \times 15}{45} = 6.667 \text{ k}\Omega$$

$$R_C = \frac{R_2 R_3}{R_1 + R_2 + R_3} = \frac{15 \times 10}{45} = 3.333 \text{ k}\Omega$$

Total resistance of the network can now be easily obtained:

$$R_t = R_A + [(R_C + 10) \| (R_B + 10)]$$
$$= 4.444 + [(3.333 + 10) \| (6.667 + 10)]$$
$$\doteq 11.85 \text{ k}\Omega$$

EXAMPLE 8-15: Compute the total resistance between points X and Y of Fig. 8-21(a).

SOLUTION: Label the circuit per Fig. 8-19 and convert the center three resistors to a delta arrangement as shown in Fig. 8-21(b):

$$R_1 = \frac{R_A R_B + R_B R_C + R_A R_C}{R_C}$$

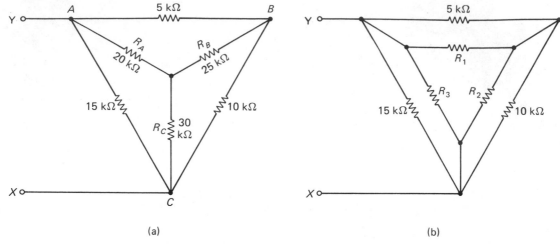

(a)   (b)

FIGURE 8-21: *Example 8-15: (a) original circuit; (b) transformed circuit.*

$$= \frac{(20 \times 25) + (25 \times 30) + (20 \times 30)}{30} = \frac{1850}{30}$$

$$= 61.667 \text{ k}\Omega$$

$$R_2 = \frac{1850}{R_A} = \frac{1850}{20} = 92.50 \text{ k}\Omega$$

$$R_3 = \frac{1850}{R_B} = \frac{1850}{25} = 74.00 \text{ k}\Omega$$

The circuit now becomes a simple series–parallel network:

$$R_t = (15 \,\|\, R_3) \,\|\, [(R_1 \,\|\, 5) + (R_2 \,\|\, 10)]$$
$$= 12.472 \,\|\, [4.625 + 9.024]$$
$$= 12.472 \,\|\, 13.65 = 6.517 \text{ k}\Omega$$

8-7. SUMMARY

In this and the preceding chapter we have presented seven methods for circuit analysis:

1. Ohm's law.

2. The loop method.

3. The nodal method.

4. Thévenin's theorem.

5. Norton's theorem.

6. Superposition.

7. Wye–delta transformation.

Let us now direct our attention to the question of when to use what method. For finding all parameters of a circuit, Ohm's law, the loop, and nodal methods are more attractive, with the loop method being used with multiple power supplies and the nodal method for circuits having many parallel branches.

The concept of Thévenin's and Norton's theorems is fundamental to electronics; every power supply, amplifier, and signal generator should be viewed as both a voltage supply in series with a resistance (Thévenin's theorem) and a current supply in parallel with a resistance (Norton's theorem). Thévenin's theorem is also a useful tool for the bench, and can be used to evaluate a single resistance or current within a circuit. Neither of these theorems is very useful for analyzing all the parameters within a circuit.

Superposition is useful for analyzing circuits with multiple supplies. Wye–delta transformations are cookbook formulas for converting a wye-resistive network to a delta-resistive network, and vice versa.

8-8. *REVIEW QUESTIONS*

8-1.   What types of circuit problems are more easily solved using the loop method?

8-2.   When developing a loop equation, what sign should be applied to a resistor whose mesh direction is opposite to the assumed current direction?

8-3.   What determines how many loop equations must be written?

8-4.   What does a minus sign mean in the final solution of a branch current?

8-5.   What types of circuit problems are more easily solved using nodal analysis?

8-6.   What determines how many nodal equations are required?

8-7.   State Thévenin's theorem.

8-8.   What are the Thévenized voltage and resistance for each of the following? Assume that each is ideal.

   (a)   A voltmeter.

   (b)   An ammeter.

   (c)   A +10-V supply.

   (d)   A 1.5-V dry cell.

   (e)   A 120-V 100-W light bulb.

   (f)   A 1-MΩ resistor.

8-9.   State the method for finding V_{th}.

8-10.   State the method for finding R_{th}.

8-11. State Norton's theorem.

8-12. How is I_n found?

8-13. How is R_n found?

8-14. How can a constant current supply be approximated using a constant voltage supply and a resistance?

8-15. What is the superposition method of circuit analysis?

8-16. What is the wye–delta method of analysis?

8-9. PROBLEMS

8-1. Compute the current through R_1 in Fig. 8-22.

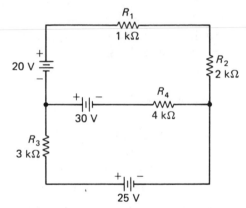

FIGURE 8-22: *Problems 8-1 and 8-18.*

8-2. Compute the voltage across R_2 in Fig. 8-23 using the loop method.

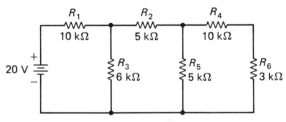

FIGURE 8-23: *Problems 8-2 and 8-7.*

8-3. A three-wire distribution system similar to Fig. 8-3 supplies ± 120 V. The positive source feeds a 75-Ω load and the negative source feeds a 10-Ω load through 500 ft of size 14 wire. What is the voltage across each of the loads at 20°C?

8-4. Compute the voltage of A with respect to B across R_3 in Fig. 8-24 using the loop method.

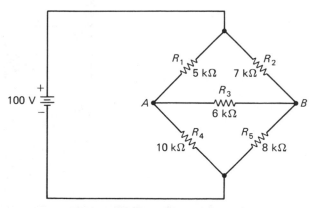

FIGURE 8-24: *Problems 8-4 and 8-9.*

8-5. Compute the voltage of *A* with respect to *B* across R_5 of Fig. 8-25 using the loop method.

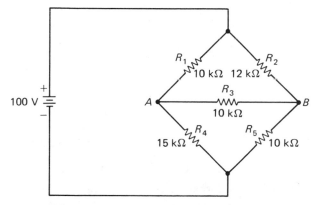

FIGURE 8-25: *Problems 8-5, 8-10, 8-13, and 8-16.*

8-6. Compute the current through R_3 of Fig. 8-26 using nodal analysis.

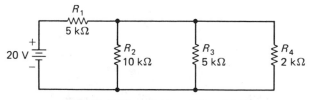

FIGURE 8-26: *Problems 8-6 and 8-14.*

8-7. Compute the current through R_5 of Fig. 8-23 using nodal analysis.

8-8. Compute the current through R_6 of Fig. 8-27 using nodal analysis.

8-9. Compute the voltage across R_5 of Fig. 8-24 using nodal analysis.

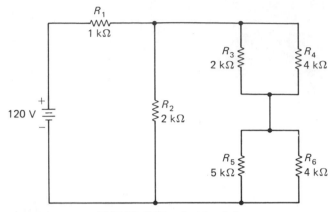

FIGURE 8-27:   *Problem 8-8.*

8-10.   Compute the voltage of A with respect to B across R_3 of Fig. 8-25 using nodal analysis.

8-11.   Compute V_{th} and R_{th} of Fig. 8-28 looking into terminals A and B.

8-12.   Change R_3 of Fig. 8-28 to 100 kΩ and compute V_{th} and R_{th} looking into terminals A and B.

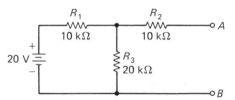

FIGURE 8-28:   *Problems 8-11 and 8-12.*

8-13.   Compute the current through R_3 of Fig. 8-25 using Thévenin's theorem.

8-14.   Compute the current through R_1 of Fig. 8-26 using Thévenin's theorem.

8-15.   Compute the current through R_4 of Fig. 8-29 using Norton's theorem.

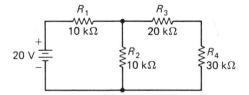

FIGURE 8-29:   *Problem 8-15.*

8-16.   Compute the current through R_3 of Fig. 8-25 using Norton's theorem.

8-17.   Compute the currents through R_1, R_2, and R_3 of Fig. 8-30 using superposition.

8-18.   Compute the currents through R_1, R_3, and R_4 of Fig. 8-22 using superposition.

8-19.   A distribution system similar to Fig. 8-18 has sources of ± 100 V, supplying a 20-Ω load from the positive source and a 150-Ω load from the negative source through

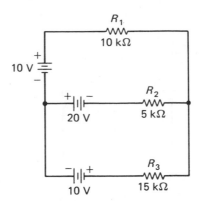

FIGURE 8-30: *Problem 8-17.*

750 ft of size 16 wire. Using superposition, compute the voltages across each of the loads.

8-20. In Fig. 8-20(a), $R_1 = 500\ \Omega$, $R_2 = 1\ k\Omega$, $R_3 = 700\ \Omega$, $R_4 = 600\ \Omega$, and $R_5 = 800\ \Omega$. Find the total resistance of the circuit using wye–delta transformation.

8-21. Find the resistance between A and B of Fig. 8-31 using wye–delta transformation.

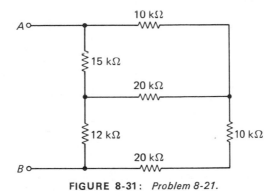

FIGURE 8-31: *Problem 8-21.*

8-10. PROJECTS

8-1. Construct the circuit shown in Fig. 8-32. Measure all the branch currents and compare with the loop-method computed values.

8-2. Construct the unbalanced bridge shown in Fig. 8-33. Measure the voltages across each of the resistors and compare these with the computed values.

8-3. Construct the circuit shown in Fig. 8-33. Vary the resistance of R_3 from $0\ \Omega$ to $1\ M\Omega$ and measure current through R_3 and voltage across R_3. Plot the result on a graph. Next, Thévenize the circuit assuming that R_3 is the load. Construct this Thévenin equivalent circuit, reconnect R_3, and vary it from $0\ \Omega$ to $1\ M\Omega$, measuring its current and voltage drop. Plot these results on the previous graph. How do they compare? What does this prove?

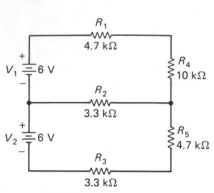

FIGURE 8-32: *Project 8-1.*

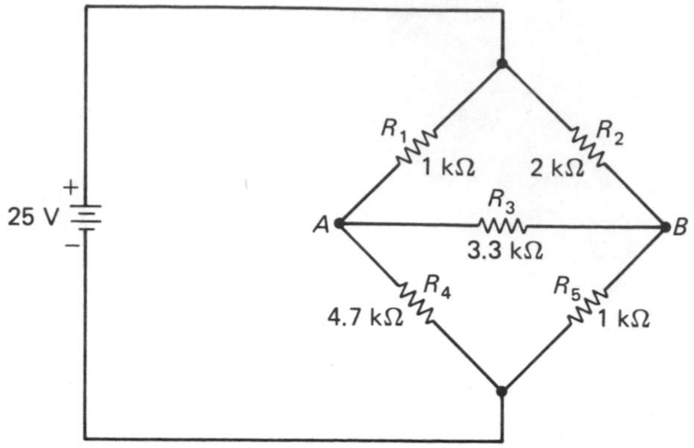

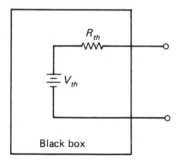

FIGURE 8-33: *Projects 8-2 and 8-3.*

8-4. Assume that you had access only to the two terminals of the black box shown in Fig. 8-34 (that is, you could not modify the circuit inside the box). Describe a procedure by which you could find V_{th} and R_{th}. Develop formulas for such a procedure.

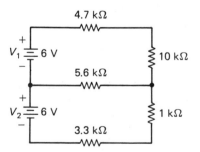

FIGURE 8-34: *Project 8-4.*

8-5. Construct the circuit shown in Fig. 8-35. Measure branch currents due to V_2 alone (V_1 removed and its former terminals shorted), V_1 alone (V_2 removed and its former terminals shorted), and with both V_1 and V_2 in the circuit. Do your results verify the superposition theorem?

FIGURE 8-35: *Project 8-5.*

8-6. Construct the circuit shown in Fig. 8-36. Measure R_{AB}, R_{BC}, and R_{AC}. Next, compute the equivalent wye circuit and construct it. Again, measure R_{AB}, R_{BC}, and R_{AC}. Do your results verify the wye–delta transformation?

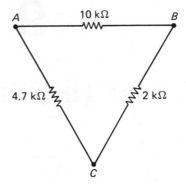

FIGURE 8-36: *Project 8-6.*

9

DIRECT-CURRENT METERS

In electricity, we are dealing with a force that is both silent and invisible. Its effects, however, are quite observable, as our chattering teeth will testify when we come in contact with a live wire. Thus, we can measure the effect of electricity, and translate this into such terms as volts, amps, and ohms. This chapter introduces us to the insides of voltmeters, ammeters, and ohmmeters, allowing us to understand how each functions, their specifications and limitations, and how we can design special meter circuits for specific applications. The sections include:

9-1. Meter types

9-2. Moving-coil meters

9-3. The voltmeter

9-4. The ammeter

9-5. The ohmmeter

9-6. The volt-ohm-milliammeter

9-7. Effects of meters upon a circuit

9-1. METER TYPES

There are two ways of categorizing meters:

1. The type of display used.
2. The source of power for the meter.

FIGURE 9-1: *Digital voltmeter (photo by Gilchrist).*

FIGURE 9-2: *Volt-ohm-milliammeter (photo by Gilchrist).*

Two types of displays are commonly used today: digital and scale, Figs. 9-1 and 9-2. The digital types require no interpretation on the part of the observer, whereas the scale types are subject to parallax errors (see Chapter 2).

Ammeters and voltmeters can be powered from three sources of energy:

1. An internal battery that supplies an internal electronic amplifier, Fig. 9-3.

2. Power from a wall receptacle that supplies an internal amplifier, Fig. 9-1.

3. The circuit that is being measured, Fig. 9-2.

The first two types amplify the signal received from the circuit under test, thus extracting very little energy from it; the third type, used only in scale-type meters, drives a moving-coil mechanism.

FIGURE 9-3: *Electronic VOM (photo by Ruple).*

9-2. MOVING-COIL METERS

Most, but not all, scale-type meters use a basic meter movement called a *d'Arsonval movement* after its inventor, Fig. 9-4. The device has a stationary, permanent magnet; its movable portion consists of a coil of wire wound around an iron core mounted upon jewel bearings. Current passing through the coil creates a magnetic field that opposes the field produced by the permanent magnet, causing the coil to rotate. An indicating needle is attached to the coil mechanism and a scale provided for readout. Figure 9-5 illustrates a practical such device called a *Weston movement*.

We shall study the interaction of magnetic fields in more detail in Chapter 10.

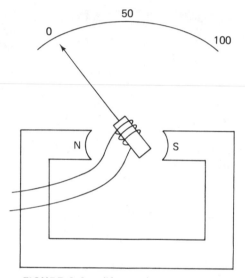

FIGURE 9-4: *d'Arsonval meter movement.*

FIGURE 9-5 *Weston movement (photo by Ruple).*

178

Specifications

Such a basic meter movement has two specifications of interest to us. Since the coil part of the meter consists of very fine wire, the meter has resistance. In addition, the manufacturer will define a meter as requiring a certain current to cause it to read the highest number on the scale. This is its full-scale current (I_{FS}). Using these two specifications, we can design a voltmeter, ammeter, or ohmmeter.

9-3. THE VOLTMETER

Since the voltmeter is placed in parallel with the circuit it is to measure, it should, ideally, have an infinite resistance. We can construct a voltmeter by merely providing a resistance in series with the basic meter movement. It will, however, require some elementary calculations.

EXAMPLE 9-1: Design a 0- to 1-V voltmeter from a 100-μA meter movement having an internal resistance of 1000 Ω.

SOLUTION: The schematic for such a meter is shown in Fig. 9-6(a). Since, with 1.00 V impressed, the meter is to read full scale, it will have 100 μA flowing through it, Fig. 9-6(b). Thus, total current is 100 μA, and total voltage is 1.00 V. Total resistance must therefore be

$$R_t = \frac{V_t}{I_{FS}} = \frac{1}{100 \times 10^{-6}} = 10 \text{ k}\Omega$$

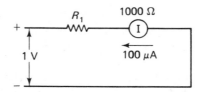

(a) Schematic

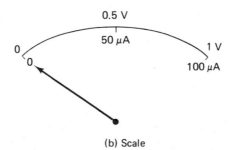

(b) Scale

FIGURE 9-6: *Voltmeter: (a) schematic; (b) scale.*

Since the meter has a resistance of 1000 Ω (R_m), the resistance of R_1 is

$$R_1 = R_t - R_m = 10,000 - 1000 = 9000 \ \Omega$$

EXAMPLE 9-2: Using the same meter movement as in Example 9-1, design a voltmeter that reads 100 V full scale (FS).

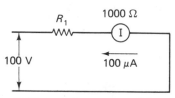

FIGURE 9-7: *100-V voltmeter.*

SOLUTION: The schematic is shown in Fig. 9-7. Calculate R_1:

$$R_t = \frac{V_t}{I_{FS}} = \frac{100}{100 \times 10^{-6}} = 1 \ M\Omega$$

$$R_1 = R_t - R_m = 1,000,000 - 1000 = 999.0 \ k\Omega$$

Note that the larger the voltage scale, the larger the total resistance of this type of voltmeter.

Ohms/Volt Rating

From the preceding two examples, we can derive a specification that is very frequently used to describe voltmeters, its *ohms/volt rating*, also called its *sensitivity*. There are two methods by which this rating can be calculated:

1. Knowing the total resistance of the meter and its voltage range, ohms/volt rating equals total resistance of the meter divided by the full-scale voltage on that range.

$$\frac{\Omega}{V} = \frac{R_t}{V_{FS}}$$

2. It is the inverse of I_{FS}:

$$\frac{\Omega}{V} = \frac{1}{I_{FS}}$$

Note that Ω/V is constant for a particular voltmeter, regardless of its voltage range.

EXAMPLE 9-3: Calculate the sensitivity of the meters in Examples 9-1 and 9-2.

SOLUTION: We shall calculate this rating three ways:

1. Total resistance and total voltage for the meter in Fig. 9-6:

$$\frac{\Omega}{V} = \frac{R_t}{V_{FS}} = \frac{10{,}000}{1} = 10 \text{ k}\Omega/\text{V}$$

2. Total resistance and voltage for the meter in Fig. 9-7:

$$\frac{\Omega}{V} = \frac{R_t}{V_{FS}} = \frac{1{,}000{,}000}{100} = 10 \text{ k}\Omega/\text{V}$$

3. From I_{FS} alone:

$$\frac{\Omega}{V} = \frac{1}{I_{FS}} = \frac{1}{100 \times 10^{-6}} = 10 \text{ k}\Omega/\text{V}$$

Note that all three are identical.

Since the sensitivity and I_{FS} are reciprocals, either can be used to calculate the series resistor required for a voltmeter:

$$R_t = \frac{V_{FS}}{I_{FS}}$$

but

$$I_{FS} = \frac{1}{\Omega/V}$$

Therefore,

$$R_t = V_{FS} \times \frac{\Omega}{V}$$

EXAMPLE 9-4: A meter has a sensitivity of 7.500 kΩ/V and a resistance of 1500 Ω. Calculate the series resistance needed for a 50-V scale.

SOLUTION: The total resistance for a 50-V range is

$$R_t = V_{FS} \times \frac{\Omega}{V} = 50 \times 7500 = 375 \text{ k}\Omega$$

$$R_1 = R_t - R_m = 375{,}000 - 1500 = 373.5 \text{ k}\Omega$$

Accuracy

The accuracy of a voltmeter is usually specified as a percent of full scale. Thus, a $\pm5\%$ accuracy on the 100-V scale would represent

$$\text{error} = \pm5\% \times 100 \text{ V} = \pm5 \text{ V}$$

It is very important to observe that this accuracy depends, not upon the voltage read, but upon the scale used. If, for example, we were to read 20 V on the 100-V scale of a 5% meter, the error would be $\pm5\%$ of 100 V, not $\pm5\%$ of 20 V. Thus, if 20 V were impressed on this voltmeter, it could read between 15 and 25 V and still be considered within its $\pm 5\%$ FS specification. This 5-V error would represent a percent of reading error of

$$\% \text{ error} = \tfrac{5}{20} \times 100 = 25\% \text{ error}$$

It is, therefore, considered good practice to use the lowest range possible on a voltmeter without exceeding the V_{FS} of that range.

Multirange Voltmeters

A single meter can be designed for a variety of voltage ranges by adding a selector switch and properly sized resistors. Figure 9-8(a) shows one possible circuit.

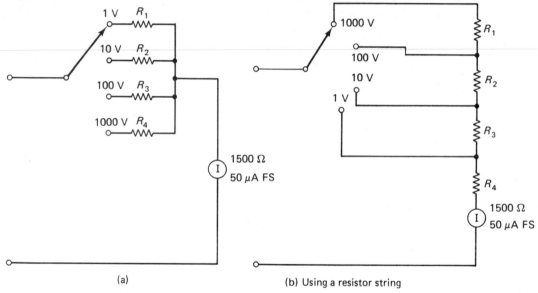

(a)	(b) Using a resistor string

FIGURE 9-8: *Multirange voltmeter:* (a) *using separate range resistors;* (b) *using a resistor string.*

EXAMPLE 9-5: Calculate the resistors necessary for the multirange voltmeter shown in Fig. 9-8(a).

SOLUTION: Compute the sensitivity of the movement:

$$\frac{\Omega}{V} = \frac{1}{V_{FS}} = \frac{1}{50 \times 10^{-6}} = 20 \text{ k}\Omega/V$$

For the 1-V scale,

$$R_1 = V_{FS} \times \frac{\Omega}{V} - R_m$$

$$= (1 \times 20{,}000) - 1500 = 18.50 \text{ k}\Omega$$

For the 10-V scale,

$$R_2 = V_{FS} \times \frac{\Omega}{V} - R_m$$

$$= (10 \times 20{,}000) - 1500 = 198.5 \text{ k}\Omega$$

For the 100-V scale,

$$R_3 = V_{FS} \times \frac{\Omega}{V} - R_m$$

$$= (100 \times 20{,}000) - 1500 = 1{,}998{,}500 \ \Omega$$

For the 1000-V scale,

$$R_4 = V_{FS} \times \frac{\Omega}{V} - R_m$$

$$= (1000 \times 20{,}000) - 1500$$

$$= 19{,}998{,}500 \ \Omega$$

Note the peculiar values of resistances computed in this example. These values would be very difficult to obtain in 1%-tolerance resistors, resulting in a rather high cost. By revising the basic circuit a bit, Fig. 9-8(b), more conventional values can be used.

EXAMPLE 9-6: Compute the resistor string values shown in Fig. 9-8(b).

SOLUTION: Although "brute-force" Ohm's law techniques can be used, the sensitivity rating makes life much simpler. For the 1-V scale,

$$R_4 = V_{FS} \times \frac{\Omega}{V} - R_m$$

$$= (1 \times 20{,}000) - 1500 = 18{,}500 \ \Omega$$

Since the difference between the 10-V and 1-V scales is 9 V:

$$R_3 = V_{DIFF} \times \frac{\Omega}{V}$$

$$= 9 \times 20{,}000 = 180 \text{ k}\Omega$$

Similarly,

$$R_2 = V_{DIFF} \times \frac{\Omega}{V}$$

$$= (100 - 10) \times 20{,}000$$

$$= 1.8 \text{ M}\Omega$$

$$R_1 = V_{DIFF} \times \frac{\Omega}{V}$$

$$= (1000 - 100) \times 20{,}000$$

$$= 18.0 \text{ M}\Omega$$

Note that each of the preceding values, except R_4, is a commonly available resistor, resulting in a much lower cost.

EXAMPLE 9-7: A 1800-Ω 100-μA meter movement is to be used as a voltmeter with ranges of 1.5, 5, 15, 50, 150, and 500 V. Design the divider string.

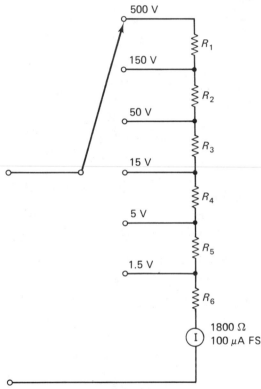

FIGURE 9-9: *Example 9-7.*

SOLUTION: The schematic is shown in Fig. 9-9.

$$\frac{\Omega}{V} = \frac{1}{I_{\text{FS}}} = \frac{1}{100 \times 10^{-6}} = 10 \text{ k}\Omega/\text{V}$$

$$R_6 = V_{\text{FS}} \times \frac{\Omega}{V} - R_m$$

$$= (1.5 \times 10{,}000) - 1800$$

$$= 13.20 \text{ k}\Omega$$

$$R_5 = V_{\text{DIFF}} \times \frac{\Omega}{V}$$

$$= (5 - 1.5) \times (10{,}000)$$

$$= 35.0 \text{ k}\Omega$$

$$R_4 = V_{\text{DIFF}} \times \frac{\Omega}{V}$$

$$= (15 - 5) \times (10,000)$$

$$= 100 \text{ k}\Omega$$

$$R_3 = V_{\text{DIFF}} \times \frac{\Omega}{V}$$

$$= (50 - 15) \times (10,000)$$

$$= 350 \text{ k}\Omega$$

$$R_2 = V_{\text{DIFF}} \times \frac{\Omega}{V}$$

$$= (150 - 50) \times (10,000)$$

$$= 1.00 \text{ M}\Omega$$

$$R_1 = V_{\text{DIFF}} \times \frac{\Omega}{V}$$

$$= (500 - 150) \times (10,000)$$

$$= 3.50 \text{ M}\Omega$$

9-4. THE AMMETER

An *ammeter* is placed in series with the circuit it is to measure and, ideally, should have zero ohms resistance. We can construct such a device from a basic meter movement by placing a resistor in parallel with the meter.

> **EXAMPLE 9-8:** Design a 1-mA FS ammeter from a meter movement having a FS current of 50 μA and a resistance of 1500 Ω.
>
> *SOLUTION:* We have two requirements that must be met simultaneously:
>
> 1. The maximum current through the movement is to be 50 μA.
> 2. With this current of 50 μA through the meter movement, the total current through the entire circuit of the meter must be 1 mA.

These two conditions can only be met by using a parallel circuit such as that shown in Fig. 9-10. The resistor placed in parallel with the meter movement is called a *shunt resistor*, for it shunts most of the current around the sensitive meter movement. Its value can be calculated as follows. By Kirchhoff's law,

$$I_t = I_R + I_m$$

Therefore,

$$I_R = I_t - I_m$$

but

$$V = I_R R_S = (I_t - I_m) R_S$$

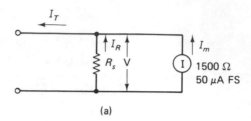

(a)

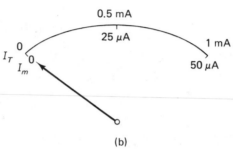

(b)

FIGURE 9-10: *Example 9-8, the ammeter: (a) schematic; (b) scale.*

and

$$V = I_m R_m$$

Equate these two:

$$I_m R_m = (I_t - I_m) R_S$$

Solve for R_S:

$$R_S = \frac{I_m R_m}{I_t - I_m} \qquad (9\text{-}1)$$

Substitute the values given:

$$R_S = \frac{I_m R_m}{I_t - I_m}$$

$$= \frac{50 \times 10^{-6} \times 1500}{1 \times 10^{-3} - 50 \times 10^{-6}}$$

$$= 78.95 \ \Omega$$

Note that, with this resistance in parallel with the 1500-Ω meter resistance, the total ammeter resistance is

$$R_A = R_S \| R_m$$

$$= \frac{1500 \times 78.95}{1500 + 78.95}$$

$$= 75.00 \ \Omega$$

EXAMPLE 9-9: Design a 1-A FS ammeter using the meter movement in Example 9-8.

SOLUTION: From Eq. (9-1),

$$R_S = \frac{I_m R_m}{I_t - I_m}$$

$$= \frac{50 \times 10^{-6} \times 1500}{1 - (50 \times 10^{-6})}$$

$$= 75.00 \text{ m}\Omega$$

Multirange Ammeters

The circuit of Fig. 9-11 could be used for a multiple-range ammeter. However, should one of the switch positions become dirty, preventing a good contact, all the current would pass through the meter movement, possibly destroying it. For this

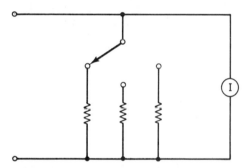

FIGURE 9-11: *Possible multirange ammeter.*

reason, the *universal*, or *Ayrton shunt* is usually used, Fig. 9-12(a). Note that the meter is always shunted, and dirt between the switch contacts would lessen meter current rather than increase it. Using Ohm's law, we can calculate each of the resistors. However, let us develop a general-purpose formula for examining the circuit. From Fig. 9-12(b), we can calculate current in the divider string, I_s:

$$I_s = I_t - I_m$$

Knowing this, we can compute the total resistance of the string, R_s:

$$I_m R_m = I_s R_s$$

for we know I_m, R_m, and R_s. Now we shall move the selector switch to the next higher range, Fig. 9-12(c). Note that the voltage between points A and B can be computed two ways:

$$V_A = I_m(R_1 + R_m)$$
$$V_A = I_s R_p$$

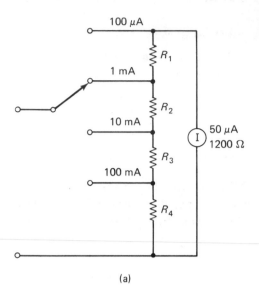

(a)

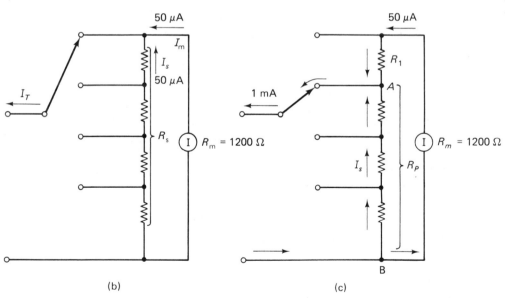

(b)   (c)

FIGURE 9-12: *Universal or Ayrton shunt: (a) schematic; (b) 100-μA range; (c) 1-mA range; (d) 10-mA range; (e) final circuit.*

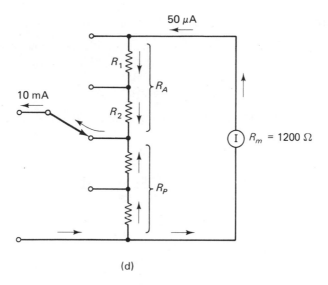

(d)

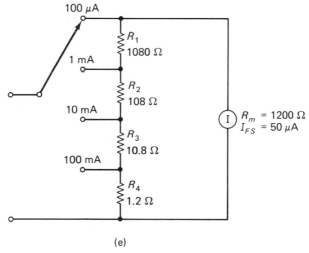

(e)

FIGURE 9-12: (*Cont.*)

Equate these two:

$$I_m(R_1 + R_m) = I_s R_p$$

Substitute $R_s - R_p$ for R_1 and $I_t - I_m$ for I_s and solve for R_p:

$$I_m(R_s - R_p + R_m) = (I_t - I_m)R_p$$

$$R_p = \frac{I_m(R_s - R_m)}{I_t} \qquad (9\text{-}2)$$

Note that the numerator is independent of the range.

EXAMPLE 9-10: Compute the Ayrton shunt resistors shown in Fig. 9-12(a).

SOLUTION: Going to the most sensitive range, 50 μA will flow through the total shunt and 50 μA through the meter movement, Fig. 9-12(b). Thus,

$$R_s = R_m = 1200 \ \Omega$$

Moving to the 1-mA range, Fig. 9-12(c), from Eq. 9-2:

$$R_p = \frac{I_m(R_s + R_m)}{I_t}$$

$$= \frac{50 \times 10^{-6}(1200 + 1200)}{1 \times 10^{-3}}$$

$$= \frac{0.120}{1 \times 10^{-3}}$$

$$= 120 \ \Omega$$

Thus,

$$R_1 = R_s - R_p$$

$$= 1200 - 120 = 1080 \ \Omega$$

Moving to the 10-mA range, Fig. 9-12(d):

$$R_p = \frac{I_m(R_s + R_m)}{I_t} = \frac{0.120}{0.01} = 12.0 \ \Omega$$

$$R_A = R_s - R_p = 1200 - 12.0 = 1188 \ \Omega$$

But

$$R_A = R_1 + R_2$$

$$R_2 = R_A - R_1 = 1188 - 1080 = 108 \ \Omega$$

Moving to the 100-mA range:

$$R_4 = \frac{I_m(R_s + R_m)}{I_t} = \frac{0.12}{0.1} = 1.200 \ \Omega$$

$$R_3 = R_s - R_1 - R_2 - R_4$$

$$= 1200 - 1080 - 108 - 1.2$$

$$= 10.80 \ \Omega$$

The finished schematic is shown in Fig. 9-12(e).

EXAMPLE 9-11: Compute the values for an Ayrton shunt with scales of 100 μA, 500 μA, 1.0 mA, and 5 mA using a 50-μA movement with a resistance of 2000 Ω.

SOLUTION: The schematic is shown in Fig. 9-13. $R_s = 2000 \ \Omega$, since on the 100-μA scale, 50 μA must flow through the meter and 50 μA through the shunt. Moving to the 500-μA position:

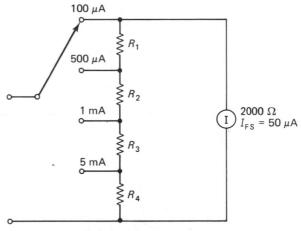

FIGURE 9-13: *Example 9-11.*

$$R_p = \frac{I_m(R_s + R_m)}{I_t} = \frac{50 \times 10^{-6}(2000 + 2000)}{500 \times 10^{-6}}$$

$$= \frac{0.200}{500 \times 10^{-6}} = 400\ \Omega$$

$$R_1 = R_s - R_p = 2000 - 400 = 1600\ \Omega$$

In the 1-mA position,

$$R_p = \frac{I_m(R_s + R_m)}{I_t} = \frac{0.200}{0.001} = 200.0\ \Omega$$

$$R_2 = R_s - R_1 - R_p = 2000 - 1600 - 200 = 200\ \Omega$$

In the 5-mA position,

$$R_p = \frac{I_m(R_s + R_m)}{I_t} = \frac{0.200}{0.005} = 40.00\ \Omega$$

$$R_4 = R_p = 40.00\ \Omega$$

$$R_3 = R_s - R_1 - R_2 - R_4 = 2000 - 1600 - 200 - 40 = 160\ \Omega$$

9-5. THE OHMMETER

The *ohmmeter* is constructed from a basic meter movement, a battery, and scaling resistors, Fig. 9-14. With leads A and B open, the meter reads 0 mA (infinite ohms on the resistance scale, indicated by the "∞" symbol). When leads A and B are shorted together, resistor R_A is adjusted to read full scale, in this case 1 mA (zero ohms on the resistance scale). If a 10-kΩ resistor were now placed between leads A and B, the meter would read half scale, since there would be twice as much resistance in the

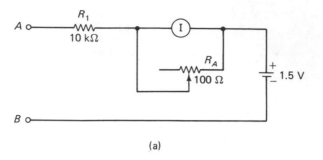

(a)

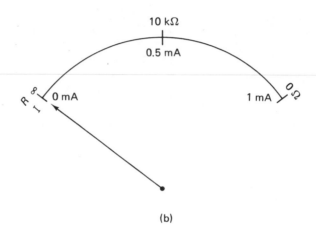

(b)

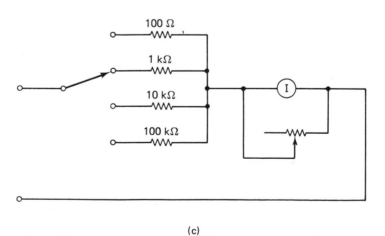

(c)

FIGURE 9-14: *Ohmmeter: (a) basic circuit; (b) resistance scale; (c) multiple-range ohmmeter.*

circuit as when the leads were shorted. The resistance scale would be marked "10 kΩ" as this point. Thus, the resistance scale is nonlinear, reading from right to left.

As the battery within the meter ages, its terminal voltage is reduced, providing erroneous resistance readings. Thus, every time the meter is used, the two leads should be shorted and R_A adjusted for the meter to read full scale. In this manner the effects of the aging, feeble battery are minimized.

Multiple resistance ranges can be provided by selecting various values of R_1, Fig. 9-14(c). Note that the resistance selected represents the half-point on the scale. That is, if the 10-kΩ resistor is selected and a 10-kΩ resistor were being measured, the meter, when properly calibrated, would read half scale.

Many ohmmeters have two batteries, a high-voltage battery (about 6 V) and a low-voltage battery (1.5 V). The high-voltage battery is switched into the circuit on the high-resistance scale to obtain enough emf to provide the 1 mA of current for a meter reading. (A 6-V battery is high voltage???)

9-6. THE VOLT-OHM-MILLIAMMETER

The *volt-ohm-milliammeter* (VOM) is one of the most common types of meters found in the electronics industry. It is a combination voltmeter, ohmmeter, and ammeter requiring no external power supply and containing a variety of ranges. Figure 9-2 illustrates the device. The function to be performed is selected using the large, central selector switch, with the test leads connected between the "+" and "−" jacks.

9-7. EFFECTS OF METERS UPON A CIRCUIT

A famous physicist once observed: "No system can be measured without disturbing that system." This is certainly true within electronics: whenever a meter is used to measure an electrical property of a circuit, it disturbs that circuit. It changes the original voltage and current from that which existed prior to attaching the meter.

This phenomenon is quite apparent in the use of an ohmmeter; prior to measurement, no current flows through the resistor under test. While making the measurement, however, current does flow. Although this current flow usually has no effect upon a resistor, there are some devices which are greatly affected. An ordinary incandescent lamp has different resistances when it is cold and when it is hot. The very act of using an ohmmeter will cause current flow through the lamp, heating the filament and changing its resistance. Thus, the reading will not represent the resistance when the lamp is cold.

Voltmeter Loading

A voltmeter also affects the circuit it is measuring. The important question we must ask, however, is whether its effect is negligible. Consider the circuit shown in Fig. 9-15(a). Here, the voltage between A and B is seen to be 12 V. However, when

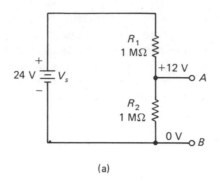

(a)

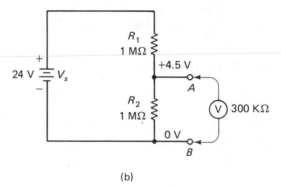

(b)

FIGURE 9-15: *Voltmeter loading: (a) original circuit; (b) connecting a voltmeter.*

a 20-kΩ/V voltmeter is added and its 15-V range selected, we note that its resistance is

$$R_m = \frac{\Omega}{V} \times V_{FS}$$

$$= 20,000 \times 15 = 300 \text{ k}\Omega$$

When this 300-kΩ meter is placed in parallel with R_2, the resultant resistance between A and B is

$$R_{AB} = R_2 \| R_m$$

$$= \frac{R_2 R_m}{R_2 + R_m} = \frac{(10^6) \times (0.3 \times 10^6)}{(10^6) + (0.3 \times 10^6)} = 230.77 \text{ k}\Omega$$

Thus, we now have a voltage across R_2 of

$$V_{AB} = \frac{R_{AB}}{R_{AB} + R_1} \times V_s = \frac{230.77}{230.77 + 1000} \times 24 = 4.500 \text{ V}$$

Whereas we had 12 V across AB without the meter, we have 4.5 V with it in the circuit.

Although we can always compute the effect of voltmeter loading as was done above, it is much more important to recognize when such a problem may exist. Consider the circuit shown in Fig. 9-16. Here, the voltage between A and B without the voltmeter is 12.000 V. However, with the voltmeter it is 11.803 V, a 1.6% error. If we are using a 5% meter, this error is obviously negligible. How, then, can we approximate our percent loading error? Read on.

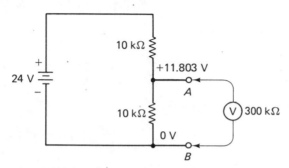

FIGURE 9-16: *Negligible voltmeter loading.*

The error of a meter can be approximated by considering the ratio of the Thévenized resistance of the circuit being measured to the resistance of the meter. Thus, if we were to look into terminals A and B of Fig. 9-16, and short the battery (mentally, of course), the Thévenized resistance of the circuit would be

$$R_{th} = 10 \text{ k}\Omega \| 10 \text{ k}\Omega = 5 \text{ k}\Omega$$

The ratio of this to meter resistance is

$$\text{ratio} = \frac{R_{th}}{R_m} = \frac{5}{300} = 0.01667 \text{ or } 1.667\%$$

Thus, our error when using this meter is about 1.7%.

EXAMPLE 9-12: Approximate the measurement error of the circuit shown in Fig. 9-17.

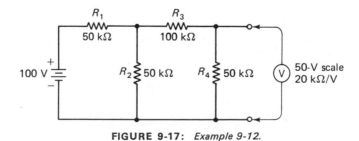

FIGURE 9-17: *Example 9-12.*

SOLUTION: We shall use approximations. Mentally shorting out the battery,

$$R_{eq} = R_1 \| R_2 = 50 \text{ k}\Omega \| 50 \text{ k}\Omega = 25 \text{ k}\Omega$$

This is in series with a 100-kΩ resistor, yielding 125 kΩ. This resultant resistance is in parallel with R_4, producing

$$R_{th} = R_{eq} \| R_4 = 125 \| 50 = 35.71 \text{ k}\Omega$$

The resistance of the meter is

$$R_m = \frac{\Omega}{\text{V}} \times V_{FS} = 20,000 \times 50 = 1 \text{ M}\Omega$$

The ratio of R_{th} and R_m is

$$\text{ratio} = \frac{R_{th}}{R_m} = \frac{35.71}{1000} = 0.0357 \text{ or } 3.57\%$$

Thus, the meter reading would be 3.57% in error, owing to the loading effect of the voltmeter.

Electronic Voltmeters

Electronic voltmeters, both digital and scale types, differ in their resistance characteristics from the voltmeters discussed above. These devices contain a power supply and amplifiers, thus extracting very little energy from the circuit they are measuring. Most have a constant resistance of 11 MΩ regardless of the scale used. Thus, when figuring their loading effect, R_m should be considered as 11 MΩ.

Ammeter Insertion Loss

Like the voltmeter, the ammeter also affects the circuit it measures. Since the ammeter has a very low resistance, its effect can be neglected in all but circuits of very low resistance. Consider the circuit shown in Fig. 9-18(a). According to Ohm's law,

$$I = \frac{V}{R} = \frac{0.1}{0.1} = 1 \text{ A}$$

If, however, an ammeter with a resistance of 50 mΩ were inserted in the circuit, the current would be

$$I = \frac{V}{R} = \frac{0.1}{0.100 + 0.050} = 0.667 \text{ A}$$

This represents a 33.3% error and is called an *insertion loss*.

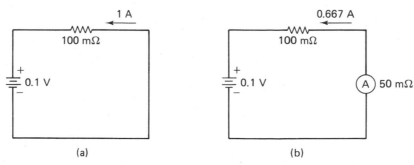

FIGURE 9-18: *Ammeter insertion loss: (a) original circuit; (b) ammeter inserted.*

It is quite easy to estimate the error caused by an ammeter. Knowing the ammeter resistance (this can be approximated by knowing most ammeters have a voltage drop of from 50 to 100 mV) and the Thévenized circuit resistance, we obtain

$$\% \text{ error} \approx \frac{R_m}{R_{\text{th}}} \times 100$$

Thus, in Fig. 9-18:

$$\% \text{ error} \approx \frac{R_m}{R_{\text{th}}} \times 100 = \frac{50}{100} \times 100 = 50\%$$

This should tell us that the error is substantial and cannot be discounted.

EXAMPLE 9-13: Approximate the insertion error of Fig. 9-19.

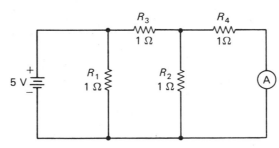

FIGURE 9-19: *Example 9-13.*

SOLUTION: The Thévenized resistance is

$$R_{\text{th}} = (R_2 \,\|\, R_3) + R_4$$

Note that when the 5-V battery is shorted, it shorts R_1.

$$R_{\text{th}} = 1.500 \ \Omega$$

A rough guess of the current would be about 3 A. Thus, assuming that we use the 5-A scale of the ammeter and it has a 50-mV drop:

$$R_m = \frac{V}{I} = \frac{50 \text{ mV}}{3 \text{ A}} = 16.67 \text{ m}\Omega$$

$$\% \text{ error} = \frac{R_m}{R_{\text{th}}} \times 100 = \frac{16.67}{1500} \times 100$$

$$= 1.11\%$$

9-8. SUMMARY

There are two basic types of displays used for meters: the scale type and the digital type. The digital type requires no interpretation, whereas the scale type can be subject to parallax errors. Moving-coil meters operate using magnetic fields, and such meter movements have the resistance and the full-scale current specified by the manufacturer. Voltmeters are constructed by inserting multiplying resistors in series with the meter movement, whereas ammeters require a resistor in parallel with the movement. An Ayrton shunt is used in ammeters to protect the meter movement while switching scales. The ohms/volt rating of a voltmeter is equal to the inverse of the full-scale current. The accuracy of meters is expressed in percent of full scale rather than percent of reading; thus, one should try to use the upper one-third of a scale for improved accuracy. An ohmmeter has an internal battery to supply current to the resistance under test, and a nonlinear scale reading from right to left. The VOM is a combination voltmeter, ohmmeter, and ammeter. Any meter affects the circuit it is measuring. The effect can be estimated by considering the Thévenized resistance of the circuit and the resistance of the meter. Electronic voltmeters usually have a constant resistance of 11 MΩ, whereas VOMs have a resistance that varies with the scale selected.

9-9. REVIEW QUESTIONS

9-1.   What two types of displays are used for meters?

9-2.   What three sources of energy are used for meters?

9-3.   What is a d'Arsonval meter movement? Describe it.

9-4.   What are the two specifications usually given for a meter movement?

9-5.   What is the resistance of an ideal voltmeter?

9-6.   How can a basic meter movement be converted to a voltmeter?

9-7.   How are the ohms/volt rating and the full-scale current rating related?

9-8.   Meter accuracy is usually specified in percent of _____.

9-9. What is the advantage of using a divider string rather than separate resistors for a multirange voltmeter?

9-10. What is the resistance of an ideal ammeter?

9-11. How can a basic meter movement be converted into an ammeter?

9-12. What is the advantage of the Ayrton shunt over the separate resistor shunt?

9-13. Name the three components necessary to construct a basic ohmmeter.

9-14. In what two ways does the basic ohmmeter scale differ from the basic voltmeter scale?

9-15. What is the purpose for a higher-voltage battery in a multirange ohmmeter?

9-16. Voltmeter _____ is the phenomenon whereby the voltage decreases when the voltmeter is connected to a circuit.

9-17. How can the effect of a voltmeter on a circuit be estimated?

9-18. What is the usual resistance of an electronic voltmeter?

9-19. How does the resistance characteristic of an electronic voltmeter differ from that of a VOM?

9-20. When an ammeter is placed in a circuit, the current decreases. This is called _____ _____ .

9-10. PROBLEMS

9-1. Design a 5-V FS voltmeter from a 1-mA 100-Ω meter movement.

9-2. Design a 10-V FS voltmeter from a 100-μA 500-Ω meter movement.

9-3. Calculate the ohms/volt rating for the meter in (a) Prob. 9-1 and (b) Prob. 9-2.

9-4. A meter has an accuracy of 5%. What is the permissible range of voltage readings on the 50-V scale when 20 V is impressed?

9-5. A meter has an accuracy of 3% and reads 26.0 V on the 50-V scale when 24.8 V is actually impressed. Is the meter within tolerance?

9-6. Using a 1-mA 100-Ω meter movement, design a multirange voltmeter having 1.5-, 10-, 20-, and 300-V scales. Use a divider string.

9-7. Using a 50-μA 1500-Ω meter movement, design a multirange voltmeter having 10-, 30-, 100-, and 750-V scales. Use a divider string.

9-8. Design a 10-mA ammeter from a 1-mA 150-Ω meter movement.

9-9. Design a 2-mA ammeter from a 100-μA 1-kΩ meter movement.

9-10. Design an ammeter having ranges of 2 mA, 50 mA, 200 mA, and 1 A using a 1-mA 100-Ω meter movement.

9-11. Design an ammeter having ranges of 100 μA, 750 μA, 20 mA, and 100 mA from a 50-μA 2000-Ω meter movement.

9-12. What is the approximate measurement error for the circuit in Fig. 9-20 if a 20-kΩ/V voltmeter were used on the 50-V scale?

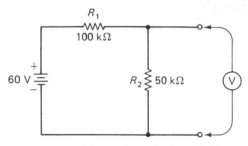

FIGURE 9-20: *Problems 9-12 and 9-13.*

9-13. What is the approximate error in Prob. 9-12 if the voltmeter were on the 150-V scale?

9-11. PROJECTS

9-1. (a) Measure the FS current of a meter movement using the diagram shown in Fig. 9-21(a). Make sure R is very large, at least 5 MΩ initially.

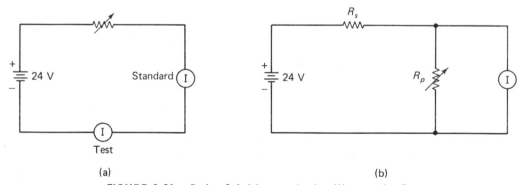

(a)   (b)

FIGURE 9-21: *Project 9-1: (a) measuring I_{FS}; (b) measuring R_m.*

(b) Measure the resistance of the meter movement according to the following procedure, Fig. 9-21(b):

CAUTION

Do not use an ohmmeter to measure meter resistance. The current supplied by the ohmmeter may permanently damage the delicate meter movement.

---

1. Open R_p and adjust R_s for a full-scale reading.

2. Connect R_p and adjust until the meter reads half scale; R_p will then equal R_m.

9-2. Construct a voltmeter with the following ranges: 1, 5, 10, and 50 V. Use a divider string.

9-3. Construct an ammeter with ranges of 5, 15, and 50 mA.

9-4. (a) Construct the circuit of Fig. 9-15 and verify meter loading with a VOM and with an electronic voltmeter.

(b) Construct the circuit of Fig. 9-16 and repeat the experiment.

9-5. Develop a formula for finding the resistance of a voltmeter using only one power supply and a known value of resistance.

9-6. Write a biographical sketch of Edward Weston.

9-7. Describe the hot-wire ammeter.

PART II

ALTERNATING CURRENT

In Part I we introduced direct current, a phenomenon whereby current flows in only one direction. However, much of the work in electronics involves *alternating current* (ac), a system in which charges reverse direction along a conductor at regular time intervals. Like the man who cannot make up his mind, much energy is expended, with charges going nowhere. However, when we harness this energy, we can send messages to the moon, transmit television into our homes, and power giant motors for our mills. In Part II we shall examine the origins of alternating current, the characteristics of it, and the components used to make up ac circuits. The chapters include:

10

MAGNETISM

Most of us have observed the curious workings of magnetism—the action of a compass, the repulsion of two magnets, the attraction of iron materials to the magnet. We even speak of our having a magnetic personality. Thus, we think of magnetism independently from electricity. However, in 1819, Hans Christian Oersted showed that electricity and magnetism are closely linked; magnetism can cause electricity and electricity causes magnetism. We in electronics are constantly faced with its effects in such varied devices as meters and memories, recorders and relays. This chapter introduces magnetism, discussing first some general external properties of magnetism, then the units we apply to these properties, and, finally, the workings of magnetism at the atomic level. The sections include:

10-1. The magnet

10-2. The electromagnet

10-3. Total magnetic units

10-4. Per-unit magnetic terms

10-5. The physics of magnetism

10-6. Applications of magnetism

10-1. THE MAGNET

The science of magnetism extends back to the dawn of history. As early as 2600 B.C. the Chinese were aware of its effects and by the second century A.D. they were guiding their ships using a magnetic compass. The lodestone (literally *leading stone*) is a magnet that occurs in nature, and is named after its north-seeking ability. It is formed from a material called *magnetite* and was the basis for this early compass.

We know now that certain metals can be made into magnets by stroking them with an existing magnet. Some retain this magnetism very well, thus possessing a high degree of retentivity. Others lose it as soon as the magnetizing force is removed, having low retentivity. Those magnets with high retentivity, such as shown in Fig. 10-1, are called *permanent magnets*.

The earth itself is a giant permanent magnet with two poles, a north pole and a south pole. When a bar magnet is suspended in air, it tends to align itself in such a manner that the north-seeking pole of the bar magnet (marked N) points to the north pole of the earth, and the south-seeking pole (marked S) points to the earth's south pole, Fig. 10-2. Thus, every magnet has two opposite poles.

The field of a magnet consists of many magnetic lines of force extending from the north pole of the magnet to the south pole, Fig. 10-3. The stronger the magnet is, the more lines of force it has. We measure these lines of force in a unit called a *weber* (Wb), there being 10^8 lines of force in 1 Wb. It is easier to remember that there are 100 lines of force in 1 microweber (μWb).

When the north pole of one magnet is placed next to the south pole of another magnet, the lines of force extend from one to the other and there is a force of attraction between the poles. We can observe this phenomenon by sprinkling iron filings on a paper placed over the magnets, Fig. 10-4. When the north pole of one magnet is placed next to the north pole of another, the two fields interact, causing a force of repulsion between the magnets, Fig. 10-5. We can summarize the actions of magnetic poles by observing: Like poles repel, unlike poles attract.

The repulsion of like poles can best be explained by recognizing that these lines of force form complete loops and will not cross one another. The attraction of unlike poles is caused by the tendency of lines of force to form the smallest closed loop possible; if two materials can be moved closer together to form a smaller loop, they will be forced closer together. Furthermore, these lines of force will travel the path of least resistance, preferring metals such as iron to air.

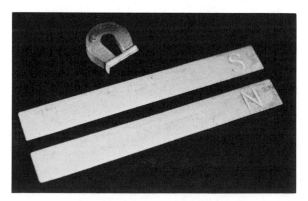

FIGURE 10-1: *Magnets (photo by Ruple).*

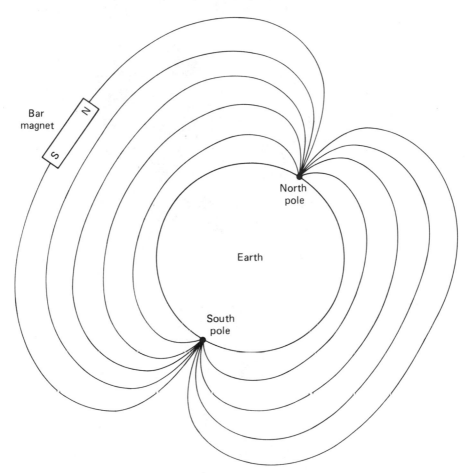

FIGURE 10-2: *Bar magnet in Earth's magnetic field.*

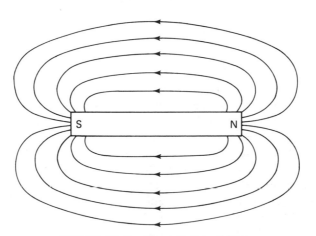

FIGURE 10-3: *Magnetic lines of force.*

207

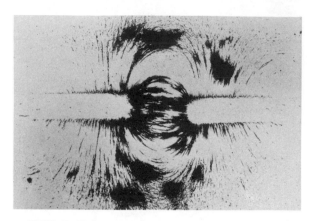

FIGURE 10-4: *Unlike poles attract (photo by Ruple).*

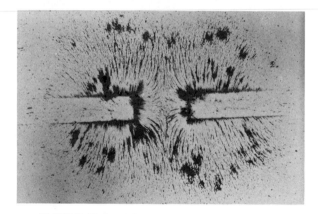

FIGURE 10-5: *Like poles repel (photo by Ruple).*

10-2. THE ELECTROMAGNET

It was in 1819 that Oersted observed that a compass was influenced by electrical current within a wire. He found that the greater the current, the greater the compass tended to align itself perpendicular to the wire. We now know that a wire carrying current generates concentric lines of magnetic force around the wire, Fig. 10-6(a).

Next, if we were to form a coil from a length of wire, the lines of force would be more concentrated in the center of the coil, Fig. 10-6(b). By adding more coils, the field can be further concentrated, forming a virtual magnet with a north pole and a south pole, Fig. 10-6(c). Note that if the coil were grasped with the left hand such that the fingers point in the direction of current flow, the thumb would point to the north pole of the coil, Fig. 10-6(d). This is called the *left-hand rule* for coils.

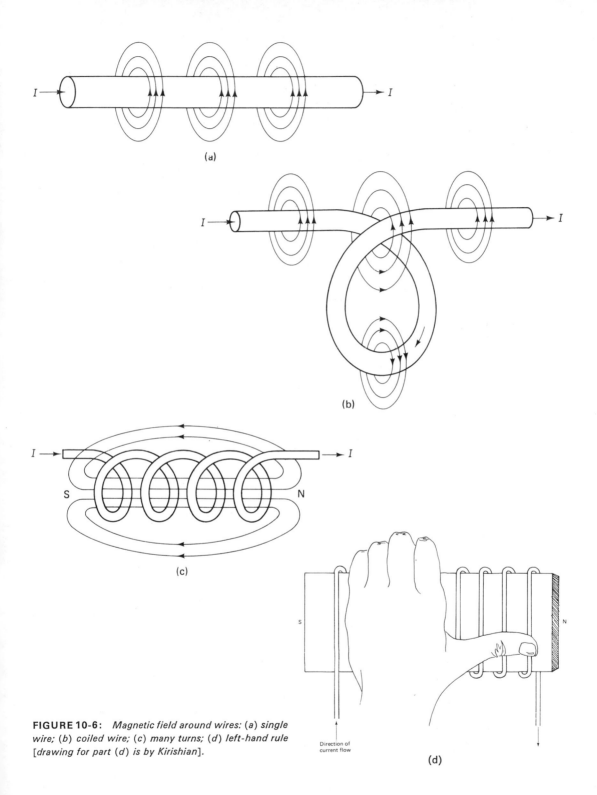

FIGURE 10-6: *Magnetic field around wires: (a) single wire; (b) coiled wire; (c) many turns; (d) left-hand rule [drawing for part (d) is by Kirishian].*

Direction of current flow

209

10-3. TOTAL MAGNETIC UNITS

From the foregoing discussion it is apparent that we can increase the number of lines of force in two ways:

1. Increase the number of turns in the coil.
2. Increase the current within the coil.

Thus, we are saying that a force causes more lines. This force is called *magnetomotive force* (*F*) and is expressed in ampere-turns (abbreviated A). Its equation is

$$F = NI$$

where *F* is the magnetomotive force, *N* the number of turns of the coil, and *I* the current within the coil in amperes.

EXAMPLE 10-1: A 100-turn coil has 50 mA flowing through it. What is the magnetomotive force produced?

SOLUTION: Solve for *F*:

$$F = NI$$
$$= 100 \times 50 \times 10^{-3} = 5 \text{ A}$$

EXAMPLE 10-2: If the current in Example 10-1 is doubled, what is the magnetomotive force produced?

SOLUTION:

$$F = NI$$
$$= 100 \times 100 \times 10^{-3} = 10 \text{ A}$$

Note that doubling the current doubles the force. Doubling the number of turns also doubles the force.

Magnetic Relationships

We have, therefore, a force (ampere-turns) producing a result (lines of force). We should also note that this force meets a resistance that, in magnetics, is called *reluctance*. These three properties are quite analogous to electricity in which a force (voltage) is opposed (resistance) in its attempt to produce a result (current). Thus:

1. Magnetomotive force (*F*) is analogous to electromotive force (*V*).
2. Lines of force (Φ) is analogous to current (*I*).
3. Reluctance ($\mathfrak{R}$) is analogous to resistance (*R*).

We can relate the three electrical properties by Ohm's law, $V = IR$. Similarly, we can relate the three magnetic units:

$$F = \Phi \mathfrak{R}$$

where F is magnetomotive force expressed in ampere-turns, Φ the magnetic lines of force (also called magnetic flux) expressed in webers, and $\mathfrak{R}$ is reluctance expressed in ampere-turns per weber. Note that reluctance does not have its own units but is expressed in terms of F and Φ.

EXAMPLE 10-3: A force of 20 ampere-turns produces a flux of 1500 μWb. What is the reluctance of the material?

SOLUTION:

$$F = \Phi \mathfrak{R}$$

Therefore,

$$\mathfrak{R} = \frac{F}{\Phi} = \frac{20}{1500 \times 10^{-6}} = 13,333 \text{ A/Wb}$$

EXAMPLE 10-4: A magnetomotive force of 10 A is applied to a medium having a reluctance of 10,000 A/Wb. What is the resultant flux?

SOLUTION: Since

$$F = \Phi \mathfrak{R},$$

$$\Phi = \frac{F}{\mathfrak{R}} = \frac{10}{10,000} = 1000 \ \mu\text{Wb}$$

There is one more definition we should introduce here, permeance. *Permeance* is the inverse of reluctance and, thus, is equivalent to conductance in electricity.

These magnetic relationships are summarized in Table 10-1. In addition we have the following equations:

$$F = NI \qquad \text{force} = \text{amperes} \times \text{turns}$$

$$F = \Phi \mathfrak{R} \qquad \text{force} = \text{flux} \times \text{reluctance}$$

TABLE 10-1: *Total magnetic units.*

Property	Symbol	Units	Electrical Equivalent
Force	F	Ampere-turns (A)	Voltage
Flux	Φ	Webers (Wb)	Current
Reluctance	$\mathfrak{R}$	Ampere-turns per weber (A/Wb)	Resistance
Permeance	$\mathcal{P}$	Webers per ampere-turn (Wb/A)	Conductance

10-4. PER-UNIT MAGNETIC TERMS

In the foregoing discussion we expressed total flux in webers. However, compare a coil 1 cm in diameter producing a flux of 1000 μWb with a coil 1 m in diameter producing the same 1000 μWb. Would not the 1-cm-diameter coil have a more dense magnetic field per unit area than the 1-m-diameter coil? Thus, in magnetics we have a unit of flux density (B) such that

$$B = \frac{\Phi}{A}$$

That is, flux density, measured in teslas (T), is equal to flux, measured in webers, divided by the area in square meters over which this flux extends.

EXAMPLE 10-5: A coil 2 cm in diameter has a magnetic flux of 20,000 μWb. What is the flux density of the magnetic field within the center of the coil?

SOLUTION: The area of the coil's center is $A = \pi r^2$, where r is 1 cm or 0.01 m. Thus,

$$B = \frac{\Phi}{A} = \frac{20,000 \times 10^{-6}}{\pi \times (0.01)^2} = 63.66 \text{ T}$$

Consider a 100-turn coil with 1 A flowing through it. Would not the field (or magnetic "push") be more intense if the coil length were 1 cm rather that 1 m? Thus, we have another unit, called *field intensity*, defined as the magnetomotive force per unit length:

$$H = \frac{F}{l}$$

where H is field intensity measured in ampere-turns per meter (A/m), F is magneto-motive force measured in ampere-turns, and l is length in meters.

EXAMPLE 10-6: A 100-turn coil is 50 mm in length and has 20 mA flowing through it. What field strength is produced by the coil?

SOLUTION: First, we must compute magnetomotive force:

$$F = NI = 100 \times 0.02 = 2 \text{ A}$$

Knowing this, we obtain

$$H = \frac{F}{l} = \frac{2}{50 \times 10^{-3}} = 40 \text{ A/m}$$

Finally, since flux density involves area and field intensity involves length, we can relate these using a term analogous to conductance, called *permeability*:

$$\mu = \frac{B}{H}$$

where μ is permeability. We can determine the units of μ by substituting the units for B and H:

$$\mu = \frac{B}{H} = \frac{\Phi/\text{m}^2}{F/\text{m}} = \frac{\Phi}{F \cdot \text{m}}$$

where m is meters. But, since $F = NI$:

$$\mu = \frac{\Phi}{NI \cdot \text{m}}$$

Thus, the units of μ are webers per ampere-turn-meter.

> **EXAMPLE 10-7**: A certain material has a μ of 5000. What field intensity is required to produce a flux density of 1 Wb/m²?
>
> *SOLUTION:* Since $\mu = B/H$,

$$B = \mu H = 5000 \times 1 = 5000 \text{ T}$$

Since the permeability of free space is $4\pi \times 10^{-7}$, we can relate the permeability of a material to that of free space. This is called its *relative permeability*. Thus,

$$\mu_r = \frac{\mu}{\mu_0}$$

where μ_0 is the permeability of free space.

Finally, we can summarize the relationships of magnetic units in Table 10-2. In addition, we have the following formulas:

$$B = \frac{\Phi}{\text{area}} \qquad H = \frac{F}{\text{length}} \qquad \mu = \frac{B}{H}$$

TABLE 10-2: *Per-unit magnetic terms.*

Property	Symbol	Units	Electrical Equivalent
Field intensity	H	Ampere-turns per meter (A/m)	Voltage
Flux density	B	Teslas (T)	Current
Permeability	μ	Wb/A·m or T·m/A	Conductance

Other Units

Although the SI system of units is the accepted one, some *cgs units* are still used within the electronic community. The gauss, in particular, is still the basis for the gaussmeter, a device used to measure flux; removing the magnetic field from a television set or recording head is still called degaussing, rather that deteslaing. This cgs (centimeter-gram-second) system of units is shown in Table 10-3.

TABLE 10-3: *Cgs system of magnetic units.*

Property	Units	Comments
Flux	Maxwells (Mx)	1 Mx = 1 line of force
Magnetomotive force	Gilbert (Gb)	1 Gb = 0.7958 ampere-turn
Flux density	Gauss (G)	1 G = 1 Mx/cm²
Field intensity	Oersted (Oe)	1 Oe = 1 Gb/cm

10-5. THE PHYSICS OF MAGNETISM

It is only recently that the physics of magnetism has been fairly well understood. First, let us note that any electron in motion creates a magnetic field about it, just as any wire carrying current has a field associated with it. In the atom, electrons are both in motion around the nucleus, and spinning much as the earth revolves on its axis. In some materials, the net effect of its electron motion results in the total magnetic field of the atom being zero, whereas in other materials, this net effect is not zero.

Magnetic Classifications

As a result of these differing electron spin characteristics, materials fall into one of four magnetic classifications: nonmagnetic, diamagnetic, paramagnetic, and ferromagnetic. *Nonmagnetic* materials are those, such as glass and wood, which do not react to a magnetic field. That is, the total magnetism due to electron spin is zero and remains zero under the influence of such a field. However, a bar of *diamagnetic* material, such as zinc, lead, copper, or oxygen, reacts very weakly to a magnetic field, aligning itself in a plane perpendicular to the impressed field. A bar of *paramagnetic* material, such as aluminum and platinum, reacts in a similar manner, aligning itself parallel to the field. Finally, those materials such as iron, nickel, and cobalt which react strongly to such a field are called *ferromagnetic* substances. These are the elements used for magnets in use today.

Domains

Physicists have determined that in these ferromagnetic substances the orbits of a group of adjacent atoms tend to align themselves such that the magnetic field of this "group" is all in the same direction. These groups, called *domains*, consist of on the order of 10^{12} atoms and arrange themselves in serpentlike configurations as illustrated in Fig. 10-7(a). When subjected to an external magnetic field, the boundaries of adjacent domains move in such a manner that domains aligned with the external field

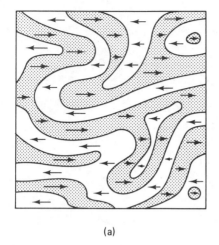

(a)

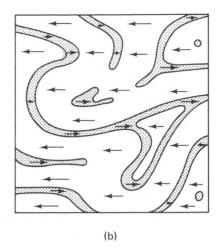

(b)

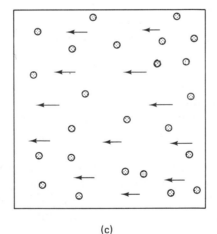

(c)

FIGURE 10-7: *Magnetic domains: (a) no external field; (b) mild external field; (c) strong external field.*

tend to grow, and those opposed to the external field tend to shrink, Fig. 10-7(b). If the intensity of the magnetic field is further increased, the opposing domains shrink further until they are just small cylinders, called *bubbles*, Fig. 10-7(c). Further increase in the external field results in these bubbles disappearing completely. When this occurs, the material is said to be in *magnetic saturation*.

These bubbles can be moved about much as a block of wood can be moved over water, and form the basic principle of the recent bubble memory developed for computer systems.

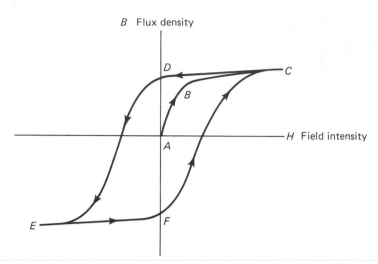

FIGURE 10-8: BH *curve.*

The BH Curve

Figure 10-8 is a graph of flux density (B) versus field intensity (H) for a magnetic material. If the material is totally demagnetized, at zero field intensity the flux density would be zero, point A on the graph. That is, with no external field, there would be no lines of force produced by the material. As field intensity is increased, the flux density increases until it reaches point B. Here the curve starts to flatten out, caused by the disappearance of opposing domains. As flux density is further increased, a point is reached in which all the domains are aligned with the field and no further flux can be obtained, point C on the graph. The material is then said to be saturated.

If we were now to reduce the field intensity to zero, we would find that the domains would not completely return to their previous alignment and the material would retain a magnetic field, point D on the graph. This magnetic field is called a *residual field*, and the measure of a material's capability of retaining a large residual magnetism is called its *retentivity*.

As we apply a field in the negative direction, we eventually reach saturation, point E. By now, reducing the field to zero, we still retain a residual magnetism, point F.

This curve we have examined is called a *hysteresis curve*, after a Greek word meaning "lagging behind." We shall find that hysteresis occurs in many electronic as well as magnetic circuits.

10-6. APPLICATIONS OF MAGNETISM

Those in electronics constantly use the properties of magnetism for performing tasks. We shall mention a few of these applications in this section.

Solenoids and Relays

The *solenoid* is a very useful device for converting electrical energy into linear motion, Fig. 10-9, and is used in such equipment as electrically operated valves, electric door locks in cars, and electric latches for door hatches in aircraft. Current flowing through the coil creates a magnetic field, causing the armature to pull sharply toward the center of the coil. This motion can then be mechanically coupled to a variety of systems.

When this mechanical action is coupled to electrical contacts, the device is called a *relay*, Fig. 10-9(c). Current through the coil creates an electromagnet out of the core, pulling the switch contacts to the closed position. Spring tension causes the contacts to open upon cessation of current.

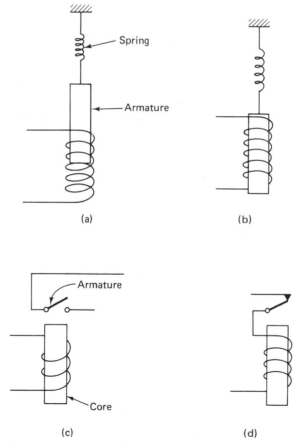

FIGURE 10-9: *Solenoids and relays: (a) solenoid de-energized; (b) solenoid energized (c) relay; (d) buzzer.*

A variation of the relay is the *buzzer*, Fig. 10-9(d). When a voltage is applied across its input leads, the coil energizes, opening the contacts. This deenergizes the coil, causing the contacts to return to their normally closed position. The mechanical action of the vibrating contacts creates the "buzz."

Meters

In the Weston meter movement, Fig. 10-10, current flowing through the coil creates a magnetic field about the core. This field opposes that of the permanent magnet, causing rotation of the coil assembly. The more current that is supplied the coil, the greater is this rotation. An indicating pointer is attached to the coil assembly and a scale provided for readout.

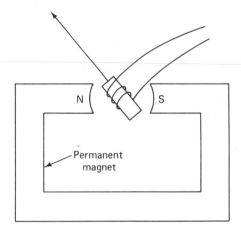

FIGURE 10-10:   *Weston meter movement.*

Recording Media

Magnetic tape recorders operate by supplying a variable current to the recording head, determined by the intensity of the signal, Fig. 10-11. The head then magnetizes particles within the thin magnetic film on the surface of the moving tape. In the reading process, the magnetic field on the tape is detected by a read head and converted back to a varying electrical signal.

Computing systems use similar equipment to record information on magnetic disks, Fig. 10-12. Shaped like a phonograph record, these disks are coated with fine magnetic particles and rotate at high speeds past recording and reading heads.

Computer Memories

Computers use small magnetic cores shaped like doughnuts to record numbers, Fig. 10-13. Current flowing in the X and Y leads in the direction indicated will create a magnetic field around the wires, magnetizing the cores in the CCW direction. This

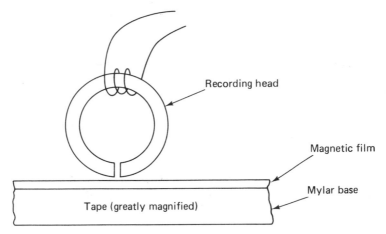

FIGURE 10-11: *Magnetic tape recording.*

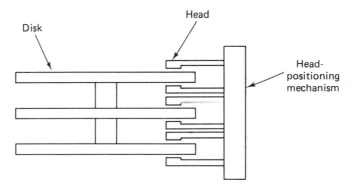

FIGURE 10-12: *Magnetic disk.*

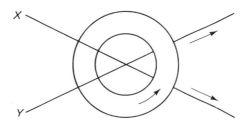

FIGURE 10-13: *Core memory.*

direction is defined to be a 1. Current flowing in the opposite direction on the X and Y leads will magnetize the core in the CW direction, defined as a zero. Thus, the core is able to store a one or a zero.

One of the newer developments is the bubble memory, where magnetic bubbles are lined up in a row, the presence of a bubble defined as a 1 and the absence of a

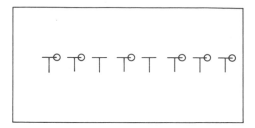

FIGURE 10-14: *Bubble memory.*

bubble a zero, Fig. 10-14. The bubbles are lined up by supplying an external magnetic field to the small T-shaped metal objects. When this field is rotated, the bubbles move from one T to the next.

10-7. SUMMARY

The effects of magnets have been known for centuries, the earth itself being one. A magnet has lines of force extending from its north pole to its south pole. Like poles repel, unlike poles attract. Current within a wire creates a magnetic field about the wire; this field becomes more intense as the wire is coiled. There are two systems of units in magnetism, both analogous to Ohm's law, summarized in Tables 10-1 and 10-2. Electron spin determines whether a material is magnetic, paramagnetic, diamagnetic, or nonmagnetic. Magnetic materials consist of small domains having a uniform field. The *BH* curve, also called a hysteresis curve, illustrates the relationship between flux density and field intensity. The electronic industry uses many magnetic devices such as relays, solenoids, meters, and storage media.

10-8. REVIEW QUESTIONS

10-1. What did Hans Christian Oersted discover?

10-2. A magnet that retains its magnetism after removing the external field is called a _____ magnet.

10-3. Every magnet contains a _____ pole and a _____ pole.

10-4. The unit of magnetic lines of force is the _____.

10-5. Like poles _____, unlike poles _____.

10-6. The direction of the magnetic field created by a coil may be found using the _____ _____ hand rule.

10-7. Magnetomotive force is measured in _____ and its symbol is _____.

10-8. Resistance to magnetomotive force is called _____ and its symbol is _____.

10-9. Magnetomotive force is analogous to _____ in electricity. Lines of force is analogous to _____ and reluctance is analogous to _____ _____.

10-10. The inverse of reluctance is _____ and its symbol is _____. This quality is analogous to _____ in electricity.

10-11. Flux density is measured in _____, which is _____ per unit _____. Its symbol is _____.

10-12. Field intensity is measured in _____ per unit _____. Its symbol is _____.

10-13. The measure of a material's ability to permit a certain field intensity is called its _____, and uses the symbol _____.

10-14. Flux density is analogous to _____ in electricity; field intensity is analogous to _____; permeability is analogous to _____.

10-15. Define relative permeability.

10-16. _____ determines whether a material is nonmagnetic, paramagnetic, diamagnetic, or magnetic.

10-17. Name two nonmagnetic materials, two paramagnetic materials, two diamagnetic materials, and two magnetic materials.

10-18. Groups of atoms aligned in the same magnetic direction are called _____.

10-19. Define saturation in terms of domains.

10-20. Small, cylindrical magnetic domains that oppose the applied field are called _____ _____.

10-21. Define saturation in terms of ability of the material to respond to a magnetizing force.

10-22. The S-shaped graph of *B* verses *H* is called a _____ curve after a Greek word meaning "lagging behind."

10-23. A coil that converts electrical motion into linear motion is called a _____.

10-24. What is the difference between a solenoid and a relay?

10-25. How does a Weston movement work?

10-26. Describe the recording process of a magnetic tape recorder.

10-27. Name two types of computer memories.

10-9. PROBLEMS

10-1. A coil having 1000 turns has 20 mA flowing through it. What is the magnetomotive force it produces?

10-2. How much current must flow through a 400-turn coil to produce a magnetomotive force of 10 ampere-turns?

10-3. A coil produces a magnetomotive force of 5 ampere-turns, resulting in a field of 100 μWb. What is the reluctance of the magnetic material?

10-4. A material has a reluctance of 1000 *NI*/Wb. How much current must flow through a 100-turn coil to produce 5000 μWb of flux?

10-5. What is the flux density in a coil 3 cm in diameter if the flux is 7500 μWb?

10-6. What is the total flux in a 2-cm diameter coil having a flux density of 3500 T?

10-7. A 200-turn coil is 25 mm in length and has 10 mA flowing through it. What field strength is produced by the coil?

10-8. What is the length of a coil with 200 turns having 20 mA flowing through it that is necessary to produce a field intensity of 100 ampere-turns per meter?

10-9. What is the permeability μ of a material if a field strength of 100 ampere-turns per meter produces a flux density of 20,000 T?

10-10. What is the relative permeability of the material in Prob. 10-9?

10-10. PROJECTS

10-1. Repeat Oersted's experiment by passing current through a wire and observing the action of a compass about the wire.

10-2. Construct an electromagnet from a wire coiled about a large nail. Vary the current through the coil and observe the strength of the field. Construct a graph of the current through the coil versus the weight of iron particles it will pick up.

10-3. This chapter has used the SI units to express magnetic relationship. What are the corresponding units in the cgs system?

10-4. Who was Gilbert and what did he discover?

10-5. Investigate how bubbles are created and destroyed in the bubble memory.

10-6. What materials are used for shielding magnetic lines of force?

11

INTRODUCTION
TO ALTERNATING CURRENT

Having laid the groundwork for the generation of alternating current by the discussion of magnetism, we shall now formally introduce the subject. We shall first study one method of generating alternating current, then we shall pick apart the resultant waveform in detail. We shall then study several methods of measuring these elusive charges, and, finally, examine the power system used within our homes. The sections include:

11-1. Generating alternating current

11-2. The sine wave

11-3. Expression of ac values

11-4. Measurement of ac values

11-5. 60-hertz power

11-1. GENERATING ALTERNATING CURRENT

There are several ways in which alternating current can be generated, but one of the most straightforward methods is that of the ac generator. If we were to move a wire through a magnetic field, Fig. 11-1, the electrons within that wire would be influenced by the magnetic field in such a manner that they would tend to flow to one end of the wire creating a voltage across the wire. The magnitude of this voltage is dependent upon how rapidly the wire cuts the magnetic field. If many lines per second are cut, a large voltage occurs; if only a few lines per second are cut, a small voltage occurs. This phenomenon is called *induction*, after Latin words meaning "lead into." Thus, the magnetic field *induces* a voltage into the wire.

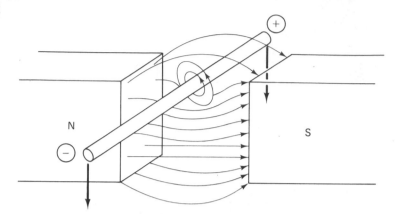

FIGURE 11-1: *Induction.*

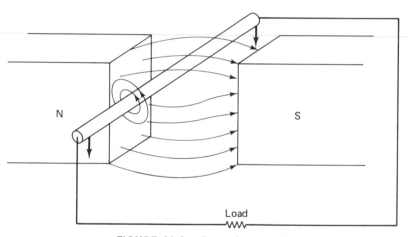

FIGURE 11-2: *Generating current.*

If we were now to connect this generating wire to a load, current would flow during the time the wire is cutting the field, Fig. 11-2. This current would, in turn, create a magnetic field around the generating wire in such a manner that the field would "bunch up" beneath the wire, tending to push it back up. Thus, the current generated within a wire flows in a direction such that its magnetic field opposes the action that created that current. If we were to push the wire down in the figure, the induced current would tend to push the wire back up. If we were to move the wire up, the induced current would tend to push the wire back down. This principle is known as *Lenz's law*.

If we were to connect a loop of wire as shown in Fig. 11-3, rotate the loop at a constant speed, and take the output from *A* and *B*, we would have a simple ac generator. The brushes contact the slip rings, removing and supplying current from the loop. Let us now examine what form this voltage takes as the loop rotates, Fig. 11-4. When the loop is as shown in Fig. 11-4(a), a maximum number of lines of force are

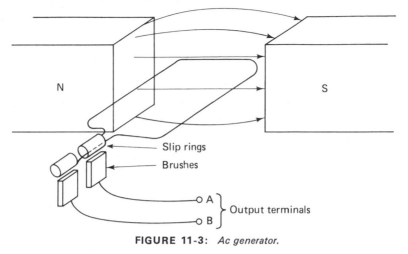

FIGURE 11-3: *Ac generator.*

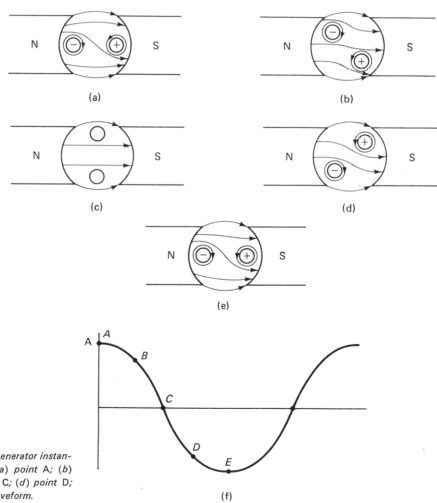

(a)

(b)

(c)

(d)

(e)

FIGURE 11-4: *Generator instantaneous voltage: (a) point A; (b) point B; (c) point C; (d) point D; (e) point E; (f) waveform.*

(f)

cut per unit time and the voltage is as shown at point *A* in Fig. 11-4(f) at the polarity shown. Note that this device is a generator and that electrons come out of the paper on the − lead and feed into the paper on the + lead.

As the loop proceeds as shown in Fig. 11-4(b), fewer lines are cut per unit time and less voltage is produced until, at Fig. 11-4(c), no lines are cut, resulting in zero volts. As rotation continues, Fig. 11-4(d), current reverses direction in the top and bottom wires and a negative voltage is produced. Finally, maximum negative voltage is produced at Fig. 11-4(e).

This waveform is called a *sinusoidal waveform*, or a *sine wave*, and will continue to be produced as long as rotation occurs. We shall analyze the sine wave in the following section.

11-2.   THE SINE WAVE

The name of the sine wave is derived from Fig. 11-5. If the vector *V* were to rotate at a uniform rate, and we were to plot *H* on a vertical axis and time on a horizontal axis, we would find it would plot the sine wave shown. This waveform has the equation

$$H = V \sin \alpha$$

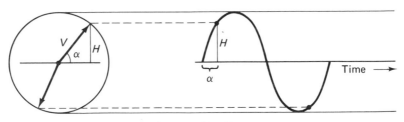

FIGURE 11-5:   *Sine wave.*

The angle α is the angle of the vector through its rotation cycle. *V* is the length of the vector; it is also the maximum positive height of the waveform, referred to as the waveform's maximum *amplitude*, and the maximum negative height of the waveform. *H* represents the height (or amplitude) of the waveform at the angle under consideration, α.

> **EXAMPLE 11-1:** What is the amplitude of a sine wave when it is 30° through its cycle, if the maximum amplitude is 12?
>
> *SOLUTION:* The equation for the sine wave is
>
> $$H = V \sin \alpha$$
>
> But *V* is the maximum amplitude and was given as 12: Therefore,

$$H = V \sin \alpha$$
$$= 12 \times \sin 30°$$
$$= 12 \times 0.5 = 6$$

EXAMPLE 11-2: At an angle of 52°, the amplitude of a waveform is 32. What is its maximum amplitude?

SOLUTION:

$$H = V \sin \alpha$$
$$32 = V \sin 52°$$
$$V = \frac{32}{\sin 52°} = 40.61$$

Thus, the maximum amplitude of the waveform is 40.61.

In electronics, we throw a few more variables into the basic sine equation. The complete equation for voltage is

$$v = V \sin (2\pi ft + \theta)$$

where v represents the instantaneous amplitude (in this case voltage) and V is the maximum amplitude (voltage) of the waveform. We shall discuss f (frequency) and θ (phase angle) separately, then return to this equation.

Frequency

Let us consider the waveform shown in Fig. 11-6. Note that one cycle of this waveform represents one complete repetition of the signal. Time has been plotted along the horizontal axis and, according to the graph, five cycles of this waveform occur

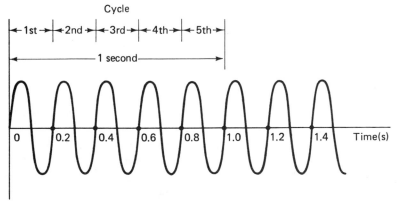

FIGURE 11-6: *Frequency.*

within 1 s. This number, five, represents the *frequency* of the waveform. We can define frequency as the number of repetitive waveforms occurring within 1 s of time. The unit of frequency is the *hertz*; thus, this signal has a frequency of 5 hertz (Hz). In electronics, signals occur all the way from 1 Hz to several GHz (10^9 Hz).

Period

Since frequency represents cycles per second, the inverse of frequency, the *period* of the waveform, is measured in seconds per cycle. Thus,

$$f = \frac{1}{P}$$

EXAMPLE 11-3: A signal has 365 cycles within 6 s. What is its frequency and period?

SOLUTION: Since frequency is cycles per second:

$$f = \frac{\text{cycles}}{\text{time}} = \frac{365}{6} = 60.83 \text{ Hz}$$

Its period is:

$$P = \frac{1}{f} = \frac{1}{60.83} = 16.44 \text{ milliseconds (ms)}$$

EXAMPLE 11-4: A signal source produces 4.69 million cycles over a period of 26.95 ms. What are the frequency and period of the waveform?

SOLUTION:

$$f = \frac{\text{cycles}}{\text{time}}$$

$$= \frac{4.69 \times 10^6}{26.95 \times 10^{-3}}$$

$$= 174.0 \text{ MHz}$$

$$P = \frac{1}{f} = \frac{1}{174.0 \times 10^6} = 5.746 \text{ ns}$$

Phase Angle

The *phase angle* is a measurement of the angle difference between two waveforms. In Fig. 11-7 waveform *A* is said to have a phase angle of 20° with reference to waveform *B*. Thus, it is said to lead *B* by 20°. One way to look at this is to pretend you are a lady (or gentleman) bug and start traveling along the horizontal axis from the extreme left toward the right. The first waveform you encounter is the waveform that leads. The second waveform you encounter is said to lag the first waveform by the difference in phase angle. Thus, the following are true statements about Fig. 11-7:

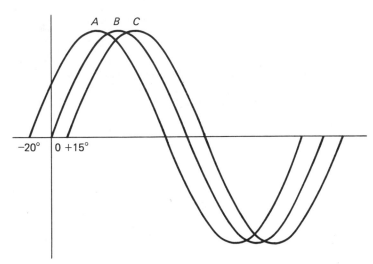

FIGURE 11-7: *Phase angle.*

1. *A* leads *B* by 20°.
2. *B* lags *A* by 20°.
3. *C* lags *A* by 35°.
4. *C* lags *B* by 15°.
5. *A* leads *C* by 35°.
6. *B* leads *C* by 15°.

Angles can be measured in both degrees and in units called *radians*. There are 2π radians in 360° (one complete rotation).

EXAMPLE 11-5: Convert the following to radians:

(a) 30°.   (b) 180°.   (c) 60°.   (d) −26°.

SOLUTION: To convert to radians, multiply degrees by $\pi/180$.

(a)   Angle $= 30 \times \dfrac{\pi}{180} = 0.5236$ rad.

(b)   Angle $= 180 \times \dfrac{\pi}{180} = 3.1416$ rad.

(c)   Angle $= 60 \times \dfrac{\pi}{180} = 1.047$ rad.

(d)   Angle $= -26 \times \dfrac{\pi}{180} = -0.4538$ rad.

EXAMPLE 11-6: Convert the following to degrees:

(a) 0.39 rad.   (b) 1.319 rad.   (c) −2.32 rad.   (d) 0.005 rad.

SOLUTION: To convert radians to degrees, multiply by 180/π.

(a) Angle $= 0.39 \times \dfrac{180}{\pi} = 22.35°$.

(b) Angle $= 1.319 \times \dfrac{180}{\pi} = 75.57°$.

(c) Angle $= -2.32 \times \dfrac{180}{\pi} = -132.9°$.

(d) Angle $= 0.005 \times \dfrac{180}{\pi} = 0.2865°$.

Angular Velocity

The general equation for a sine wave is

$$v = V \sin (2\pi f t + \theta)$$

The term $2\pi f$ represents the number of radians our rotating vector must pass through in 1 s. This is called the *angular velocity* of the waveform and is represented by the symbol ω. Note the following analysis of units:

$$\omega = 2\pi f$$
$$= \frac{2\pi \text{ radians}}{\text{cycle}} \times \frac{f \text{ cycles}}{\text{second}}$$
$$= 2\pi f \frac{\text{radians}}{\text{second}}$$

Thus, if we multiply radians per cycle by cycles per second, we obtain radians per second, which, indeed, represents an angular velocity.

If we were now to multiply this $2\pi f$ by t (time), we would have

$$\frac{\text{radians}}{\text{seconds}} \times \text{seconds} = \text{radians}$$

Observe, therefore, that $2\pi f$ represents radians per second, t represents seconds, and $2\pi f t$ represents an angle, radians. Since we can only find the sine of angles and we must find the sin $(2\pi f t + \theta)$, then $2\pi f t$ must represent an angle.

The Sine-Wave Equation

Now, having discussed each part of the equation in detail, let us put it together. The basic waveform for voltage at any time t is represented as

$$v = V \sin(2\pi ft + \theta)$$

where: $v =$ instantaneous amplitude in volts at time t
$V =$ maximum amplitude, in volts
$f =$ frequency, in hertz
$t =$ time, in seconds
$\theta =$ the phase angle, in radians or degrees
$2\pi f =$ the angular velocity, in radians/second
$2\pi ft =$ an angle in radians at a time, t

EXAMPLE 11-7: The equation for a certain waveform is

$$v = 160 \sin(366t + 15°)$$

What is this waveform's (a) maximum amplitude, (b) frequency, (c) period, (d) phase angle, (e) angular velocity, and (f) graph?

SOLUTION:

(a) The waveform's maximum amplitude is 160.

(b) Since the 366 represents $2\pi f$:

$$2\pi f = 366$$

$$f = \frac{366}{2\pi} = 58.25 \text{ Hz}$$

(c) $P = 1/f = 1/58.25 = 17.17$ ms.

(d) The phase angle is $+15°$ with reference to the signal, $v = 160 \sin(366t)$.

(e) The angular velocity is 366 rad/s.

(f) The waveform is shown in Fig. 11-8.

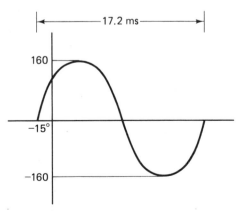

FIGURE 11-8: *Example 11-7.*

EXAMPLE 11-8: The equation for a certain current is

$$i = 26 \sin (2620t - 36°)$$

What is the current's (a) maximum amplitude, (b) frequency, (c) period, (d) phase angle, (e) angular velocity, and (f) graph?

SOLUTION:

(a) The maximum amplitude is 26 A.

(b) $\qquad \omega = 2\pi f$

$\qquad 2620 = 2\pi f$

$$f = \frac{2620}{2\pi} = 417.0 \text{ Hz.}$$

(c) $P = 1/f = 1/417.0 = 2.398$ ms.

(d) $\theta = -36°$.

(e) $\omega = 2620$ rad/s.

(f) The graph is shown in Fig. 11-9.

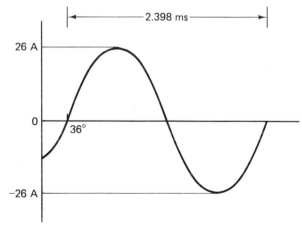

FIGURE 11-9: *Example 11-8.*

EXAMPLE 11-9: Find the current at $t = 30 \ \mu$s for the following waveform:

$$i = 16 \sin (360,000t)$$

SOLUTION: Merely substitute for t:

$$i = 16 \sin (360,000t)$$
$$= 16 \sin 10.800 \text{ rad}$$
$$= -15.695 \text{ A}$$

EXAMPLE 11-10: A certain voltage waveform is 20 V at 200 μs after $t = 0$. If its frequency is 1 kHz, what is the maximum amplitude of the signal?

SOLUTION: Note that we are given the instantaneous value v and must solve for V. Unless otherwise specified, phase angle should be assumed zero.

$$v = V \sin 2\pi ft$$

$$20 = V \sin (2 \times \pi \times 10^3 \times 200 \times 10^{-6})$$

$$V = \frac{20}{\sin 1.257 \text{ rad}} = 21.03 \text{ V}$$

Harmonics

A *harmonic* is a multiple of a given, fundamental, frequency. Thus, the second harmonic of 1 kHz is 2 kHz, the third harmonic 3 kHz, the tenth harmonic 10 kHz. The first harmonic is considered the fundamental frequency. All repetitive waveforms, regardless of shape, can be thought of as the sum of an infinite number of harmonics. For example, a square wave can be constructed from odd-numbered harmonics added in proper proportions, Fig. 11-10. The equation for this waveform is

$$v = V_m(\sin \omega t + \tfrac{1}{3} \sin 3\omega t + \tfrac{1}{5} \sin 5\omega t + \tfrac{1}{7} \sin 7\omega t + \dots)$$

The fundamental is the term $\sin \omega t$. Each harmonic is indicated by the number preceding the ω. Thus, the seventh harmonic of the fundamental is $\sin 7\omega t$. This seems reasonable since its angular velocity, 7ω, is seven times faster than that of the

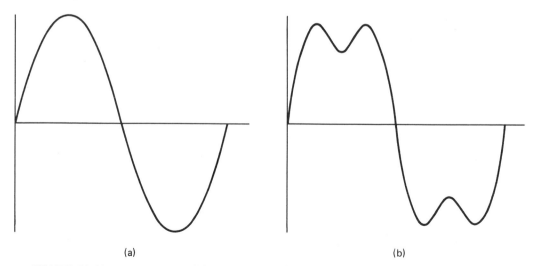

(a)   (b)

FIGURE 11-10: *Square wave: (a) fundamental; (b) fundamental and third harmonic; (c) fundamental, third and fifth harmonics; (d) final square wave.*

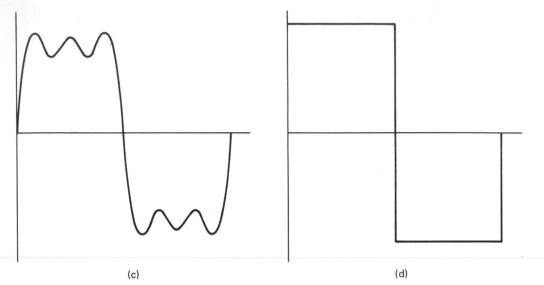

(c)   (d)

FIGURE 11-10:   *Continued.*

fundamental, ω. Note that the fractions in the equation indicate the maximum amplitude of the third harmonic is $\frac{1}{3}$ that of the fundamental; the amplitude of the fifth harmonic is $\frac{1}{5}$ that of the fundamental; the amplitude of the nth harmonic is $1/n$ that of the fundamental, n being any odd integer.

Thus, the square wave is the sum of an infinite number of odd harmonics, each of proper amplitude. Other nonsinusoidal, repetetive waveforms are merely the sum of harmonics with the proper amplitude, frequency, and phase angle.

Wavelength

The signal from a radio or television station is an ac waveform that travels at approximately the speed of light, 300 Mm/s. If we were to take a snapshot of the waveform at a particular time (and we could see a radio wave), we would find that the distance from zero degrees on one cycle to zero degrees on the next cycle would be 300 m for a 1-MHz signal. This distance is called the waveform's *wavelength*, and may be computed according to the formula

$$v = f\lambda$$

where v is the velocity in meters/second at which the waveform travels, f is its frequency in hertz, and λ is its wavelength in meters. Note that, for radio waves, v is 300×10^6 m/s.

> **EXAMPLE 11-11:** What is the wavelength of a radio station operating on a frequency of 570 kHz?

SOLUTION: Since $v = f\lambda$,

$$\lambda = \frac{v}{f} = \frac{300 \times 10^6}{570 \times 10^3} = 526.3 \text{ m}$$

EXAMPLE 11-12: The velocity of a sound wave through dry air at 20°C is 344 m/s. What is the wavelength of an 800-Hz tone as it travels through this medium?

SOLUTION: Since $v = f\lambda$,

$$\lambda = \frac{v}{f} = \frac{344}{800} = 43.0 \text{ cm}$$

11-3. EXPRESSION OF AC VALUES

When we speak to one another about a waveform we can specify its frequency. But how can we specify the amplitude of a waveform if it is constantly changing? We shall see that there are four characteristics of amplitude we can identify: peak amplitude, peak-to-peak amplitude, average amplitude, and rms amplitude.

Peak Measurements

There are two descriptions of amplitude involving the word "peak." The first, the *peak amplitude*, is the amplitude from the 0-V line to the maximum positive peak of the waveform. Thus, for Fig. 11-11(a) it would be 100 V.

The second, called *peak to peak* (p-p), is the amplitude from the signal's maximum negative excursion to its maximum positive excursion, in this case 200 V. Thus, the peak amplitude of the waveform in Fig. 11-11(a) is 100 V, the peak-to-peak amplitude, 200 V. In a similar manner, the peak amplitude of the waveform in Fig. 11-11(b) is 5 V, and its p-p amplitude, 10 V.

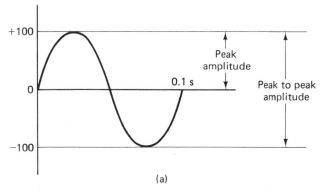

(a)

FIGURE 11-11: *Peak amplitudes: (a) sine wave; (b) rectangular waveform.*

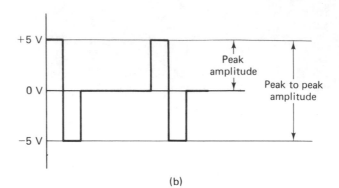

(b)

FIGURE 11-11: *Continued.*

Average Amplitude

The average amplitude of a waveform tells us something about its long-term characteristics. To find the average of the waveform in Fig. 11-12(a):

1. Find the total area underneath the waveform for one period.
2. Divide this by the period.

Thus, we have merely to find the area indicated by the shaded portion, and divide this by the period. We can total it as follows:

Portion	Height	Width	Area
A	10	1	10
B	10	1	10
C	5	1	5
D	0	1	0
Total			25

Since the total width is 4, the average is

$$\text{average} = \tfrac{25}{4} = 6.25$$

Thus, the average is 6.25 V.

> **EXAMPLE 11-13:** Find the average voltage of the waveform in Fig. 11-12(b).
>
> *SOLUTION:* Since the formula for the triangle, *A*, is
>
> $$A_A = \tfrac{1}{2}lh$$
> $$= \tfrac{1}{2} \times 3 \times 15 = 22.50$$

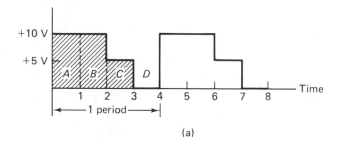

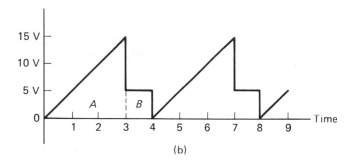

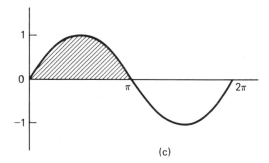

FIGURE 11-12: *Average of a waveform: (a) rectangular waveform; (b) Example 11-13; (c) sine wave.*

The area of the rectangle is

$$A_B = lh = 1 \times 5 = 5$$

The total area is

$$A_t = A_A + A_B = 22.5 + 5 = 27.5$$

The average is equal to the area divided by the period

$$V_{av} = \frac{A_t}{P} = \frac{27.5}{4} = 6.875 \text{ V}$$

The method for finding the average of a sinusoidal waveform is identical to that of the rectangular and the triangular waveforms. Note, however:

1. The average over the complete waveform is zero. We shall average only one-half cycle.

2. Finding the area underneath a sine wave is no easy trick.

It turns out that the area of the shaded part of Fig. 11-12(c) is, by calculus, equal to 2.000. Thus, the average is

$$V_{av} = \frac{area}{period} = \frac{2}{\pi} = 0.6366$$

Thus, we can state that, for sinusoidal waveforms only, the average amplitude is equal to the peak amplitude multiplied by 0.6366:

$$V_{av} = 0.6366 \times V_{pk}$$

EXAMPLE 11-14: Find the average current for the following sine waves:

(a) A peak current of 20 A.

(b) A p-p current of 45.6 A.

(c) A p-p current of 20.6 mA.

SOLUTION:

(a) $V_{av} = 0.6366 \times V_{pk} = 0.6366 \times 20 = 12.73$ A.

(b) Since the peak current is one-half the p-p current:

$$V_{av} = 0.6366 \times \frac{V_{p\text{-}p}}{2}$$

$$= 0.6366 \times \frac{45.6}{2}$$

$$= 14.51 \text{ A}$$

(c) $V = 0.6366 \times \dfrac{V_{p\text{-}p}}{2}$

$$= 0.6366 \times \frac{20.6}{2}$$

$$= 6.557 \text{ V}$$

Root-Mean-Square

The *root-mean-square* (rms), or *effective*, value of an ac voltage is that value which would cause the same heating effect as an identical dc voltage. Thus, 10-V ac rms across a 10-kΩ resistor will produce the same heat as 10 V dc. This is, by far, the most common measurement of an ac waveform, for it relates ac to dc.

Rms means, literally, the root of the mean of the square. It can be found mathematically by:

1. Squaring the waveform.

2. Finding the mean (average) of this squared waveform.

3. Taking the square root of the result.

> **EXAMPLE 11-15:** Find the average and rms of the voltage waveform shown in **Fig. 11-13(a)**.
>
> *SOLUTION:* The average is:

Portion	Area ($l \times h$)
A	5
B	2
C	20
D	0
Total	27

$$V_{av} = \frac{\text{area}}{\text{period}} = \frac{27}{5} = 5.400 \text{ V}$$

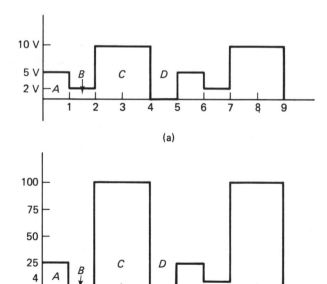

(a)

(b)

FIGURE 11-13: *Rms of a rectangular waveform: (a) original waveform; (b) squared waveform.*

The rms requires that the waveform first be squared, Fig. 11-13(b). This means that every vertical point must be squared. Next the mean (average) of the square is computed:

Portion	Area $(l \times h)$
A	25
B	4
C	200
D	0
Total	229

$$h_{av} = \frac{\text{area}}{\text{period}} = \frac{229}{5} = 45.80$$

Thus, the mean of the square is 45.80. We must now find the root of the mean of the square (rms):

$$V_{rms} = \sqrt{45.80} = 6.768 \text{ V}$$

This means that the waveform is 6.768 V rms and will have the same heating effect as a 6.768-V dc source. Note that the average is not the same as the rms, but is 5.400 V for this waveform.

EXAMPLE 11-16: Find the average and rms of the waveform in Fig. 11-14(a).

SOLUTION: We shall find the average of the absolute current and shall assign all currents positive.

$$V_{av} = \frac{(10 \times 1) + (0 \times 1) + (5 \times 2) + (0 \times 1)}{5}$$

$$= 4.000 \text{ V}$$

For rms, the waveform is squared, Fig. 11-14(b), then averaged:

$$h_{av} = \frac{(100 \times 1) + (0 \times 1) + (25 \times 2) + (0 \times 1)}{5} = 30.00$$

$$V_{rms} = \sqrt{30} = 5.477 \text{ V}$$

The rms of a sinusoidal waveform is found in the same manner, Fig. 11-15(a). First, the waveform is squared, Fig. 11-15(b). The shaded area turns out to be $\pi/2$. Thus,

$$h_{av} = \frac{\pi/2}{\pi} = 0.5000$$

$$V_{rms} = \sqrt{0.5000} = 0.7071$$

Thus, the rms of a sinusoidal waveform is 0.7071 times its peak value.

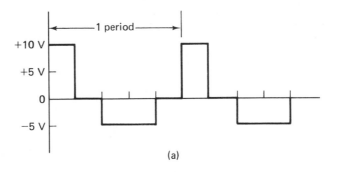

(a)

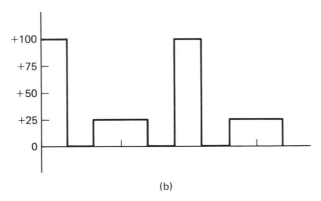

(b)

FIGURE 11-14: *Example 11-16: (a) original waveform; (b) squared waveform.*

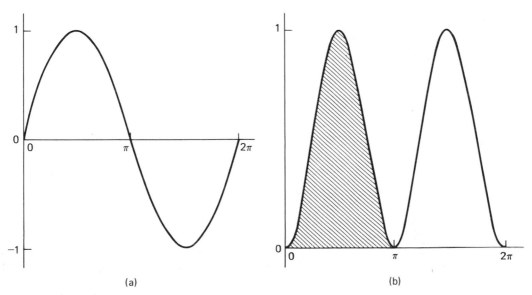

(a)   (b)

FIGURE 11-15: *Rms of sine wave: (a) sine wave; (b) sine-squared waveform.*

EXAMPLE 11-17: Compute the rms value for the following sinusoidal waveforms:

(a)   A voltage with a peak value of 29.6 V.

(b)   A current with a p-p value of 26.31 mA.

SOLUTION:

(a)
$$V_{rms} = 0.7071 \times V_{pk}$$
$$= 0.7071 \times 29.6 = 20.93 \text{ V}$$

(b)
$$I_{rms} = 0.7071 \times V_{pk}$$
$$= 0.7071 \times \frac{26.31}{2} = 9.302 \text{ mA}$$

11-4.   MEASUREMENT OF AC VALUES

As in direct current, there are several instruments used to measure alternating current. There are three types of ac voltmeters:

1. Digital voltmeters.

2. Electronic voltmeters (scale type).

3. VOM.

All three meters are calibrated to read sinusoidal waveforms in rms. However, their specifications should be examined closely for the frequency range each can handle accurately and for their reaction to nonsinusoidal waveforms.

The d'Arsonval movement is commonly used for ac VOMs by providing a circuit that converts the alternating current to direct current and displays the result. However, several other types are used, such as iron vane and thermocouple.

The Oscilloscope

One of the most useful instruments used for analyzing ac waveforms is the oscilloscope, Fig. 11-16(a). With it, not only can the amplitude be measured, but its actual waveshape can be ascertained. The instrument contains a cathode ray tube (CRT) similar to a TV picture tube, in which an electron beam is focused onto a phosphor-coated screen, Fig. 11-16(b), causing light to be emitted. The beam is made to travel at a uniform rate from left to right, Fig. 11-16(c), by the time base generator. Thus, the horizontal axis represents time. The waveform to be observed is connected to the vertical input, and, when it goes positive, will cause the beam to deflect upward on the face of the CRT; when the input goes negative, the beam will be deflected downward. Consequently, the beam "writes" the waveform onto the screen of the CRT.

Controls are provided for varying horizontal beam traveling speed, vertical amplifi-

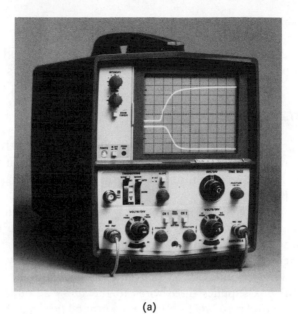

(a)

(b)

Time base
generator

Vertical
input

Vertical
amplifier

(c)

FIGURE 11-16: *The oscilloscope: (a) photograph (Courtesy, Tektronix, Inc.); (b) cathode ray tube (CRT); (c) block diagram.*

cation, and synchronization of the waveform with the time base generator. Thus, the oscilloscope rates number one in its value as an ac analysis tool.

11-5. 60-HERTZ POWER

The power used to supply our homes is 120/240 V ac, Fig. 11-17. That is, there is 120 V ac between neutral and each hot lead, and 240 V between the two hot leads. This is because the waveform of one hot lead is 180° out of phase with the waveform of the other hot lead.

Power is supplied to our homes from a transformer on a pole that converts the high voltage (1200 V and up) to 120/240 V. A wire is run down the pole from the neutral lead to earth ground. Three wires are fed into the service entrance (SE) box, and the neutral is again connected to earth ground, usually a cold water pipe. These are the only two points that earth grounding is permitted.

The SE box supplies three wires to the 120-V ac outlets: a hot lead (insulated black),[1] a neutral (insulated white), and a ground (green or bare). When a small appliance is plugged into the outlet, the chassis ground of the appliance contacts the ground lead of the outlet, grounding the appliance. Note that under ordinary conditions no current flows through the ground lead. If, however, the hot lead inside the small appliance were to accidentally short to the chassis, this hot lead would be grounded, thus tripping the circuit breaker. Without the ground lead the appliance would be hot, and anyone touching it would receive a severe shock. Therefore, this ground lead is of supreme importance in protecting our lives.

The circuits fed from the SE box are usually balanced. That is, one hot lead feeds about the same load as the other. Some appliances, such as electric stoves and dryers, require 240 V ac. These are fed as shown, with one circuit breaker in each hot lead. Equipment grounding is supplied by connecting the neutral lead of the appliance to the chassis of the stove or dryer. This is an exception to the rule that the neutral is never used for equipment grounding.

When switches are wired, the switch is always connected in series with the hot lead. Note that no fuses, switches, or circuit breakers are ever permitted in the ground lead. They may be used only if the hot lead is opened simultaneously with the neutral by means of mechanically ganged switches or breakers.

11-6. SUMMARY

Alternating current can be generated by rotating a loop of wire in a magnetic field. The result is a sine wave. The general formula for a sine wave is $v = V \sin (2\pi ft + \theta)$, where v is the instantaneous voltage at time t, V the peak voltage, f the frequency, $2\pi f$ the angular velocity, and θ the phase angle. Period is the inverse of frequency.

---

[1]This is its usual color. However, the code merely states that it must not be white or green. Thus, red and blue are sometimes used to differentiate between hot leads.

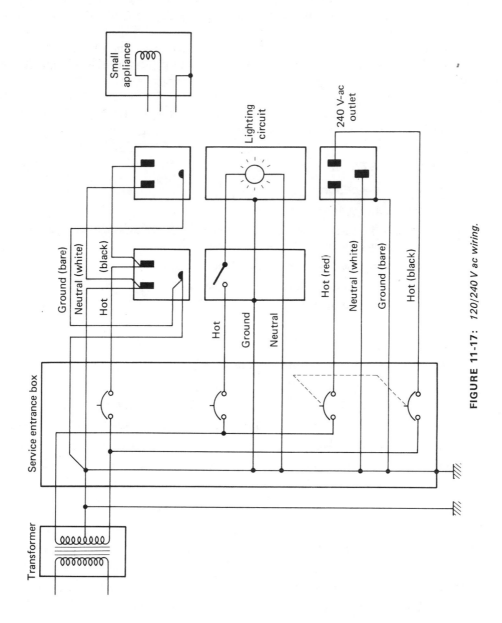

FIGURE 11-17: *120/240 V ac wiring.*

245

Harmonics are multiples of a fundamental frequency. Any repetitive waveform can be considered as a sum of harmonics of proper amplitude and phase. Wavelength is the distance from one point on the waveform to the next, identical point on the waveform.

Ac voltages are measured in peak, peak to peak, average, and rms. The rms voltage is the effective voltage and produces the same heating effect as a dc voltage of equal value.

The oscilloscope can provide a visual picture of an ac waveform. The power within our homes is 60 Hz 120/240 V ac, supplying both 120- and 240-V loads. Proper grounding within 60-Hz circuits can prevent electrical shock and save lives.

11-7. REVIEW QUESTIONS

11-1. What is induction?

11-2. What is Lenz's law?

11-3. What is the name of the waveshape that is generated by a loop of wire rotating at a constant velocity in a magnetic field?

11-4. Define frequency. What is its unit of measurement?

11-5. Define period. How is it related to frequency? What is its unit of measurement?

11-6. What is a phase angle?

11-7. In the equation $v = V \sin (2\pi ft + \theta)$, what is each of the following?

(a) v. (b) f. (c) t. (d) $2\pi f$. (e) θ.

11-8. How many degrees are there in a radian?

11-9. What is a harmonic?

11-10. What is the velocity of an electromagnetic waveform in free space?

11-11. What is the difference between peak measurement and peak-to-peak measurement?

11-12. Describe the method of finding the average of a repetitive waveform.

11-13. To find the average of a sine wave, multiply the peak value by _____.

11-14. To find the average of a sine wave, multiply the p-p value by _____.

11-15. Describe the method of finding the rms for any repetitive waveform.

11-16. To find the rms of a sine wave, multiply its peak value by _____.

11-17. To find the rms of a sine wave, multiply its p-p value by _____.

11-18. What advantage does an oscilloscope have over a meter in measurement of ac values?

11-19. What does the horizontal axis of an oscilloscope trace represent?

11-20. What does the vertical axis of an oscilloscope trace represent?

11-21. At what two points within a 60-Hz power system may the neutral be connected to earth ground?

11-22. What is the color of the hot lead within a 60-Hz power system? The neutral lead? The ground lead?

11-23. What is the purpose of the ground lead?

11-24. What is the effect of having an open ground lead upon the safety of the equipment? Why?

11-25. In what lead is the switch located in a lighting circuit?

11-26. Under what conditions may a circuit breaker or switch be connected in the neutral lead?

11-8. PROBLEMS

11-1. What is the amplitude of a sine wave when it is 25° through its cycle if its maximum amplitude is 26?

11-2. At an angle of 105° through its cycle, a sinusoidal waveform's amplitude is 105. What is its maximum amplitude?

11-3. A sinusoidal waveform goes through four complete cycles in 223 ms. What is the waveform's frequency? Its period?

11-4. A sinusoidal waveform goes through half of its cycle in 45.9 ns. What is the waveform's frequency? Its period?

11-5. Convert the following to radians:

    (a) 25.9°.   (b) 107°.   (c) −65°.

11-6. Convert the following to degrees:

    (a) 0.065 rad.   (b) 1.92 rad.   (c) 45 rad.

11-7. The waveform for a certain voltage is $v = 65 \sin (4567t + 23°)$. What is the waveform's: (a) maximum amplitude, (b) angular velocity, (c) frequency, (d) period, (e) phase angle, (f) graph?

11-8. What is the voltage of the waveform in Prob. 11-7 at 0.5 ms after $t = 0$?

11-9. A waveform has the equation $i = 0.023 \sin (45{,}000t + 65°)$. What is the waveform's: (a) maximum amplitude, (b) angular velocity, (c) frequency, (d) period, (e) phase angle, (f) graph?

11-10. What is the current in Prob. 11-9 at 1 μs after $t = 0$?

11-11. A certain sinusoidal voltage waveform is at 10 V in 20 ms after $t = 0$. What is its maximum voltage if its frequency is 10 Hz?

11-12. A certain sinusoidal current waveform is at 2.5 mA in 3 μs after $t = 0$. What is its maximum current if its frequency is 700 kHz?

11-13. (a) What is the fourth harmonic of 1.00 MHz? (b) The 17th harmonic?

11-14. The horizontal scanning frequency on a television set is 15,750 Hz. (a) What is its third harmonic? (b) Its 13th harmonic?

11-15. What is the wavelength of a radio wave in free space if its frequency is 194 MHz?

11-16. The velocity of a sound wave in Lake Geneva is 1435 m/s at a temperature of 9°C. What is the wavelength of an 860-Hz tone?

11-17.   What is the average and rms of the waveform shown in Fig. 11-18?

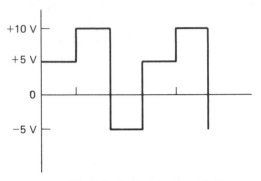

FIGURE 11-18: *Problem 11-17.*

11-18.   A voltage has the equation $v = 24 \sin (3400t + 27°)$. What are its peak, p-p, average, and rms values of the waveform?

11-19.   A sinusoidal waveform has an amplitude of 675 V p-p. What are its average and rms values?

11-20.   A sinusoidal waveform has an average of 23.9 V. What are its peak, p-p, and rms values?

11-9. PROJECTS

11-1.   Obtain a strong magnet and a sensitive microammeter. Move a wire through the magnet's field and observe the current generated. Note the difference in moving the wire up and down.

11-2.   Study the subject of Lissajous figures and compare two frequencies using this method on an oscilloscope. Compare the ratios 1 : 1, 1 : 2, 1 : 3, 1 : 4 , 1 : 5, and 1 : 6. Next, reverse the horizontal and vertical inputs to the scope and repeat the experiment.

11-3.   Obtain a spectrum analyzer and examine a square wave, noting the strengths of the various harmonics.

11-4.   Connect a sine wave of 1 kHz to an audio amplifier that feeds a speaker. Note its sound. Next, connect a 1-kHz square wave and note its sound. What does your ear tell you about the harmonics?

11-5.   What is the mathematical expression of a sawtooth waveform in terms of sine waves and harmonics? A triangular waveform?

11-6.   Feed a 100-Hz sine wave into an oscilloscope and measure its p-p value. Compare this with the rms value read on a VOM. How does the measured relationship between p-p and rms compare with the mathematical relationship?

11-7.   A three-way switching circuit enables a light to be controlled from two switches. A four-way switch enables a light to be controlled from three or more switches. What is the schematic of the three-way circuit and the four-way 60-Hz power circuit?

11-8.   Write a short biographical sketch of Gustav Hertz.

12

PHASORS

Space vehicles, men on other planets, "put your phasors on stun"—the term "phasor" conjures up all sorts of extraterrestrial experiences in our imaginations. Although these experiences have a certain fascination about them, we in electronics are concerned with another type of phasor, one that will assist us in understanding the voltage and current relationships within an ac circuit. In this chapter we shall examine phasors and how they can be mathematically manipulated so that in subsequent chapters we can apply the technique to the solution of ac problems. The sections include:

12-1. Phasors and sine waves

12-2. The complex field

12-3. Rectangular coordinates

12-4. Polar coordinates

12-1. PHASORS AND SINE WAVES

In Chapter 11 it was shown that a sine wave can be represented as the vertical component of a rotating vector. Thus, in Fig. 12-1 where two sine waves are represented, the equations for the waveforms are

$$H_A = V \sin \theta$$
$$H_B = V \sin (\theta - 45°)$$

Note that when waveform B is 23° through its cycle, waveform A is $(23° + 45°)$ or 68° through its cycle. Thus, we can represent the relationship between these two

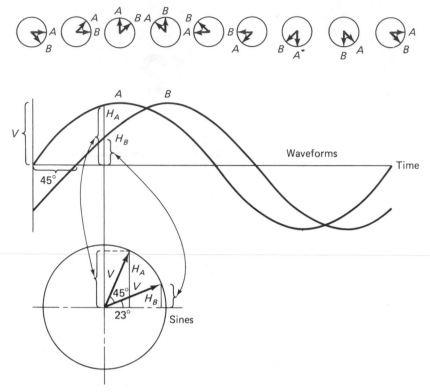

FIGURE 12-1: *Instantaneous phase and amplitude.*

waveforms as two rotating vectors, as shown at the top of the diagram. Vectors such as these that are used to represent phase are called *phasors*. Note that the lengths of the phasors and the phase angle between them is constant along the entire waveform.

These phasors can also be used to represent additive waveforms. In Fig. 12-2 waveforms A and B have been added to obtain the result. This relationship can also be represented as three rotating phasors: A, B, and $A + B$. Again, the length of each phasor and the phase relationship between them is constant along the entire waveform.

RMS Values

Since both length and relative phase are a constant along a waveform, phasors can be used to represent rms values of voltage, current, and ac resistance (called *impedance*). However, to effectively use the system of phasors, we must first study the mathematical operations of phasors: how they are expressed and how they are added, subtracted, multiplied, divided, and raised to a power. The rest of this chapter is devoted to this objective.

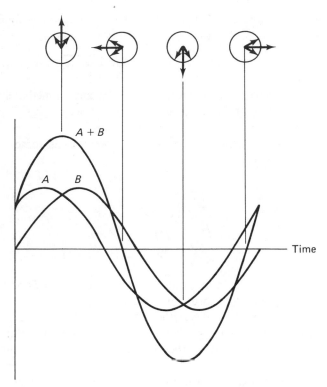

FIGURE 12-2: *Adding sine waves.*

12-2. THE COMPLEX FIELD

Since phasor algebra is borrowed from a system scientists call a complex field, let us consider its beginnings. It seems that, every time mathematicians run into an obstacle, they define a new number system. Man started out in his early history by defining a 1, 2, and so forth, enabling him to count dinosaurs, saber-toothed tigers, and wives. Then, someone subdivided a mammoth and found no way to represent the result, so fractions were introduced. This worked very well until someone tried to represent temperatures below zero, resulting in the negative number system. However, consider what happens when we try to take the square root of several numbers:

$$\sqrt{25} = \pm 5$$
$$\sqrt{2} = \pm 1.414$$
$$\sqrt{1} = \pm 1$$
$$\sqrt{0} = 0$$
$$\sqrt{-1} = \text{TILT}$$

So, mathematicians decided to call $\sqrt{-1}$ an *imaginary number* and define it as *i*. Thus, we could take the square root of -100 as follows:

$$\sqrt{-100} = \sqrt{-1}\sqrt{100} = \pm 10\sqrt{-1} = \pm 10i$$

Next, we in electronics borrowed the system and immediately ran into a problem: to us, *i* means current. So, instead of calling it *i*, we call it *j*. Therefore, to us, $\sqrt{-100}$ $= \pm 10j$, not $\pm 10i$.

It turns out that this *j* (also called a *j factor*) has some very interesting characteristics. First, note that, since $j = \sqrt{-1}$,

$$j = \sqrt{-1}$$
$$j^2 = (\sqrt{-1})(\sqrt{-1}) = -1$$
$$j^3 = j(j^2) = -\sqrt{-1}$$
$$j^4 = (j^2)(j^2) = (-1)(-1) = +1$$

Thus, we will never have an occasion to use a power of *j* above 1, a trait that will be very useful in reducing expressions containing *j*.

Additionally, *real numbers* (those that are not imaginary) and *j* factors can be placed upon a graph. In Fig. 12-3 we have defined numbers along the horizontal axis as being real, and those along the vertical axis as imaginary numbers. Thus, a number 5 units high on the vertical axis would be called *j*5 and represent $5\sqrt{-1}$. A number 3 down on the vertical axis would be called $-j3$ and represent $-3\sqrt{-1}$. A number 4

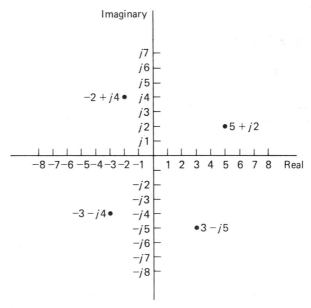

FIGURE 12-3: *Complex plane.*

to the right on the horizontal axis is just a $+4$, while one 2 to the left of the origin on the horizontal axis is known as -2. This graph, therefore, represents both real and imaginary numbers.

We should also note that each point is represented as the sum of a real and an imaginary number. A point on the graph 5 to the right of the origin and 2 up from the origin is called $5 + j2$, the 5 representing the real part of the number and the $+j2$ representing the imaginary part of the number. Other points have been noted on the graph. Since each point represents the sum of a real part and an imaginary part, such numbers are called *complex numbers*, and the entire graph is called a *complex plane*.

EXAMPLE 12-1: Express the following as complex numbers:

(a)   A number 45 places to the right of the origin, and 6 units below the origin.

(b)   A number 39.63 to the left of the origin and 63.19 above it.

(c)   A number 29.21 to the left of the origin and 92.17 units below it.

SOLUTION:

(a) $45 - j6$.   (b) $-39.63 + j63.19$.

(c) $-29.21 - j92.17$.

Complex numbers are considered as vectors or phasors, the tail originating at the origin and the head pointing to the specific point within the complex field, Fig. 12-4(a). Thus, to us in electronics, $5 + j2$ represents a phasor, as do $2 + j5$, $-6 + j6$, $-2 - j5$, and $3 - j4$. This system of representing phasors is called *rectangular coordinates*.

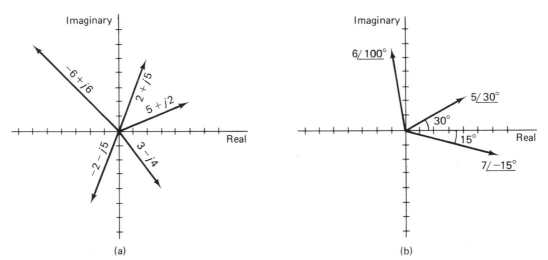

FIGURE 12-4: *Phasors in the complex field: (a) rectangular coordinates; (b) polar coordinates.*

There is, however, another way we can express these phasors: we can state their length and angle with respect to the horizontal in a system called *polar coordinates*. For example, a phasor 5 units long that has an angle of 30° above the horizontal is called 5/30°, Fig. 12-4(b). We shall be manipulating phasors in both polar and rectangular coordinate systems.

12-3. RECTANGULAR COORDINATES

As noted in the previous paragraph, rectangular coordinates can be used to specify any phasor in the form of $A + jB$, where A represents the distance from the origin along the horizontal axis and B represents the distance from the origin along the vertical axis.

Adding Phasors

Phasors can be added both graphically and numerically. To add two phasors graphically, merely form a parallelogram from the given phasors.

EXAMPLE 12-2: Graphically find the sum (also called the *resultant*) of phasors AB and AC in Fig. 12-5(a).

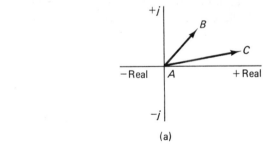

(a)

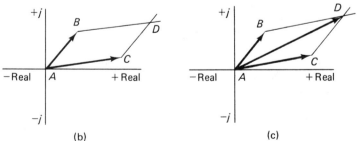

(b)   (c)

FIGURE 12-5: *Example 12-2, phasor addition: (a) given problem; (b) drawing parallels; (c) resultant.*

SOLUTION: First, a line is drawn parallel to AC, passing through B, Fig. 12-5(b). A line is then drawn parallel to AB through C. The head of the resultant phasor is located at the intersection of these lines, Fig. 12-5(c). Thus, in phasor notation,

$$\overrightarrow{AB} + \overrightarrow{AC} = \overrightarrow{AD}$$

The symbol "$\rightarrow$" identifies that line segment as a phasor.

There is a second way of graphically specifying phasor addition. Rather than connect the tails of the phasors together, they can be shown in a head-to-tail configuration, Fig. 12-6(a). The addition is actually the same as in the previous paragraph if you observe that BC is parallel to AD and AB parallel to DC, Fig. 12-6(b). The resultant is $\overrightarrow{AC}$.

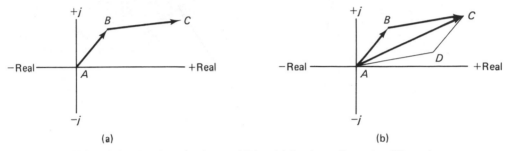

(a)   (b)

FIGURE 12-6: *Head-to-tail phasor addition: (a) head-to-tail notation; (b) resultant.*

Although graphical addition is very easy to perform, mathematical addition is even easier. Addition of phasors expressed in rectangular coordinates requires that the real numbers be summed and the j factors summed.

EXAMPLE 12-3: Add the following phasors:

   (a) $3 + j4,\ 5 - j6$.

   (b) $-4 - j6,\ -10 - j1$.

SOLUTION:

   (a)

$$3 + j4$$
$$5 - j6$$
Resultant $8 - j2$

Note that the sum of $j4$ and $-j6$ is $-j2$. Graphically, this means that we traveled north 4 units on the j axis, then south 6 units, resting at $-j2$.

   (b)

$$-4 - j6$$
$$-10 - j1$$
Resultant $-14 - j7$

Subtracting Phasors

Graphically, subtracting phasor **M** from phasor **N** can be analyzed as follows:

$$\mathbf{P} = \mathbf{M} - \mathbf{N}$$

Thus,

$$\mathbf{M} = \mathbf{P} + \mathbf{N}$$

Consequently, we can deduce that **M** is a resultant, **N** is one of its component phasors, and we must find the second component phasor. Thus, in Fig. 12-7, we are given **M** and **N** and must find **P**. We can find it by completing the parallelogram as shown.

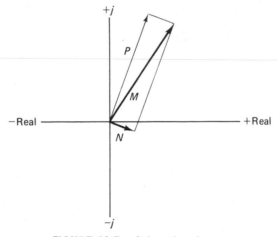

FIGURE 12-7: *Subtracting phasors.*

To subtract one phasor from the other mathematically, merely subtract the real parts and the imaginary parts.

> **EXAMPLE 12-4:** Subtract $-6 + j20$ from $12 - j6$.
>
> *SOLUTION:*

$$
\begin{array}{r}
12 - \ j6 \\
- \ (-6 + j20) \\
\hline
18 - j26
\end{array}
$$

Subtracting -6 from 12 yields 18 (remember, this is algebraic subtraction). Subtracting $j20$ from $-j6$ yields $-j26$. Note that subtracting j factors is the same as subtracting real numbers, except for the inclusion of j in the result.

> **EXAMPLE 12-5:** Subtract $-6 - j22$ from $-26 + j6$.

SOLUTION:

$$-26 + j6$$
$$- (-6 - j22)$$
$$\overline{-20 + j28}$$

Multiplying Phasors

Before we analyze how to multiply, let us look into the "what." It turns out this little quantity we have been using, j, has some other very interesting properties. Note that multiplying by j results in rotation of a phasor by 90°. Consider the number 1 (Fig. 12-8):

$$1 = 1$$
$$1 \times j = j1$$
$$j1 \times j = j^2 1 = (-1)(1) = -1$$
$$-1 \times j = -j1$$
$$-j1 \times j = -j^2 1 = -(-1)(1) = 1$$

Thus, each time we have multiplied by j we have rotated the phasor 90°. We shall make use of this property extensively in reduction of phasor expressions.

Next, let us multiply the phasor $3 + j6$ by j:

$$j(3 + j6) = j3 + j^2 6 = (-1)(6) + j3 = -6 + j3$$

It can be shown by geometry that the phasors $3 + j6$ and $-6 + j3$ are also 90° apart. Thus, we can summarize:

The effect of multiplying any phasor by j is to rotate the phasor CCW by 90°.

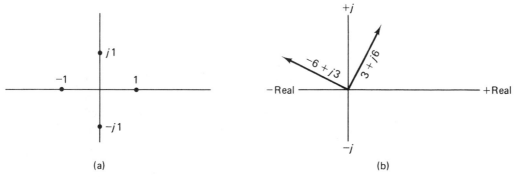

(a)   (b)

FIGURE 12-8: *Multuplying by j: (a) unity; (b) multiplying 3 + j6 by j.*

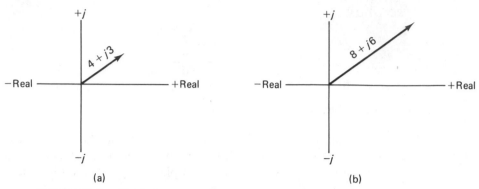

FIGURE 12-9: *Multiplying a phasor by a scalar: (a) original phasor; (b) multiplied by 2.*

Let us now examine the effect of multiplying a phasor by a scalar. A scalar is a number that has magnitude but no direction. It contrasts with a vector that has both magnitude and direction.

Consider the phasor shown in Fig. 12-9(a). To multiply this phasor by the scalar 2, merely multiply both the real part by 2 and the imaginary part by 2. Thus,

$$(4 + j3) \times 2 = 8 + j6$$

Note that the effect as shown in Fig. 12-9(b) is merely to lengthen the phasor.

Next, let us combine the unity *j* factor with a scalar and note the effect. Multiplication is performed as if the *j* were an ordinary algebraic literal.

$$(3 + j4) \times (j2) = j6 + j^2 8 = -8 + j6$$

It turns out that this rotates the phasor by 90° and extends its length by a factor of 2, Fig. 12-10.

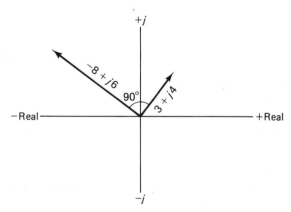

FIGURE 12-10: *Multiplying a phasor by j2.*

We can now summarize the effect of scalar and j-factor multiplication:

1. Multiplying a phasor by a unity j factor rotates the phasor 90°.
2. Multiplying a phasor by a scalar changes the length of the phasor.
3. Multiplying by a nonunity j factor rotates the phasor by 90° and changes its length.

We have just one more case to examine, that of multiplying a phasor by a phasor. The feat is accomplished by treating each phasor as an ordinary binomial and performing the multiplication.

EXAMPLE 12-6: Multiply the phasors $3 - j1$ and $2 + j3$.

$$
\begin{array}{r}
3 - j1 \\
2 + j3 \\
\hline
j9 - j^2 3 \\
6 - j2 \\
\hline
6 - j7 - j^2 3
\end{array}
$$

but $-j^2 3 = 3$. Therefore, the solution is

$$6 - j7 + 3 = 9 - j7$$

The result is graphed in Fig. 12-11. Note that changes in both magnitude and direction are reflected in the result.

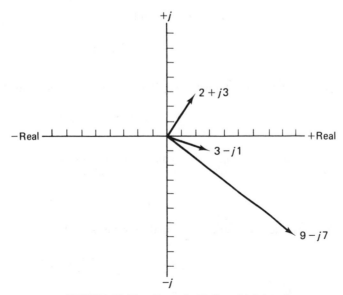

FIGURE 12-11: *Example 12-6, multiplying phasors.*

EXAMPLE 12-7: Multiply $-3 + j6$ and $-7 - j8$.

$$
\begin{array}{r}
-3 + j6 \\
-7 - j8 \\
\hline
j24 - j^2 48 \\
21 - j42 \\
\hline
\end{array}
$$

$$21 - j18 - j^2 48 = 48 + 21 - j18 = 69 - j18$$

Dividing Phasors

Division of phasors expressed in rectangular coordinates can be done by multiplying both numerator and denominator by a function of the denominator called a *conjugate*. A conjugate of a phasor is the original phasor with its *j*-factor sign reversed. Table 12-1 lists some examples. Now let us use this method in a few examples.

TABLE 12-1: *Conjugates.*

Phasor	Conjugate
$3 - j6$	$3 + j6$
$-4 + j12$	$-4 - j12$
$-6 - j23$	$-6 + j23$

EXAMPLE 12-8: Divide $6 - j4$ by $2 + j6$.

SOLUTION: The problem in fraction form is

$$\frac{6 - j4}{2 + j6}$$

Next, multiply both numerator and denominator by the conjugate of the denominator:

$$\frac{6 - j4}{2 + j6} \cdot \frac{2 - j6}{2 - j6} = \frac{12 - j36 - j8 + j^2 24}{4 - j12 + j12 - j^2 36} = \frac{-12 - j44}{40}$$

Note that multiplying the denominator by its conjugate results in the *j* factor dropping out. This is similar to the theorem in algebraic factoring:

$$(a + b)(a - b) = a^2 - b^2$$

Next, we must leave the answer in the form of a real number and an imaginary number:

$$\frac{-12 - j44}{40} = -\frac{12}{40} - j\frac{44}{40} = -0.3 - j1.1$$

This, then, is the final answer.

EXAMPLE 12-9: Divide $12.6 - j22.9$ by $-24.9 - j12.9$.

SOLUTION:
Original problem:

$$\frac{12.6 - j22.9}{-24.9 - j12.9}$$

Multiplying by conjugates:

$$\frac{12.6 - j22.9}{-24.9 + j12.9} \cdot \frac{-24.9 - j12.9}{-24.9 - j12.9} = \frac{-313.74 + j162.54 + j570.21 - j^2 295.41}{620.01 + 166.41}$$

$$= \frac{-18.33 + j732.75}{786.42}$$

$$= -0.02331 + j0.9318$$

Note that there is no need to multiply real by j factor in the denominator, for it will just drop out.

We have now discussed addition, subtraction, multiplication, and division of phasors using rectangular coordinates. Next, we shall do the same using polar notation.

12-4. POLAR COORDINATES

As was stated earlier, we can also express a phasor by stating its length and its angle referred to the horizontal. In Fig. 12-12 the phasor $3 + j4$ can also be expressed as $5\underline{/53.13°}$. That is, it has a length of 5 units and its angle with reference to the horizontal is $53.13°$. This is a very useful phasor form. However, we must be able to relate it to the rectangular form.

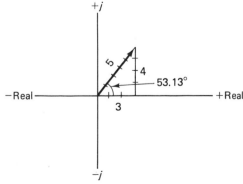

FIGURE 12-12: *Polar coordinates.*

Polar-to-Rectangular Conversion

Many calculators can convert directly from polar to rectangular coordinates. In addition, polar to rectangular conversion is easily accomplished by observing from Fig. 12-13 that:

1. Real $= V \cos \alpha$.
2. Imaginary $= V \sin \alpha$.

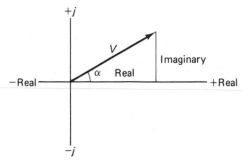

FIGURE 12-13: *Polar–rectangular relationships.*

EXAMPLE 12-10: Convert the following phasors to rectangular coordinates:

(a) $120\underline{/30°}$.

(b) $60\underline{/212°}$.

SOLUTION:

(a) Real $= V \cos \alpha$
$= 120 \cos 30°$
$= 120 \times 0.8660$
$= 103.9$
Imaginary $= V \sin \alpha$
$= 120 \sin 30°$
$= 120 \times 0.5000$
$= 60.00$
Therefore, $120\underline{/30°} = 103.9 + j60.00$.

(b) Real $= V \cos \alpha$
$= 60 \cos 212°$
$= -50.88$
Imaginary $= V \sin \alpha$
$= 60 \sin 212°$
$= -31.80$
Therefore, $60\underline{/212°} = -50.88 - j31.80$.

Rectangular-to-Polar Conversion

Many calculators allow direct rectangular to polar conversion. The process can be described mathematically by the following formulas:

$$\alpha = \tan^{-1}\left(\frac{\text{imaginary}}{\text{real}}\right)$$

$$V = \frac{\text{imaginary}}{\sin \alpha}$$

This procedure can easily be done on any calculator with trigonometric functions.

EXAMPLE 12-11: Convert the following to polar coordinates:

    (a) $3 - j4$.

    (b) $26 + j63$.

SOLUTION:

    (a) To find the angle:

$$\alpha = \tan^{-1}\left(\frac{\text{imaginary}}{\text{real}}\right)$$

$$= \tan^{-1}\left(\frac{-4}{3}\right)$$

$$= \tan^{-1}(-1.333)$$

$$= -53.13°$$

To find the phasor length:

$$V = \frac{\text{imaginary}}{\sin \alpha}$$

$$= \frac{-4}{\sin(-53.13°)}$$

$$= 5.000$$

Thus, $3 - j4 = 5.000 \underline{/-53.13°}$.

    (b) $\alpha = \tan^{-1}\left(\dfrac{\text{imaginary}}{\text{real}}\right)$

$$= \tan^{-1}\left(\frac{63}{26}\right)$$

$$= 67.57°$$

$$V = \frac{\text{imaginary}}{\sin \alpha}$$

$$= \frac{63}{\sin(67.57°)}$$

$$= 68.15$$

Thus, $26 + j63 = 68.15 \underline{/67.57°}$.

Addition and Subtraction

Phasors in polar form can only be added or subtracted if their angles are the same, in which case the lengths are combined and the angle remains the same.

EXAMPLE 12-12: Perform the following phasor operations:

(a) $3\underline{/25°} + 6\underline{/25°}$.

(b) $7\underline{/-35°} - 2\underline{/-35°}$.

SOLUTION:

(a) To add, merely add the phasor lengths, keeping the angle the same:

$$3\underline{/25°} + 6\underline{/25°} = (3 + 6)\underline{/25°} = 9\underline{/25°}$$

(b) To subtract, merely subtract the phasor lengths, keeping the angle the same:

$$7\underline{/-35°} - 2\underline{/-35°} = (7 - 2)\underline{/-35°} = 5\underline{/-35°}$$

If the angles are not identical, the phasors must be converted into rectangular form, and then added or subtracted.

EXAMPLE 12-13: Perform the following phasor operations:

(a) $4\underline{/26°} + 9\underline{/85°}$.

(b) $6\underline{/-23°} - 7\underline{/64°}$.

SOLUTION:

(a) Since the angles are not identical, the phasors must be converted to rectangular form, then added:

$$4\underline{/26°} = 3.595 + j1.753$$
$$9\underline{/85°} = 0.784 + j8.966$$
$$\text{Sum:} \quad 4.379 + j10.719$$

Converting back to polar: $11.58\underline{/67.78°}$. Thus, $4\underline{/26°} + 9\underline{/85°} = 11.58\underline{/67.78°}$.

(b)
$$6\underline{/-23°} = 5.523 - j2.344$$
$$7\underline{/64°} = 3.069 + j6.292$$
$$\text{Subtracting:} \quad 2.454 - j8.636$$

Converting back to polar: $8.978\underline{/-74.13°}$. Thus, $6\underline{/-23°} - 7\underline{/64°} = 8.978\underline{/-74.13°}$.

Multiplication

To multiply polar phasors, the lengths are multiplied and the angles added:

$$(A\underline{/\alpha_1})(B\underline{/\alpha_2}) = (A \cdot B)\underline{/\alpha_1 + \alpha_2}$$

Note that we are essentially using the same rules for the angles of these phasors as we would for exponents in algebra. That is, in algebra the exponents are added during multiplication. The reason for adding these angles is that they are, in fact, exponents. This can be shown by the following gymnastics:

The $\cos \alpha$ is represented by the series

$$\cos \alpha = 1 - \frac{\alpha^2}{2!} + \frac{\alpha^4}{4!} - \frac{\alpha^6}{6!} + \cdots$$

The $\sin \alpha$ is represented by the series

$$\sin \alpha = \alpha - \frac{\alpha^3}{3!} + \frac{\alpha^5}{5!} - \frac{\alpha^7}{7!} + \cdots$$

But $\epsilon^{j\alpha}$ is represented by the series

$$\epsilon^{j\alpha} = 1 + j\alpha - \frac{\alpha^2}{2!} - j\frac{\alpha^3}{3!} + \frac{\alpha^4}{4!} + j\frac{\alpha^5}{5!} - \frac{\alpha^6}{6!} - j\frac{\alpha^7}{7!} + \cdots$$

Therefore, $\epsilon^{j\alpha} = \cos \alpha + j \sin \alpha$. However, $V\underline{/\alpha} = V(\cos \alpha + j \sin \alpha)$. Therefore, $V\underline{/\alpha} = V\epsilon^{j\alpha}$, and α is, indeed, an exponent.

Now let us try this multiplication procedure on a few problems.

EXAMPLE 12-14: Multiply the following:

(a) $3\underline{/40°}$ by $26\underline{/-25°}$.

(b) $3\underline{/-60°}$ by $5\underline{/-20°}$.

SOLUTION:

(a) $3\underline{/40°} \times 26\underline{/-25°} = (3 \times 26)\underline{/40° - 25°} = 78\underline{/15°}$.

(b) $3\underline{/-60°} \times 5\underline{/-20°} = (3 \times 5)\underline{/-60° - 20°} = 15\underline{/-80°}$.

Division

Division of polar phasors is accomplished by dividing the phasor lengths and subtracting the angles (as you would exponents).

EXAMPLE 12-15: Divide the following phasors:

(a) $26\underline{/25°}$ by $33\underline{/-75°}$.

(b) $-33\underline{/-88°}$ by $27\underline{/45°}$.

SOLUTION:

(a) $\dfrac{26\underline{/25°}}{33\underline{/-75°}} = \left(\dfrac{26}{33}\right)\underline{/25° - (-75°)} = 0.788\underline{/100°}$

(b) $\dfrac{-33\underline{/-88°}}{27\underline{/45°}} = \left(\dfrac{-33}{27}\right)\underline{/(-88° - 45°)} = -1.222\underline{/-133°}.$

Exponentiation

Polar phasors can be raised to any power by raising the phasor length to the desired power and multiplying the angle by the power. Note that, again, the angle is treated as an exponent.

$$(V\underline{/\alpha})^n = V^n\underline{/n\alpha}$$

EXAMPLE 12-16: Perform the following:

(a) $(12\underline{/25°})^3$.

(b) $(6\underline{/-30°})^{2.6}$.

SOLUTION:

(a) $(12\underline{/25°})^3 = 12^3\underline{/25° \times 3} = 1728\underline{/75°}$

(b) $(6\underline{/-30°})^{2.6} = 6^{2.6}\underline{/-30° \times 2.6} = 105.5\underline{/-78.0°}$

12-5. SUMMARY

Phasors can be used to represent both magnitude and phase of ac electrical quantities. They are applicable to both instantaneous and rms values. Phasors consist of a real component, shown horizontally on a complex graph, and an imaginary component (labeled j) shown vertically on the complex graph. The j is equal to $\sqrt{-1}$; thus j^2 is equal to -1. Multiplying a phasor by j rotates the angle by 90°, whereas multiplying by a scalar merely changes its length. Phasors can be expressed in both rectangular and polar coordinates. Table 12-2 summarizes the operations that can be performed in the two coordinate systems. Phasors can also be added graphically by the parallelogram method. The angle in polar coordinates is, in fact, an exponent and is treated as such in all operations.

TABLE 12-2: *Phasor operations.*

	Add	*Subtract*	*Multiply*	*Divide*	*Powers*
Polar	No[a]	No[a]	Yes	Yes	Yes
Rectangular	Yes	Yes	Yes	Yes	No[b]

[a] If the angles are identical, addition and subtraction can be performed in polar.

[b] Exponentiation can be done in rectangular for integer powers.

12-6. REVIEW QUESTIONS

12-1. What is the formula for finding the length of the vertical component of a rotating vector?

12-2. The phase angle between two rotating vectors of the same velocity (increases) (decreases) (remains the same) as time increases.

12-3. What is the difference between a phasor and a vector?

12-4. In a complex field, the horizontal axis represents _____ numbers and the vertical axis represents _____ numbers.

12-5. The operator j has what numerical value?

12-6. Name the two coordinate systems for phasors.

12-7. Graphical addition is possible using a _____ for phasors.

12-8. What is the effect of multiplying a phasor by j?

12-9. What is the effect of multiplying a phasor by 3?

12-10. What is the effect of multiplying a phasor by $j3$?

12-11. What is the value of j^2?

12-12. What is a conjugate of a phasor?

12-13. How is rectangular phasor division performed?

12-14. What do the two numbers in a rectangular phasor represent?

12-15. What do the two numbers in a polar phasor represent?

12-16. Under what conditions may polar phasors be added?

12-17. Of the five operations of addition, subtraction, multiplication, division, and exponentiation, which can always be performed within the polar system? Within the rectangular system?

12-7. PROBLEMS

12-1. Compute:

(a) j^7.   (b) j^{13}.   (c) j^{239}.

12-2. Express the following as a complex number:

    (a) A number 23.8 places to the left of the origin and 87.3 units above it.

    (b) A number 59.7 units above the origin and 186 units to its right.

12-3. Graphically find the sum of the phasors in Prob. 12-2.

12-4. Algebraically add:

    (a) $6 - j5$ and $3 - j9$.

    (b) $13.3 - j45.9$ and $-94.2 + j58.3$.

    (c) $111.3 + j456.9$ and $-105.0 + j567.8$.

12-5. Subtract the following phasors:

    (a) $3 + j9$ from $4 - j7$.

    (b) $17 - j16$ from $-9 - j20$.

    (c) $459.3 + j749.2$ from $450.3 - j620.1$.

12-6. Multiply the following phasors:

    (a) $1 - j9$ and $5 + j7$.

    (b) $15 + j45$ and $21 - j60$.

    (c) $-45.7 - j39.6$ and $-60.4 - j50.5$.

12-7. Divide using rectangular operations:

    (a) $5 - j8$ by $3 + j4$.

    (b) $-6 + j7$ by $-3 - j5$.

    (c) $13.3 + j45.9$ by $-13.5 + j57.3$.

12-8. Convert to rectangular coordinates:

    (a) $12\underline{/45°}$.   (b) $21\underline{/67°}$.   (c) $45.9\underline{/-74.3°}$.

12-9. Convert to polar coordinates:

    (a) $3 - j7$.   (b) $4 + j9$.   (c) $38.4 - j73.1$.

12-10. Obtain the polar sum:

    (a) $73.9\underline{/45°}$ and $39.4\underline{/45°}$.

    (b) $35.7\underline{/39°}$ and $57.2\underline{/-21.3°}$.

    (c) $743\underline{/85°}$ and $830\underline{/21.7°}$.

12-11. Obtain the polar difference:

    (a) $45\underline{/69°}$ from $94\underline{/69°}$.

    (b) $147\underline{/56°}$ from $256\underline{/-56°}$.

    (c) $14.6\underline{/-34°}$ from $56.7\underline{/13.3°}$.

12-12. Multiply:

    (a) $5\underline{/21°}$ by $7\underline{/56°}$.

    (b) $14.5\underline{/-33°}$ by $84.3\underline{/79.4°}$.

    (c) $793\underline{/39.1°}$ by $839\underline{/88.8°}$.

12-13. Divide:

    (a) $73.9\underline{/45°}$ by $39.4\underline{/45°}$.

    (b) $35.7\underline{/39°}$ by $57.2\underline{/-21.3°}$.

    (c) $793\underline{/39.1°}$ by $256\underline{/-56.4°}$.

12-14. Find the following:

    (a) $(3+j6)^2$.   (b) $(4+j3)^{3.6}$.   (c) $(3.5\underline{/4.39°})^{2.5}$.

    (d) $\sqrt{4-j6}$.   (e) $\sqrt[3]{45.9\underline{/57°}}$.

12-8. PROJECTS

12-1. Draw sinusoidal graphs of the voltages $V_1 = 13.6 \sin t$ and $V_2 = 24.0 \sin (t + 26°)$. Next, graphically add the two waveforms and measure the phase angle between this result and V_1. Also measure its peak value. Next, mathematically add the phasors $13.6\underline{/0°}$, representing V_1, and $24.0\underline{/26°}$, representing V_2 and compare this result with the measured values.

12-2. Explain the meaning of a circle and a 45° line in a Lissajous pattern. How may other phase angles be measured using this method? How accurate is it? Measure several phase angles in this manner.

12-3. In terms of vectors, why does it require more tension at points A and B of Fig. 12-14 to make the wire more nearly horizontal? Can it ever be made perfectly horizontal? Why?

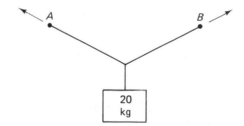

FIGURE 12-14: *Project 12-3.*

13

INDUCTANCE

In Chapter 10 we learned that current flowing through a wire generates a magnetic field of concentric flux lines. In Chapter 11 we discussed generating electricity by moving a wire within a magnetic field. We shall now combine these concepts as we study the subject of inductance, a property of a device that opposes any change in current flow. We shall first consider the physical and mathematical characteristics of inductance, then the devices on the market used to obtain it. We shall also consider some common circuit configurations having inductors. Finally, we shall examine the transformer, a device used to change ac from one voltage to another. The sections include:

13-1. Induced currents

13-2. Inductors

13-3. Inductors in series

13-4. Inductors in parallel

13-5. Mutual inductance

13-6. Transformers

13-7. *RL* time constants

13-1. INDUCED CURRENTS

In our simple ac generator of Chapter 11, we observed that, by moving a wire in a magnetic field, we could produce an electromotive force (emf). If we can move a wire in a stationary magnetic field and cause this emf, does it not seem logical that we should be able to move the magnetic field and keep the wire stationary and produce this voltage? Let us examine this point further.   .

Inductance Between Two Wires

Figure 13-1(a) illustrates a configuration in which neither wire B nor wire A has any current. Consider what happens when the current in B is raised from zero to some finite value. Its magnetic lines of force grow outward, cutting wire A and causing an emf to be generated in the direction shown, Fig. 13-1(b). This action is known as

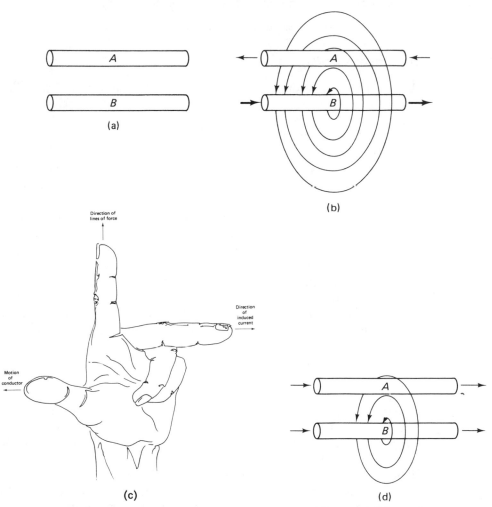

FIGURE 13-1: *Induction in a wire: (a) initial conditions; (b) current in B increasing; (c) left-hand rule for generators (drawing by Kirishian); (d) current decreasing.*

induction, and wire *B* is said to *induce* an emf into wire *A*. The direction of the induced current can be predicted by using a rule called the *left-hand rule* for generators, Fig. 13-1(c). Holding the left hand as shown:

1. The thumb points in the direction of the motion of the conductor.

2. The forefinger points in the direction of the magnetic lines of force.

3. The middle finger points in the direction of the induced current.

Let us now decrease the current in *B* back to zero, Fig. 13-1(d). The lines of force cut *A* in the opposite direction, causing an emf of opposite polarity to be generated.

From these two illustrations, we can state the following:

Any moving magnetic field that cuts a wire will generate an emf.

Inductance in Coils

Let us now assume these same two wires are part of a coil, Fig. 13-2(a). In our discussion of direct current we learned that we could study the effects of two batteries within a circuit by studying the currents produced by each separately, then adding to obtain the total currents. We shall apply this principle of superposition as we examine the currents and voltages of these two wires within the coil.

If we were to apply an external voltage, V_A, in order to increase the current through wire *B* to a value I_L and observe its effect upon *A*, we find that the lines of force from *B* cut *A*, generating a voltage V_g that opposes the applied voltage and a current I_g that opposes the increased coil current. This opposing voltage is called a *counter emf*, for it opposes any increase in applied voltage.

In a similar manner, if we were to increase the current through *A* from zero to a value, I_L, a counter emf will be generated across *B*, opposing the applied voltage, V_A. By the principle of superposition, these two effects are additive in the coil. That is, increasing the coil current in *A* causes a voltage that opposes the applied voltage, and increasing current in *B* causes a voltage that opposes the applied voltage. Therefore, increasing the current in both *A* and *B* simultaneously, as would actually happen within the coil, will cause a voltage that opposes the applied voltage, Fig. 13-2(d). This counter emf is caused by the expanding magnetic field cutting the wires of the coil.

Next, if we were to assume a dc current flow through the coil and suddenly decrease this current, we can note the effect of the collapsing field. If the expanding field created an emf in opposition to the applied voltage, the collapsing field generates an emf in the opposite direction, one that aids the applied voltage. Thus, decreasing coil current causes a force that tends to prevent this decrease.

Since increasing the current within the coil produces a voltage that tends to prevent that increase, and decreasing the current within the coil produces a voltage that tends to prevent that decrease, we can summarize inductance as follows:

Inductance is the property of a circuit that opposes any change in current flow.

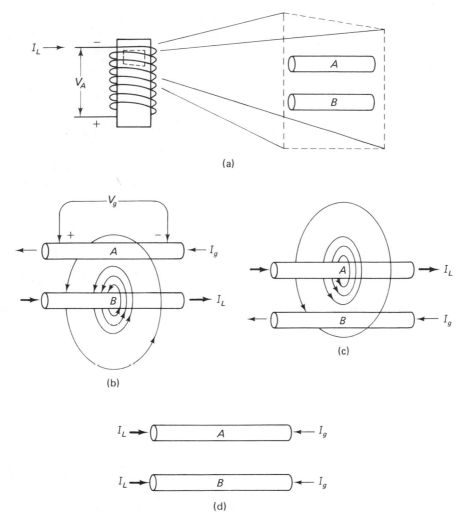

FIGURE 13-2: *Induced coil currents: (a) two adjacent coil wires; (b) effect of* B *upon* A*; (c) effect of* A *upon* B*; (d) net effect of induction.*

Let us compare this with the property of resistance. Whereas inductance opposes any change of current flow, resistance opposes any current flow. Thus, one is dynamic, the other is static. Resistance applies to both ac and dc circuits, but the inductive effect only occurs if the current changes, as occurs in an ac circuit.

A Mathematical Description

Let us now compare a slow increase in coil current with a rapid increase. The slow increase will generate a few wire-cutting lines of force per second, whereas a very rapid increase will generate many such lines of force per second. Since generated voltage depends upon how many lines of force per second are cutting a wire, counter emf is determined by how rapidly current is increased in the coil. Thus, the more rapid the increase, the greater the opposition to that increase. We can relate this effect in the formula

$$V = L\frac{di}{dt}$$

where V is the voltage; L the magnitude of inductance measured in henries (H), di represents "a small change in current," and dt indicates "a small change in time." The term di/dt can be thought of as a change in current per unit time and is expressed in amperes per second. Thus, if the current were increasing by 1 A/s ($di/dt = 1$), and inductance were 1 H, the voltage would be

$$V = L\frac{di}{dt} = 1 \times 1 = 1 \text{ V}$$

This can be visualized by picturing an increase of 1 A/s, Fig. 13-3(a), resulting in a continual generation of flux lines, Fig. 13-3(b).[1] These lines move outward from wire A, cutting wire B at a constant rate resulting in 1 V being generated across wire B, Fig. 13-3(c).

Note that, according to the formula, the faster current increases per second, the greater the voltage generated at the coil's terminals. Further, a decrease in current per unit time will result in a negative voltage being generated.

EXAMPLE 13-1: What emf will be generated by a 1-mH coil if its current increases at a constant rate of 10 mA every 20 μs?

SOLUTION:

$$V = L\frac{di}{dt} = 1 \times 10^{-3} \times \frac{10 \times 10^{-3}}{20 \times 10^{-6}} = 0.500 \text{ V}$$

Thus, the voltage at the terminals of the coil will be 500 mV.

EXAMPLE 13-2: A certain 100-mH coil has a decrease in current of 10 mA/s. What voltage will be generated at its terminals?

SOLUTION:

$$V = L\frac{di}{dt} = 100 \times 10^{-3} \times \frac{-10 \times 10^{-3}}{1} = -1.000 \text{ mV}$$

The minus sign is affixed to di to represent a decreasing rate.

---

[1]As noted in Chapter 10, flux versus current will plot as a hysteresis curve. However, if the change in magnetomotive force is kept small, the change in flux will be linear with a change in current.

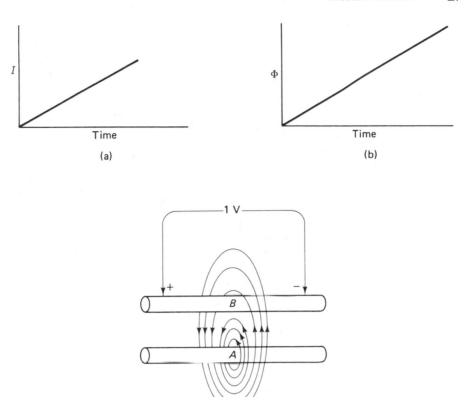

FIGURE 13-3: *Constantly increasing current: (a) current; (b) flux lines; (c) induction.*

Effect of Sine Waves

Since most of ac work involves sine waves, let us examine what happens when a coil is subjected to a sinusoidal waveform. Its reaction is more easily understood by assuming a sinusoidal current existing within the coil and examining the resultant voltage waveform, Fig. 13-4.

As the current increases from point *A*, the magnetic lines of flux are being rapidly generated, resulting in a high rate of lines of force cutting the coil, generating a high emf, as shown. At point *B*, the current is fixed; thus the flux field is fixed and no lines of force are cutting the coil. This results in zero volts being generated.

As the current decreases from point *B* to point *C*, the magnetic field collapses and lines of force cut the coil in the opposite direction, resulting in a negative voltage. Note that the greatest rate of collapse occurs at point *C*, resulting in the greatest negative voltage.

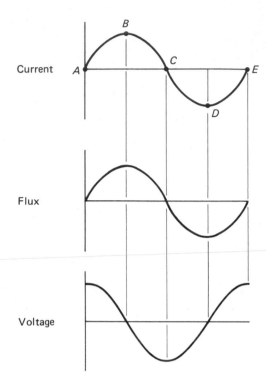

FIGURE 13-4: *Sine waves in a coil.*

At point *D* the current has reached its maximum negative value; however, the field is no longer expanding and no lines of force cut the coil. Thus, zero volts is generated. The cycle is then completed as the current moves to point *E*, zero amps.

Let us now examine the relationship between the voltage and current waveforms. Note that:

In an inductive curcuit, the voltage leads the current by 90°.

We shall find that this is one of the fundamental principles within ac circuits.

13-2. INDUCTORS

An inductor can be described as a coil of wire wrapped about a center, called a *core*. Commonly referred to as coils, inductors, or chokes (after their ability to restrict or choke ac currents), they come in various sizes and shapes, each designed for a particular application. In this section we shall examine the physical characteristics of inductors: the factors that affect inductance, the electrical imperfections that each has, and the types of devices found in electronics today.

Factors Affecting Inductance

The easiest way of describing the physical factors affecting inductance is with an equation. For a coil in which the length is at least 10 times the diameter:

$$L = \frac{\mu N^2 A}{l}$$

where L is inductance in henries, μ the permeability of the core material, N the number of turns of the coil, A the cross-sectional area of the coil in square meters, and l the length of the coil in meters. Let us examine each of these factors in detail.

Inductance is proportional to the permeability of the core material. Thus, with a permeability of 1.257×10^{-6}, free space makes a much poorer core than steel, which has a permeability of 4.0×10^{-3}. In fact, an inductor of steel would have over 3000 times more inductance than one of air built to the same dimensions. This is because the steel core would conduct over 3000 times more lines of force than the air at a given field intensity, resulting in 3000 times as many lines of force cutting the wires of the coil. Counter emf, and therefore inductance, would be 3000 times as much.

Inductance is proportional to the square of the number of turns of a coil. Thus, a coil of 200 turns has four times the inductance of a coil of 100 turns, keeping other factors constant. Doubling the number of turns affects inductance in two ways:

1. The field intensity will be doubled. That is, there will be twice the number of lines of force available for cutting wires. Remember, magnetomotive force is equal to turns times current: $F = NI$. Doubling the turns doubles the mmf.

2. Doubling the number of turns will double the number of turns that each line of force may cut. Thus, instead of a particular line cutting 100 wires when the field collapses, it will cut 200 wires, resulting in increased counter emf and a greater inductance.

The combination of these two effects results in inductance increasing as the square of the number of turns.

Inductance is proportional to the cross-sectional area of the inside of the coil, assuming that the core material completely fills the center. Thus, a core material with a cross-sectional area of 2 cm² has twice the inductance of a coil with a 1-cm² area. This becomes evident when the reluctance of a larger cross-sectional area is compared with that of a smaller area, Fig. 13-5. The larger area has less reluctance and is, there-

FIGURE 13-5: *Effect of cross-sectional area on inductance.*

fore, able to conduct more lines of force than the smaller area for a given mmf. With more lines passing through the larger area, the inductor has more lines of force available to cut the coil wires, resulting in increased counter emf. Thus, the larger area has the larger inductance.

Inductance is inversely proportional to length. That is, keeping other factors constant, a 2-cm length of coil will have one-half the inductance of a 1-cm coil. This can be understood by realizing that the 2-cm coil has twice the reluctance of the 1-cm coil, and therefore "resists" lines of force twice as much. Since fewer lines are available for cutting coil wires, less counter emf is produced and the coil has less inductance.

Thus, there are four factors that affect inductance:

1. Inductance increases as permeability increases.

2. Inductance increases as the square of the number of turns.

3. Inductance increases as cross-sectional area increases.

4. Inductance decreases as coil length increases.

Each of these factors affects the inductance of a coil. Next, we shall examine some of the imperfections possessed by coils.

Inductor Losses

Ideally, all the energy that goes into a coil should be converted into magnetic lines of force as the field expands. Then, when the field collapses, this energy should be returned to the source. Thus, the inductor should act merely as a temporary energy storehouse. However, some of this energy gets lost along the way to the storehouse, resulting in inefficiencies within the inductor. These losses can be summarized as:

1. Copper losses.

2. Eddy-current losses.

3. Hysteresis losses.

Copper losses occur because the wire used for a coil contains resistance. Current flowing through the wire of the coil causes an I^2R power loss and an increase in the temperature of the device. Since copper loss is due to a resistance, the coil can be considered as a pure inductance in series with a pure resistance. Thus, a purely inductive inductor is impossible and we must settle for a compromise.

Assume that an inductor has an iron core and the coil has an ac current passing through it. Iron is, of course, an excellent conductor and lines of force from the coil are continually cutting the core. This results in current flow within the core and, since the iron has a resistance, causes another I^2R loss and another increase in temperature. These currents are called *eddy currents*, for they flow in circles—eddys—within the core, perpendicular to the magnetic field, Fig. 13-6. They can also be represented mathematically as a resistance in series with a pure inductance.

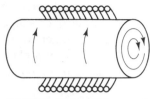

FIGURE 13-6 *Eddy currents.*

When the core of an inductor has been magnetized in one direction, then the current reduced to zero, the domain walls do not return to their neutral position. On the contrary, they must be pushed back to their original position by providing a current in the opposite direction of the magnetizing current. Thus, additional energy must be fed into the system in order for it to go to neutral. This additional energy is in the form of a power loss and is called a *hysteresis loss*, for it is caused by the magnetic hysteresis of the core. It cannot be recovered and is given off as heat to the surrounding air. Thus, we have a third element of resistance that must be mathematically placed in series with our perfect inductor.

In summary, the inductor can be thought of as pure inductance in series with three resistances, each representing a power loss: (1) copper loss, (2) eddy-current loss, and (3) hysteresis loss. We shall now study how these losses are reduced by careful design of the inductor.

Types of Coils

Figure 13-7 illustrates several common types of inductors, together with their electrical symbols. Each consists of a coil of wire wound around a core of either iron, powdered iron, or air.

Iron-core inductors are used where large values of inductances are required at relative low frequencies, such as the 60 Hz used in dc power supplies. Values ranging from 10 mH to tens of henries are common. Rather than build the core from a single

Powdered
iron core   Iron core   Air core

(a)   (b)

FIGURE 13-7: *Inductors: (a) photographs (photos by Ruple); (b) symbols.*

block of iron and encounter high eddy currents, laminated sheet steel is used. The sheets are first stamped into the proper shape, then shellacked with an insulating coating, then laminated (glued) together to obtain the desired thickness, Fig. 13-8. Since eddy currents in solid iron flow in a plane perpendicular to the magnetic field, they are effectively prevented using this lamination method.

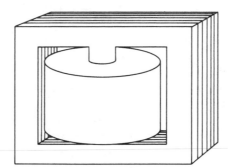

FIGURE 13-8:   *Iron-core inductor.*

Iron-core inductors used in power supplies are required to pass considerable direct current through them, resulting in a large, fixed magnetic field. Since an iron core can handle only a certain flux density, the ac field cannot be as large with this dc-generated flux as when no dc is present. Thus, inductance is reduced at higher dc currents. This phenomenon is known as the *swinging effect* and explains why an 8-H coil rated at 50 mA dc will actually have 12 to 15 H of inductance with no dc present. Its rated inductance assumes that rated dc is present.

At higher frequencies, eddy currents become so severe in laminated iron-core inductors as to make them ineffective. Thus, cores are formed from powdered materials. Two types are common: iron and ferrite. Those of iron are formed from a mix of powdered iron and an insulating binder, then pressed into shape. Thus, eddy-current paths are greatly reduced.

For more critical applications, ferrite materials are used. Formed from compounds of nickel or cobalt, iron, and oxygen, these substances have up to 1 million times the dc resistance of iron and yet have permeabilities approaching that of iron. When powdered, mixed with an insulating binder, and pressed into shape, they have extremely low eddy currents, both because of the small paths available for the current and the high resistances offered by these paths.

Some of these powdered iron and ferrite inductors have a core that may be screwed up or down as shown in Fig. 13-9 to change the inductance. Values up to 100 mH are common.

Air-core inductors are used where small inductance is required at high frequencies. Eddy currents are completely eliminated, resulting in very little power loss. Values from 0.10 μH to 100 μH are available.

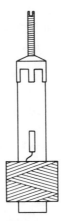

FIGURE 13-9: *Adjustable powdered iron-core inductor.*

Active Inductors

Any circuit that has the property of inductance can be considered an inductor. The most common by far is the coil. However, certain amplifier circuits can be constructed so that the voltage leads the current by 90° for sinusoidal waveforms, and any change in current is opposed. These circuits have no coils but do, in fact, exhibit the property of inductance.

13-3. INDUCTORS IN SERIES

When two inductors are connected in series, the total inductance is the sum of the individual inductances, assuming no flux interaction between the coils, Fig. 13-10.

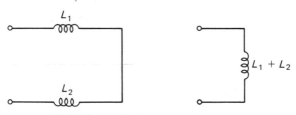

FIGURE 13-10: *Inductances in series.*

Thus,

$$L_t = L_1 + L_2 + L_3 + \ldots$$

where L_t is total inductance in henries, and L_1, L_2, and L_3 are the values of the individual inductors, also measured in henries.

EXAMPLE 13-3: Three inductors are connected in series: a 100-mH coil, a 1-H coil, and a 50-mH coil. What is the total inductance?

SOLUTION: The total inductance is the sum of the individual inductances:

$$L_t = L_1 + L_2 + L_3$$
$$= 0.100 + 1.000 + 0.050$$
$$= 1.150 \text{ H}$$

EXAMPLE 13-4: Inductances of 20, 50, 70, and 200 μH are connected in series. What is the total inductance?

SOLUTION:

$$L_t = L_1 + L_2 + L_3 + L_4$$
$$= 20 + 50 + 70 + 200$$
$$= 340 \ \mu\text{H}$$

13-4. INDUCTORS IN PARALLEL

When inductors are connected in parallel, they add in a manner similar to resistances in parallel (Fig. 13-11):

$$\frac{1}{L_t} = \frac{1}{L_1} + \frac{1}{L_2} + \frac{1}{L_3} + \frac{1}{L_4} + \cdots$$

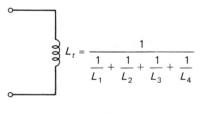

(a)

(b)

FIGURE 13-11: *Inductors in parallel:* (a) *original circuit;* (b) *equivalent circuit.*

EXAMPLE 13-5: Three inductors, of 5, 7, and 10 H, are connected in parallel. What is the total inductance of the circuit?

SOLUTION: Since they are in parallel, we must use the inverse equation:

$$\frac{1}{L_t} = \frac{1}{L_1} + \frac{1}{L_2} + \frac{1}{L_3}$$

$$= \frac{1}{5} + \frac{1}{7} + \frac{1}{10}$$

$$= 0.4429$$

$$L_t = 2.258 \text{ H}$$

EXAMPLE 13-6: What inductance must be placed in parallel with a 50-μH coil to make a total of 35 μH?

$$\frac{1}{L_t} = \frac{1}{L_1} + \frac{1}{L_2}$$

$$\frac{1}{35} = \frac{1}{50} + \frac{1}{L_2}$$

$$0.02857 = 0.02000 + \frac{1}{L_2}$$

$$L_2 = 116.7 \ \mu\text{H}$$

13-5. MUTUAL INDUCTANCE

Two inductors that are placed close enough so that the flux of one coil cuts the wires of the second coil are said to have *mutual inductance*. If these coils are connected in series, this mutual inductance can either add to the total inductance or subtract from this inductance. Figure 13-12 illustrates two possible connections of two inductors. If the mutual inductance of the first circuit is such that it opposes current flow, the mutual inductance is additive. However, when the coil is rewired as in the second circuit, the mutual inductance will aid current flow and is subtractive. Thus, the formula is

$$L_t = L_1 + L_2 \pm 2M$$

where L_t is total inductance in henries, L_1 and L_2 are the inductances of the individual coils, and M is the mutual inductance, also measured in henries.

This mutual inductance will be dependent upon how many flux lines from L_1 cut L_2 and vice versa, and is expressed using the formula

$$M = k\sqrt{L_1 L_2}$$

where k represents a factor called the *coefficient of coupling*, ranging in value from zero to 1.00. With 100% of L_1 flux lines linking L_2, k is 1.00; with 50% of L_1 flux lines linking L_2, k is 0.50; with no linking taking place, k is 0.00.

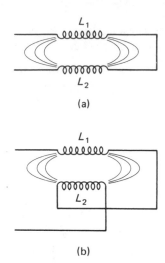

FIGURE 13-12: *Mutual inductance:* (*a*) *circuit 1;* (*b*) *circuit 2.*

EXAMPLE 13-7: A 1-H coil and a 500-mH coil are connected in series such that 50% of the lines of one inductor cut the other. The mutual inductance is such that it increases current flow. Find the total inductance.

SOLUTION: We must first find the mutual inductance, M. Since the mutual linkage is 50%, $k = 0.50$.

$$M = k\sqrt{L_1 L_2}$$
$$= 0.5\sqrt{1.00 \times 0.5}$$
$$= 0.3536 \text{ H}$$

Since this mutual inductance increases current flow, it must be subtracted from the sum.

$$L_t = L_1 + L_2 - 2M$$
$$= 1.00 + 0.500 - 2(0.3536)$$
$$= 0.7928 \text{ H} = 792.8 \text{ mH}$$

EXAMPLE 13-8: Two coils 50 mH and 250 mH are connected in series such that there is a 10% mutual linkage that opposes current flow. Find the total inductance.

$$M = k\sqrt{L_1 L_2}$$
$$= 0.1\sqrt{50 \times 250}$$
$$= 11.18 \text{ mH}$$
$$L_t = L_1 + L_2 + 2M$$
$$= 50 + 250 + 2(11.18)$$
$$= 322.4 \text{ mH}$$

Note that, because the linkage opposed current, it was additive.

One final, interesting point: assume that two identical coils are connected in series such that 100% linkage occurs,[2] aiding the inductance. This would be the case if a coil has its number of turns doubled. Thus,

$$L_t = L + L + 2M$$

but

$$M = k\sqrt{L \times L}$$
$$= 1.0 \times L = L$$

Thus,

$$L_t = L + L + 2L = 4L$$

This shows that doubling the number of turns in a coil quadruples its inductance, as indicated in Section 13-2.

Parallel Circuits

Mutual inductance can also be present in a parallel circuit if inductors are close enough for flux of one to link the second. The inductance for two inductors in parallel then becomes

$$L_t = \frac{L_1 L_2 - M^2}{L_1 + L_2 + 2M}$$

where M is in such a direction that it aids current flow. If it opposes current flow, M must have a negative sign.

> **EXAMPLE 13-9:** A 5-mH coil is in parallel with a 4-mH coil, and 40% of the flux of one links the other in such a direction that it opposes current flow. What is the total inductance of the circuit?
>
> *SOLUTION:* We must first calculate M:
>
> $$M = k\sqrt{L_1 L_2}$$
> $$= 0.4\sqrt{5 \times 4}$$
> $$= 1.789 \text{ mH}$$

Since M opposes current flow, we must assign it as negative. We can now calculate total inductance:

$$L_t = \frac{L_1 L_2 - M^2}{L_1 + L_2 + 2M}$$
$$= \frac{5 \times 4 - (-1.789)^2}{5 + 4 + 2(-1.789)}$$
$$= 3.098 \text{ mH}$$

[2]This could only occur if the coils were not increased in physical size.

13-6. TRANSFORMERS

In the previous section we discussed the effect of mutual inductance between two coils, noting that a current through one coil could affect the current in a second coil. The iron-core transformer uses this capability and is formed by winding two or more separate coils on the same magnetic frame, Fig. 13-13, creating a coefficient of coupling near 1.00. The primary is connected to an ac voltage source which induces a voltage into the secondary. Thus, the voltage of the primary is "transformed" into the secondary voltage. The transformer is capable of providing voltages on the secondary that are larger than, equal to, or less than the primary. If the voltage on the secondary is greater than the voltage on the primary, the transformer is said to be a *step-up transformer*. If the secondary voltage is less than the primary voltage, the unit is said to be a *step-down transformer*.

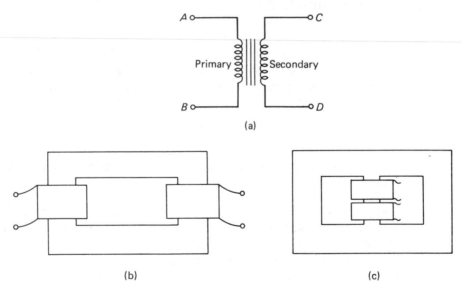

(a)

(b)   (c)

FIGURE 13-13: *Iron-core transformers:* (a) *schematic;* (b) *core-type frame;* (c) *shell-type frame.*

Loading

But how does the secondary current affect the primary? Examine Fig. 13-14(a) closely, noting that it assumes that the primary current is increasing. Under such conditions, by the left-hand rule for generators (Section 13-1), terminal D will be positive with respect to terminal C. However, since terminals C and D are open, no secondary current will flow and the primary voltage source will see only the inductance of the coil and a small amount of coil resistance.

Now, we shall short terminals C and D together and observe the consequences. Secondary current will flow, generating a magnetic field about the secondary. But

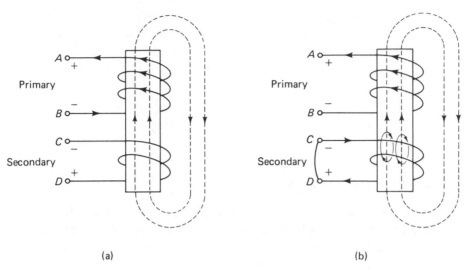

FIGURE 13-14: *Transformer loading: (a) secondary open; (b) secondary shorted.*

note the direction of this magnetic field: it opposes the field generated by the primary coil. Thus, current in the secondary causes a reduction in the magnetic field of the core, which, in turn, reduces the inductance seen by the primary source. Therefore, whatever happens in the secondary is reflected into the primary. We shall see this in the equations that follow.

Voltage and Turns Ratio

The primary and secondary voltages are related to the turns ratio by the equation

$$\frac{V_p}{V_s} = \frac{N_p}{N_s}$$

where V_p is the ac voltage applied to the primary, V_s the ac voltage generated at the secondary terminals, N_p the number of turns of the primary coil, and N_s the number of turns of the secondary coil.

> **EXAMPLE 13-10:** A transformer has 100 V applied to its primary. What is the voltage across its secondary if the primary is wound with 250 turns of wire and the secondary with 450 turns?
>
> *SOLUTION:* The secondary voltage is proportional to the turns ratio:
>
> $$\frac{V_p}{V_s} = \frac{N_p}{N_s}$$
>
> $$\frac{100}{V_s} = \frac{250}{450}$$
>
> $$V_s = 180 \text{ V}$$

EXAMPLE 13-11: What voltage must be provided to the primary of a transformer for its secondary to read 75 V? The primary has 2000 turns and the secondary 1600 turns.

SOLUTION:

$$\frac{V_p}{V_s} = \frac{N_p}{N_s}$$

$$\frac{V_p}{75} = \frac{2000}{1600}$$

$$V_p = 93.75 \text{ V}$$

Current and Turns Ratio

If we assume that a transformer is 100% efficient (it can easily be above 95% in a loaded iron-core transformer), the input power must equal the power delivered to the load, Fig. 13-15. Thus,

$$P_p = P_s$$

$$V_p I_p = V_s I_s$$

$$\frac{V_p}{V_s} = \frac{I_s}{I_p}$$

But

$$\frac{V_p}{V_s} = \frac{N_p}{N_s}$$

Therefore,

$$\frac{N_p}{N_s} = \frac{I_s}{I_p}$$

and current is inversely proportional to the turns ratio. This means that a step-up transformer will step up the voltage and step down the current, whereas a step-down transformer will step down the voltage and step up the current. Obviously, this is necessary if input power is to equal output power.

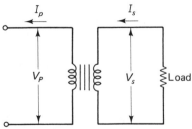

FIGURE 13-15: *Transformer voltages and currents.*

EXAMPLE 13-12: A transformer with a 400-turn primary and a 600-turn secondary has 1 A in its primary. What is its secondary current?

SOLUTION:

$$\frac{N_p}{N_s} = \frac{I_s}{I_p}$$

$$\frac{400}{600} = \frac{I_s}{1}$$

$$I_s = 0.6667 \text{ A} = 666.7 \text{ mA}$$

EXAMPLE 13-13: A transformer must supply secondary current of 25 mA at a voltage of 30 V. If the primary has 1000 turns and the secondary 3000 turns, what must the primary voltage and current be?

SOLUTION: Solve for primary voltage:

$$\frac{V_p}{V_s} = \frac{N_p}{N_s}$$

$$\frac{V_p}{30} = \frac{1000}{3000}$$

$$V_p = 10 \text{ V}$$

Solve for primary current:

$$\frac{N_p}{N_s} = \frac{I_s}{I_p}$$

$$\frac{1000}{3000} = \frac{25}{I_p}$$

$$I_p = 75 \text{ mA}$$

Note that voltage was stepped up from 10 V to 30 V and that current was stepped down from 75 mA to 25 mA.

Matching Loads

We can multiply the voltage and current equations for transformers, to obtain

$$\left(\frac{N_p}{N_s}\right)^2 = \frac{V_p/I_p}{V_s/I_s}$$

The term V_p/I_p is the ac resistance looking into the primary, known as the *primary impedance*, and is expressed as Z_p. The term V_s/I_s is the ac resistance of the load, known as the *load impedance*, Z_s. Both impedances are expressed in ohms and should be considered as resistances for the purpose of this discussion. Because of this relationship, by properly selecting a transformer, any source impedance can be matched to any load impedance to obtain the maximum power transfer from the source to the load.

EXAMPLE 13-14: A transformer is used to drive an 8-Ω speaker, Fig. 13-16. What is the impedance as seen by the source if the primary has 1000 turns and the secondary 50 turns?

SOLUTION:

$$\left(\frac{N_p}{N_s}\right)^2 = \frac{Z_p}{Z_s}$$

$$\left(\frac{1000}{50}\right)^2 = \frac{Z_p}{8}$$

$$Z_p = 3.2 \text{ k}\Omega$$

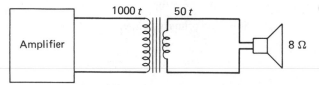

FIGURE 13-16: *Impedance matching.*

From this example we can note an amplifier with a source impedance (or resistance) of 3.2 kΩ could easily match the 3.2 kΩ of the transformer. However, if it were connected directly to the load of 8 Ω, the amplifier could not deliver optimum power, for its impedance would not match the speaker impedance. Matching impedances is one of the most common uses of transformers in electronics.

Lead Identification

Sometimes it is necessary to know the relative phase of the secondary with respect to that of the primary. To indicate this, some schematics place a dot at one of the input leads and a dot at one of the output leads, Fig. 13-17. This indicates that when lead *A* is positive with respect to *B*, lead *C* will be positive with respect to *D*.

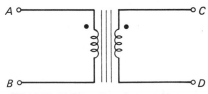

FIGURE 13-17: *Transformer phasing.*

Transformer leads may be color-coded as shown in Table 13-1 to indicate their function.

TABLE 13-1: *Transformer color code.*

Color	Meaning
Black	Primary
Red	High-voltage secondary
Green	Amplifier filament winding
Yellow	Rectifier filament winding

Autotransformers

The autotransformer is a transformer with only one winding, Fig. 13-18. It can be used as a step-up or a step-down transformer, and obeys the equations for isolated coil transformers. In the figure, if *A* and *B* are the input terminals, there would be 800 turns on the primary (600 plus 200) and 200 on the secondary. Some autotransformers have a sliding contact to vary the number of secondary turns.

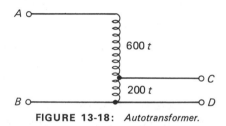

FIGURE 13-18: *Autotransformer.*

The autotransformer has the very serious disadvantage of providing an output that is referenced to the input. In the dual-coil type, Fig. 13-19, touching an output wire and ground would not result in an electrical shock. However, if you were to touch point *C* and any other terminal on the autotransformer in Fig. 13-18, you could receive a shock. Thus, the dual-coil type provides isolation from the ac line for electronic equipment.

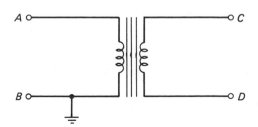

FIGURE 13-19: *Isolated coil transformer grounding.*

Cores

Transformers can be purchased in the same types of cores as previously described for inductors and have the same types of losses.

13-7. RL *TIME CONSTANTS*

In our definition of the inductor, we stated that it tends to prevent any change in current flow. Consider what happens when we close switch S_1 in Fig. 13-20. Before the switch is closed, the current through L_1 is zero. Thus, when the switch closes, the inductor (by our definition) tries to prevent the current from increasing. It does this by generating a counter emf in opposition to V_s. Therefore, at a time immediately after the switch is closed, V_L is 10 V and I is 0 mA, as shown in Fig. 13-21. With 0 mA flowing through R_1, the voltage across the resistor is 0 V.

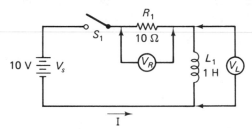

FIGURE 13-20: *dc in an RL circuit.*

However, as time progresses, the current does increase, resulting in a voltage drop, V_R, across the resistor, and a drop of V_L across the inductor. Note that the sum of V_L and V_R must at all times be 10 V, the supply voltage. Finally, when sufficient time has progressed, the inductor will have a steady 1 A through it, generating a static magnetic field, resulting in no counter emf, a drop of 0 V across L_1, and a drop of 10 V across R_1.

This curve is known as a *time-constant curve* and V_L can be computed as

$$V_L = V_s \epsilon^{-Rt/L} \tag{13-1}$$

where V_L is the instantaneous voltage across the inductor, V_s the power-supply voltage, ϵ the value 2.71828, R the resistance of R_1 in ohms, L the inductance of L_1 in henries, and t the time in seconds. Using this equation, we can calculate the voltage at any time t.

> **EXAMPLE 13-15:** Compute the voltage across L_1 and R_1 200 ms after closing switch S_1 in Fig. 13-20.
>
> *SOLUTION:* To find V_L, we write
>
> $$V_L = V_s \epsilon^{-Rt/L}$$
> $$= 10 \epsilon^{(-10 \times 0.2)/1}$$
> $$= 1.353 \text{ V}$$

292

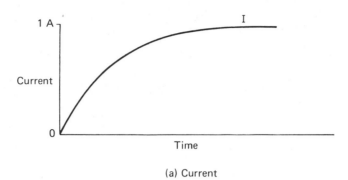

(a) Current

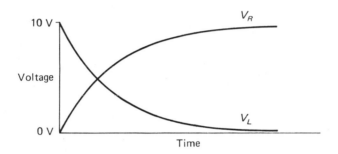

(b) Voltage

FIGURE 13-21: R/L *time constants:* (a) *current;* (b) *voltage.*

Since the supply voltage is 10 V,

$$V_s = V_R + V_L \tag{13-2}$$

$$10 = V_R + 1.353$$

$$V_R = 8.647 \text{ V}$$

Two other equations are of interest. We can find the voltage across the resistor by the equation

$$V_R = V_s(1 - \epsilon^{-Rt/L})$$

This equation can be derived by combining Eq. 13-1 with Eq. 13-2. In addition, the equation for the current through the circuit is

$$I = I_m(1 - \epsilon^{-Rt/L})$$

where I is the instantaneous current and I_m the maximum possible current at an infinite time ($I_m = V_s/R$).

EXAMPLE 13-16: An inductor of 3 mH is in series with a resistance of 10 kΩ. What will be the current through and voltage across the resistor 20 ns after 20 V is applied to the circuit?

SOLUTION:

$$V_R = V_s(1 - \epsilon^{-Rt/L})$$
$$= 20[1 - \epsilon^{-(10^4 \times 20 \times 10^{-9})/(3 \times 10^{-3})}]$$
$$= 20(1 - \epsilon^{-0.06667})$$
$$= 1.290 \text{ V}$$

The maximum current is

$$I_m = \frac{V}{R} = \frac{20}{10} = 2 \text{ mA}$$

Thus, the instantaneous current is

$$I = I_m(1 - \epsilon^{-Rt/L})$$
$$= 0.002(1 - \epsilon^{-0.06667})$$
$$= 129.0 \text{ } \mu\text{A}$$

Let us now consider what happens when the switch has been closed for a while and then is opened. The field about the inductor collapses at an extremely rapid rate, generating a very high voltage across the inductor—much higher than the supply voltage. This voltage can inflict a painful shock to anyone who happens to be touching the terminals of the inductor, and is aptly called an *inductive kick*. It is, of course, only present for a moment, until the field completely collapses.

13-8. SUMMARY

Any moving magnetic field that cuts a wire will generate an emf. In coils, this is a counter emf that opposes any change in current flow. Inductance is the property of a circuit that opposes any change in current flow and is measured in henries. It is best defined by the equation

$$v = L\frac{di}{dt}$$

When a sinusoidal waveform is applied to an inductor, the voltage leads the current by 90°. Inductance of coils depends upon the number of turns, the coil cross-sectional area, and the core material. Iron, powdered iron, powdered ferrite, and air are common core materials. Three types of losses in an inductor reduce its effectiveness: copper loss, caused by the resistance of the wire; eddy-current losses, caused by current flowing within the core; and hysteresis loss, caused by failure of the domain walls to completely return to their neutral position when the mmf is removed. Inductors connected in series can simply be added, but those connected in parallel must be added using the inverse equation. Mutual inductance, caused by linking of one coil with a second coil, can either add to or subtract from the total inductance, depending upon how the coils are connected.

Transformers are available to step up voltage (and thereby step down current) or step down voltage (and thereby step up current). The output voltage is directly proportional to the turns ratio, whereas the output current is inversely proportional to the turns ratio. Transformers are very effective in matching a source impedance to a differing load impedance. When dc is applied to a series *RL* circuit, voltage across the resistor and current within the circuit build up gradually. However, when power is removed, very large voltages appear briefly across the inductor.

13-9. REVIEW QUESTIONS

13-1. Define inductance.

13-2. Give the formula that relates instantaneous current and voltage to inductance.

13-3. Theoretically, what would the instantaneous voltage across a coil be if current increased 1 mA in zero seconds?

13-4. Explain in terms of flux lines why a high frequency produces a greater inductive reactance than a low frequency.

13-5. How are current and voltage related with respect to phase in the sinusoidal waveforms across an inductor?

13-6. Name four factors that determine the inductance of a coil.

13-7. Why is the iron in a core powdered or laminated?

13-8. Name and describe three types of losses found in core materials.

13-9. Why might a coil rated at 8 H measure 15 H?

13-10. What is the formula for finding the total inductance of inductors connected in series? In parallel?

13-11. What is mutual inductance?

13-12. What is the coefficient of coupling?

13-13. What effect does drawing secondary current have upon primary current within a transformer?

13-14. Explain in terms of power why a step-up transformer steps down the current.

13-15. How might the turns ratio of a transformer be measured without actually counting the turns?

13-16. Why is a dual-coil transformer inherently safer than an autotransformer?

13-17. Why doesn't current in a series *RL* circuit build up at a linear rate?

13-18. Why do high voltages appear across an inductor when the dc is removed?

13-1. What emf will be generated by a 10-mH coil if its current increases at a constant rate of 20 mA every 10 μs?

13-2. What emf will be generated by a 200-mH coil if its current increases at a constant rate of 1 mA every 5 s?

13-3. What is the total inductance of a series circuit made up of the following inductances: 125, 35, 60, and 75 mH?

13-4. What is the total inductance of a circuit with 1-H, 5-H, 600-mH, and 4-H coils connected in series?

13-5. What is the total inductance of 5-, 10-, and 12-H coils connected in parallel?

13-6. What is the total inductance of 200-mH, 500-mH, 1.6-H, and 2.2-H coils connected in parallel?

13-7. A 200-mH coil is connected in series with a 160-mH coil in such a manner that there is 55% flux linkage in opposition to current flow. What is the total inductance of the circuit?

13-8. A 75-μH coil is connected in series with a 125-μH coil in such a manner that there is 10% flux linkage aiding current flow. What is the total inductance?

13-9. A 150-mH coil is connected in parallel with a 220-mH coil such that 30% linkage occurs in a direction opposing current flow. What is the total inductance?

13-10. A 1.6-mH coil is connected in parallel with an 800-μH coil such that 70% linkage occurs in a direction favoring current flow. What is the total inductance of the circuit?

13-11. What voltage must be applied to the primary of a transformer in order for its secondary to read 100 V if its primary has 400 turns and its secondary 675 turns?

13-12. The secondary of a transformer reads 125 V and has 780 turns. What must its primary supply if it has 1200 turns?

13-13. If the primary current in Prob. 13-11 is 20 mA, what is its secondary current?

13-14. If the secondary current in Prob. 13-12 is 45 A, what is its primary current?

13-15. A transformer must supply 20 V at 1 A to a load. If the primary has 900 turns and the secondary 1250 turns, what must the primary voltage and current be?

13-16. The primary of a transformer draws 15 A at 25 V. What is the secondary current and voltage if the primary has 750 turns and the secondary 50 turns?

13-17. An audio transformer must supply a 16-Ω speaker from a 100-kΩ source. What must the ratio of the primary turns to the secondary turns be?

13-18. A transformer supplies a 1-kΩ load. What is the impedance looking into the primary if the primary-to-secondary turns ratio is 12:1?

13-19. A 200-mH inductor and an 800-Ω resistor are connected in series. What will the voltages across the resistor and inductor be 300 μs after energized from a 100-V source?

13-20. A 500-mH coil is connected in series with a 10-kΩ resistor. What will the voltages across the elements and the total current be 100 μs after energizing the circuit from a 25-V source?

13-21. A 60-μH coil and a 51-kΩ resistor are connected in series. What will the voltage across the elements and the current be 500 ps after energized from a 20-mV source?

13-11. PROJECTS

13-1. Connect an 8-H 50-mA coil in series with a voltage supply and a resistor that will limit the current to about 30 mA. Connect an oscilloscope across the coil and observe what happens as the circuit is opened, then closed. Explain your results.

> #### CAUTION
> *A coil can supply very high voltages at its terminals under these conditions.*

13-2 Measure the inductance of several coils on an impedance bridge and compare with the rated values.

13-3. Compare the phase of the voltage across an inductor with the phase of the current on an oscilloscope. Do the measured results agree with the theoretical results? *Hint:* Current can be observed by connecting a small resistor in series with the inductor and measuring the voltage drop across it.

13-4. Wind a coil of wire, then move it in a strong magnetic field and measure the voltage at the coil terminals with a sensitive meter. Compare your results with the left-hand rule for generators.

13-5. Investigate the various methods of winding coils and compare the advantages and disadvantages of each.

13-6. Measure the inductance of three coils. Connect them in series and measure the results. Then connect them in parallel and measure. Do the measurements agree with the theoretical values?

13-7. Determine the turns ratio on a transformer by measurement.

13-8. Connect a transformer such that it is near the rated current. Vary the secondary current and observe the primary current.

14

INDUCTIVE REACTANCE

In Chapter 13 we studied the property of inductance and the inductor itself. In this chapter we shall study the inductor within a specific environment, the sinusoidal world. We shall consider how it offers opposition to voltage, and how it reacts in series and parallel circuits. Finally, we shall consider a series circuit of resistance and inductance under the influence of sine waves. The sections include:

14-1. Inductive reactance

14-2. Reactances in series

14-3. Reactances in parallel

14-4. Series *RL* circuits

14-5. Ohm's law in series reactive circuits

14-1. INDUCTIVE REACTANCE

Inductive reactance is the amount of opposition an inductor offers to sinusoidal current flow and is measured in ohms. Whereas inductance is a fixed, physical property, inductive reactance is that property applied to the rms values of a sine wave. The formula for inductive reactance is

$$X_L = 2\pi f L$$

where X_L is the inductive reactance in ohms, f the frequency in hertz, and L the inductance in henries.

EXAMPLE 14-1: A 250-mH inductor has what reactance at 60 Hz? At 600 Hz? At 6 MHz?

SOLUTION: For 60 Hz,

$$X_L = 2\pi fL = 2 \times \pi \times 60 \times 0.25 = 94.25 \ \Omega$$

For 600 Hz,

$$X_L = 2\pi fL = 2 \times \pi \times 600 \times 0.25 = 942.5 \ \Omega$$

For 6 Mhz,

$$X_L = 2\pi fL = 2 \times \pi \times 6.0 \times 10^6 \times 0.25 = 9.425 \ \text{M}\Omega$$

Note that as frequency goes up, so does reactance.

EXAMPLE 14-2: A 25-μH inductor has what reactance at 250 kHz?

SOLUTION:

$$X_L = 2\pi fL = 2 \times \pi \times 250 \times 10^3 \times 25 \times 10^{-6} = 39.27 \ \Omega$$

It should be emphasized that these are not fake ohms but will really impede the flow of sinusoidal electron traffic as a resistor impedes its flow. If the X_L of Example 14-2 were placed in series with a 1-V source, a current would flow:

$$I = \frac{V}{X_L} = \frac{1}{39.27} = 25.46 \ \text{mA}$$

Thus, this X_L does, indeed, impede current flow, Fig. 14-1(a), much as a 39.27-Ω resistor would impede it, Fig. 14-1(b).

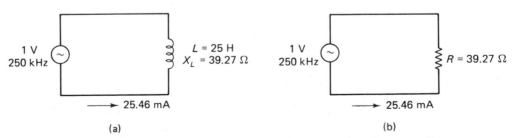

FIGURE 14-1: *Effect of inductive reactance: (a) inductor impeding current flow; (b) resistor impeding current flow.*

Reactance versus Frequency

From the inductive reactance equation, $X_L = 2\pi f L$, we can observe that reactance is highly dependent upon frequency. This can be explained by remembering that a high-frequency current generates a high rate of lines of flux cutting the coil, resulting in a large counter emf, which, in turn, greatly opposes current flow. Thus, the higher the frequency, the higher the reactance, and the smaller the current. Figure 14-2 illustrates how reactance varies with frequency for a 1-H inductor. Note that, as frequency decreases, reactance decreases. From the formula we can see that at dc (a frequency of 0 Hz),

$$X_L = 2\pi f L = 2 \times \pi \times 0 + 1 = 0\,\Omega$$

Thus, at dc, the inductor offers no opposition to current flow and, theoretically, if it were connected to a voltage source, an infinite amount of current would flow. As a practical matter, however, the dc resistance of the wire would prevent this from occurring.

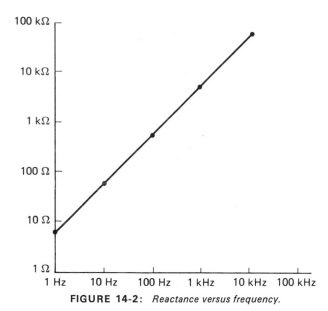

FIGURE 14-2: *Reactance versus frequency.*

14-2. REACTANCES IN SERIES

Since inductances placed in series are additive, we can derive the effect of inductive reactances in series as follows:

$$L_t = L_1 + L_2 + L_3 + \ldots$$

300

Multiplying both sides by $2\pi f$,

$$2\pi f L_t = 2\pi f L_1 + 2\pi f L_2 + 2\pi f L_3 + \cdots$$

But $2\pi f L = X_L$. Therefore,

$$X_{Lt} = X_{L1} + X_{L2} + X_{L3} + \cdots$$

and inductive reactances in series are additive.

EXAMPLE 14-3: Four inductors, of 3, 6, 5, and 10 mH, are placed in series. What is their total inductive reactance at 5 kHz?

SOLUTION: We shall obtain the solution in two ways:

METHOD 1: Find the total inductance, then compute the reactance.

$$\begin{aligned}
L_t &= L_1 + L_2 + L_3 + L_4 \\
&= 3 + 6 + 5 + 10 \\
&= 24 \text{ mH} \\
X_{Lt} &= 2\pi f L_t \\
&= 2 \times \pi \times 5 \times 10^3 \times 24 \times 10^{-3} \\
&- 754.0 \ \Omega
\end{aligned}$$

METHOD 2: Find the individual reactances, then add them:

$$\begin{aligned}
X_{L1} &= 2\pi f L_1 = 2 \times \pi \times 5000 \times 0.003 = 94.25 \ \Omega \\
X_{L2} &= 2\pi f L_2 = 2 \times \pi \times 5000 \times 0.006 = 188.50 \ \Omega \\
X_{L3} &= 2\pi f L_3 = 2 \times \pi \times 5000 \times 0.005 = 157.08 \ \Omega \\
X_{L4} &= 2\pi f L_4 = 2 \times \pi \times 5000 \times 0.010 = 314.16 \ \Omega \\
X_{Lt} &= X_{L1} + X_{L2} + X_{L3} + X_{L4} \\
&= 94.25 + 188.50 + 157.08 + 314.16 \\
&= 754.0 \ \Omega
\end{aligned}$$

Note that both methods obtain identical answers, for both inductance and inductive reactance are additive.

EXAMPLE 14-4: A coil with an inductive reactance of 1 kΩ is placed in series with a second coil of 50 mH. What is the total reactance at 1 kHz?

SOLUTION: We cannot add inductance directly to reactance; therefore, computing the reactance of the 50-mH coil,

$$X_{L2} = 2\pi f L_2 = 2 \times \pi \times 10^3 \times 0.05 = 314.16 \ \Omega$$

The total reactance is,

$$X_{Lt} = X_{L1} + X_{L2} = 1000 + 314.16 = 1314\ \Omega$$

Phasor Relationships

Since, in a reactive circuit, the voltage waveform leads the current waveform by 90°, we represent voltage as $V\underline{/\theta}$ and current as $I\underline{/\theta - 90°}$ in phasor notation, where V and I are the rms voltage and rms current, respectively. This enables us not only to determine magnitudes, but the phase differences. Inductive reactance is denoted as $X_L\underline{/90°}$, whereas resistance is $R\underline{/0°}$, for voltage and current differ by a phase angle of 90° in an inductive circuit and by 0° in a purely resistive circuit. Thus, we can apply Ohm's law in ac circuits if we are careful to observe the proper phase relationships.

EXAMPLE 14-5: An inductor of 8 H is placed in series with a voltage source of 120-V 60-Hz power. What is the current?

SOLUTION: We must first find inductive reactance:

$$X_L = 2\pi fL = 2 \times \pi \times 60 \times 8 = 3016\underline{/90°}\ \Omega$$

Note that the "$\underline{/90°}$" was affixed since this is inductive reactance. We shall now assume the 120 V is $120\underline{/0°}$ and compute the current:

$$I = \frac{V}{X_L} = \frac{120\underline{/0°}}{3.016\underline{/90°}} = 39.79\underline{/-90°}\ \text{mA}$$

Note that, by carrying phase angles along with the computations, current phase is automatically computed.

EXAMPLE 14-6: Recompute Example 14-5 assuming that the source is $120\underline{/45°}$ V.

SOLUTION:

$$I = \frac{V}{X_L} = \frac{120\underline{/45°}}{3.016\underline{/90°}} = 39.79\underline{/-45°}\ \text{mA}$$

Note that, with a voltage at 45° and a current of $-45°$ phase angle, voltage still leads current by 90° in this purely inductive circuit.

EXAMPLE 14-7: Coils of 100 mH and 65 mH are placed in series with a 50-V 400-Hz source. What are the total reactance and current of the circuit and the voltage drops across each of the inductors (Fig. 14-3)?

SOLUTION: The individual reactances are

$$X_{L1} = 2\pi fL = 2 \times \pi \times 400 \times 0.1 = 251.3\underline{/90°}\ \Omega$$
$$X_{L2} = 2\pi fL = 2 \times \pi \times 400 \times 0.065 = 163.4\underline{/90°}\ \Omega$$

The total reactance is the sum of the individual reactances:

$$X_{Lt} = X_{L1} + X_{L2} = 251.3\underline{/90°} + 163.4\underline{/90°} = 414.7\underline{/90°}\ \Omega$$

Current, according to Ohm's law, is total voltage divided by total reactance. We shall assume a phase angle of 0° for the voltage:

$$I = \frac{V}{X_{Lt}} = \frac{50\underline{/0°}}{414.7\underline{/90°}} = 120.6\underline{/-90°}\ \text{mA}$$

The voltage drops across each of the inductors are

$$V_1 = IX_{L1} = (0.1206\underline{/-90°})(251.3\underline{/90°}) = 30.30\underline{/0°}\ \text{V}$$
$$V_2 = IX_{L2} = (0.1206\underline{/-90°})(163.4\underline{/90°}) = 19.70\underline{/0°}\ \text{V}$$

Note that, according to Kirchhoff, total voltage is the sum of the individual voltage drops:

$$V_t = V_1 + V_2 = 30.30\underline{/0°} + 19.70\underline{/0°} = 50.00\underline{/0°}\ \text{V}$$

This checks with the given supply voltage.

EXAMPLE 14-8: In Fig. 14-3, L_1 is 20 mH and has a current of $20\underline{/-35°}$ mA at a frequency of 10 kHz. Find V_t if L_2 is 36 mH.

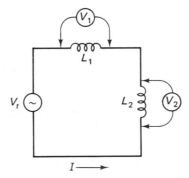

FIGURE 14-3: *Example 14-7.*

SOLUTION: We must first find the individual, then the total reactance:

$$X_{L1} = 2\pi f L = 2 \times \pi \times 10^4 \times 0.02 = 1.257\underline{/90°}\ \text{k}\Omega$$
$$X_{L2} = 2\pi f L = 2 \times \pi \times 10^4 \times 0.036 = 2.262\underline{/90°}\ \text{k}\Omega$$
$$X_t = X_{L1} + X_{L2} = 1.257\underline{/90°} + 2.262\underline{/90°} = 3.519\underline{/90°}\ \text{k}\Omega$$

Knowing the current, we can now find the total voltage:

$$V_t = IX_t = (20\underline{/-35°})(3.519\underline{/90°}) = 70.38\underline{/55°}\ \text{V}$$

Note that current and voltage are still out of phase by 90°.

14-3. REACTANCES IN PARALLEL

Reactances in parallel are treated the same way as resistors in parallel, using the inverse equation:

$$\frac{1}{X_t} = \frac{1}{X_1} + \frac{1}{X_2} + \frac{1}{X_3} + \cdots \qquad (14\text{-}1)$$

EXAMPLE 14-9: Three reactances, 1, 2, and 3 kΩ, are connected in parallel, Fig. 14-4. What is the total reactance?

SOLUTION:

$$\frac{1}{X_t} = \frac{1}{X_1} + \frac{1}{X_2} + \frac{1}{X_3}$$

$$= \frac{1}{j1} + \frac{1}{j2} + \frac{1}{j3}$$

$$= -j1 - j0.5 - j0.3333$$

$$= -j1.8333$$

$$X_t = 545.5\underline{/90°} \text{ m}\Omega$$

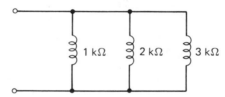

FIGURE 14-4: *Inductive reactances in parallel.*

EXAMPLE 14-10: An inductor of 350 mH is placed in parallel with inductive reactances of 8 and 12 kΩ at a frequency of 5 kHz. What is the total reactance of the circuit?

SOLUTION: We must first find the reactance of the 350-mH coil:

$$X_L = 2\pi fL = 2 \times \pi \times 5000 \times 0.35 = 10.996\underline{/90°} \text{ k}\Omega$$

We can now apply Eq. (14-1):

$$\frac{1}{X_t} = \frac{1}{X_1} + \frac{1}{X_2} + \frac{1}{X_3}$$

$$= \frac{1}{j10.996} + \frac{1}{j8} + \frac{1}{j12}$$

$$= -j0.09095 - j0.125 - j0.08333$$

$$= -j0.2993$$

$$X_t = 3.341\underline{/90°} \text{ k}\Omega$$

Current and Voltage

Using Ohm's law and Kirchhoff's laws, we can calculate the branch currents of a parallel reactive circuit. All work should be done in phasors, however.

EXAMPLE 14-11: Compute the branch currents and total current for Fig. 14-5.

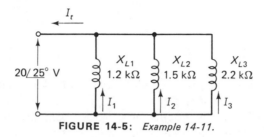

FIGURE 14-5: *Example 14-11.*

SOLUTION: Applying Ohm's law to each branch:

$$I_1 = \frac{V}{X_{L1}} = \frac{20/25°}{1.2/90°} = 16.67/{-65°} \text{ mA}$$

$$I_2 = \frac{V}{X_{L2}} = \frac{20/25°}{1.5/90°} = 13.33/{-65°} \text{ mA}$$

$$I_3 = \frac{V}{X_{L3}} = \frac{20/25°}{2.2/90°} = 9.091/{-65°} \text{ mA}$$

Computing total current,

$$I_t = I_1 + I_2 + I_3$$
$$= 16.67/{-65°} + 13.33/{-65°} + 9.091/{-65°} = 39.09/{-65°} \text{ mA}$$

Incidently, since we know total current and total voltage, we can compute total reactance:

$$X_t = \frac{V}{I_t} = \frac{20/25°}{39.09/{-65°}} = 511.6/90° \text{ } \Omega$$

EXAMPLE 14-12: Two inductors, 750 mH and 1.2 H, are connected in parallel across a 2-V 1-kHz source. Compute current through each element and the total current.

SOLUTION: We must first find the reactance of each coil:

$$X_{L1} = 2\pi f L_1 = 2 \times \pi \times 10^3 \times 0.75 = 4.712/90° \text{ k}\Omega$$
$$X_{L2} = 2\pi f L_2 = 2 \times \pi \times 10^3 \times 1.2 = 7.540/90° \text{ k}\Omega$$

The current through each branch can now be computed:

305

$$I_1 = \frac{V}{X_{L1}} = \frac{2\underline{/0°}}{4.712\underline{/90°}} = 424.4\underline{/-90°} \ \mu A$$

$$I_2 = \frac{V}{X_{L2}} = \frac{2\underline{/0°}}{7.540\underline{/90°}} = 265.3\underline{/-90°} \ \mu A$$

The total current is the sum of the individual branch currents:

$$I_t = I_1 + I_2 = 424.4\underline{/-90°} + 265.3\underline{/-90°} = 689.7\underline{/-90°} \ \mu A$$

14-4. SERIES RL CIRCUITS

In this section we shall consider ac circuits that have both resistance and inductive reactance connected in series. However, we must first define the term "impedance." *Impedance* (Z) is the total opposition to ac current flow and consists of reactance and resistance. Expressed in ohms, impedance is the phasor sum of reactance and resistance as shown in Fig. 14-6. Since inductive reactance is always a phasor at 90° and resistance a phasor at 0°, the sum of an inductive reactance and a resistance will be an impedance at some angle between 0° and 90°.

Reactance and resistances connected in series are *phasor-additive*. That is, their values cannot be added directly, but they must be summed as phasors.

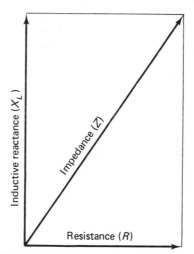

FIGURE 14-6: *ac phasor relationships.*

EXAMPLE 14-13: An inductor with a reactance of 500 Ω is in series with a 680-Ω resistor. What is the total impedance?

SOLUTION: The inductive reactance always has a phase angle of 90°, and the resistance, 0°. Thus,

$$Z = R + jX_L = 680 + j500 = 844.0\underline{/36.33°} \ \Omega$$

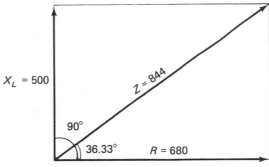

FIGURE 14-7: *Example 14-13.*

Thus, the impedance is 844.0 Ω at a phase angle of 36.33°. The phasor diagram is shown in Fig. 14-7.

EXAMPLE 14-14: A 100-μH coil is in series with a 470-Ω resistor at a frequency of 100 kHz. What is the total impedance?

SOLUTION: We must first find the inductive reactance:

$$X_L = 2\pi fL = 2 \times \pi \times 10^5 \times 100 \times 10^{-6} = 62.83\underline{/90°}\ \Omega$$

We can now find the total impedance:

$$Z = R + jX_L = 470 + j62.83 = 474.2\underline{/7.614°}\ \Omega$$

EXAMPLE 14-15: A 500-, 250-, and 600-μH coil are connected in series with three series resistors with values of 1.2 kΩ, 6.8 kΩ, and 750 Ω. What is the total impedance of the circuit at 1.4 MHz?

SOLUTION: The resistors are in series and can be lumped:

$$R_t = R_1 + R_2 + R_3$$
$$= 1.2 + 6.8 + 0.75$$
$$= 8750\underline{/0°}\ \Omega$$

In a similar manner, series inductors are additive:

$$L_t = L_1 + L_2 + L_3$$
$$= 500 + 250 + 600$$
$$= 1350\ \mu H$$

Next, we must compute the total reactance:

$$X_L = 2\pi fL_t = 2 \times \pi \times 1.4 \times 10^6 \times 1350 \times 10^{-6}$$
$$= 11{,}875\underline{/90°}$$

Find the impedance:

$$Z = R + jX_L = 8750 + j11{,}875 = 14.75\underline{/53.62°}\ \text{k}\Omega$$

A theoretically perfect inductor will only have inductance and, according to the formula $X_L = 2\pi fL$, would offer no resistance to dc. The coils that are used as inductors, however, are formed by wrapping many turns of wire around a core. Since this wire has resistance, the coil can be thought of as a pure inductor in series with a resistor that represents the wire resistance.

EXAMPLE 14-16: A 5-H coil has a dc resistance of 500 Ω. What is its impedance at 260 Hz?

SOLUTION:

$$X_L = 2\pi fL = 2 \times \pi \times 260 \times 5 = 8.168\underline{/90°}\ \text{k}\Omega$$

But it has a resistance of 500 Ω. Therefore,

$$Z = R + jX_L = 500 + j8168 = 8183\underline{/86.50°}\ \Omega$$

Thus, the coil is imperfect, for its phase angle is not quite 90°.

14-5. OHM'S LAW IN SERIES REACTIVE CIRCUITS

As long as all the calculations are done in phasor algebra, all the techniques used in dc analysis are applicable to ac analysis.

EXAMPLE 14-17: A 2-mH coil is in series with a 100-Ω resistor across a 1-V 10-kHz source. Find the total impedance, the current, and the voltages across the coil and the resistance.

SOLUTION: We must first find inductive reactance:

$$X_L = 2\pi fL = 2 \times \pi \times 10^4 \times 2 \times 10^{-3} = 125.7\underline{/90°}\ \Omega$$

Solve for impedance:

$$Z = R + jX_L = 100 + j125.7 = 160.6\underline{/51.49°}\ \Omega$$

Solve for the current:

$$I = \frac{V}{Z} = \frac{1\underline{/0}}{160.6\underline{/51.49}} = 6.227\underline{/-51.49°}\ \text{mA}$$

Solve for the voltage across the inductor:

$$V_L = IX_L = (6.227 \times 10^{-3}\underline{/-51.49})(125.7\underline{/90})$$
$$= 782.7\underline{/38.51°} \text{ mV}$$

Solve for the voltage across the resistor:

$$V_R = IR = (6.227 \times 10^{-3}\underline{/-51.49})(100\underline{/0})$$
$$= 622.7\underline{/-51.49°} \text{ mV}$$

Just as a check, add V_R and V_L:

$$V_s = V_R + V_L = 622.7\underline{/-51.49°} + 782.7\underline{/38.51°}$$
$$= (387.7 - j487.3) + (612.5 + j487.3)$$
$$= 1000 + j0 \text{ mV} = 1.000\underline{/0°} \text{ V}$$

Thus, Kirchhoff's law triumphs again. Figure 14-8 illustrates the phasor diagram. Note that the voltages vectorally add to obtain V_s. Note also that current is in phase with voltage across the resistor, but out of phase by 90° with the voltage across the inductor. You will also observe that the phase angle between current and voltage is the same as that between resistance and impedance, except that the sign is opposite.

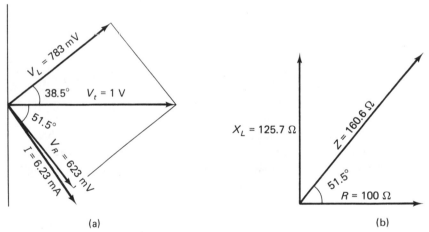

(a)   (b)

FIGURE 14-8: *Example 14-17, voltages and currents in ac circuits: (a) voltage and current phasors; (b) impedance phasors.*

EXAMPLE 14-18: Find the supply voltage V_s in Fig. 14-9.

SOLUTION: We must first find X_L:

$$X_L = 2\pi fL = 2 \times \pi \times 75 \times 10^3 \times 0.01 = 4.712 \text{ k}\Omega$$

We can now find the current through the inductor:

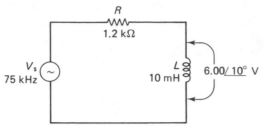

FIGURE 14-9:   *Example 14-18.*

$$I = \frac{V}{X_L} = \frac{6.00\underline{/10°}}{4.712\underline{/90°}} = 1.273\underline{/-80°} \text{ mA}$$

The total impedance is

$$Z = R + jX_L = 1.200 + j4.712 = 4.862\underline{/75.71°} \text{ k}\Omega$$

Thus, the total voltage is

$$V_s = IZ = (1.273\underline{/-80°})(4.862\underline{/75.71°}) = 6.189\underline{/-4.29°} \text{ V}$$

14-6. SUMMARY

Inductive reactance is the opposition inductance offers to sinusoidal voltages and is measured in ohms. It is computed using the equation $X_L = 2\pi f L$ and, thus, increases linearly with frequency. Reactances expressed as phasors follow the same rules as resistors and are additive when in series; when in parallel the inverse equation must be used. Inductive reactance is always considered at 90° as a phasor to that of resistance, which is considered as 0°. Phasor addition of resistance and reactance is called impedance. All the analysis techniques used in dc apply to ac as long as phasor algebra is used.

14-7. REVIEW QUESTIONS

14-1.   Give the formula for finding inductive reactance.

14-2.   If frequency goes up, what happens to inductance? To inductive reactance?

14-3.   How is total inductive reactance found for inductive reactances in series? In parallel?

14-4.   What phase angle is assumed for an inductive reactance? For resistance?

14-5.   What is the range of impedance phase angles for a series *RL* circuit?

14-8. PROBLEMS

14-1. (a) Find the inductive reactance of a 100-mH coil at 10 Hz. (b) At 100 Hz. (c) At 1 kHz.

14-2. (a) Find the inductive reactance of a 5-H coil at 10 Hz. (b) At 50 Hz. (c) At 120 Hz.

14-3. At what frequency does a 500-mH coil have an inductive reactance of 1 kΩ?

14-4. At what frequency does a 250-μH coil have a reactance of 1 MΩ?

14-5. What is the total reactance of the following series connected coils at 1 kHz? 700 mH, 1.2 H, 3.6 H.

14-6. Three coils are connected in series: a 5-mH coil, a 12-mH coil, and an 850-μH coil. What is the total inductive reactance at 25 kHz?

14-7. A coil with an inductive reactance of 2 kΩ is placed in series with a second coil of 75 mH. What is the total reactance at 1200 Hz?

14-8. Repeat Prob. 14-7 for a frequency of 10 kHz.

14-9. Three coils are connected in parallel: a 5-mH coil, a 12-mH coil, and a 7.5-mH coil. What is the total inductive reactance of the circuit at 45 kHz?

14-10. A coil with an inductive reactance of 4 kΩ is placed in parallel with a second coil of 100 mH. What is the total reactance at 2 kHz?

14-11. Repeat Prob. 14-10 for 20 kHz.

14-12. An inductor of 500 mH is placed in series with a 60-V 400-Hz supply. What is the current through the inductor?

14-13. A 60-μH coil has 1.0/$-20°$-mA current through it. What is the voltage across it if the frequency is 8 MHz?

14-14. Two coils, 50 mH and 60 mH, are connected in series across a 5-V 500-Hz supply. What are the currents through the coils and the voltages across each?

14-15. Three coils, 150, 200, and 600 mH, are connected in series with a supply of 1 V at 2 kHz. What is the total current and what are the voltages across each of the coils?

14-16. Two coils, 1.5 and 3.6 mH, are connected in series with 2.5/$-60°$ mA flowing through them. What is the supply voltage and what is the voltage across each of the coils if the frequency is 2.5 kHz?

14-17. What is the impedance of a circuit with a 50-mH coil in series with a 10-kΩ resistor at a frequency of 25 kHz?

14-18. What is the impedance of a series circuit with the following circuit elements at 40 kHz: 10-kΩ resistor, 50-kΩ resistor, 10-mH inductor, and 15-mH inductor?

14-19. A 10-H coil has a dc resistance of 250 Ω and is in series with a 750-Ω resistor at a frequency of 60 Hz. What is the total impedance of the circuit?

14-20. A 500-mH coil is in series with a 5.1-kΩ resistor and a supply of 2.5 V at 6 kHz. What are the voltages across the coil and the resistor?

14-21.   Coils of 40 mH and 55 mH are in series with resistors of 680 Ω and 1.5 kΩ and a supply voltage of 6 V at 3 kHz. What are the voltage drops across each of the elements?

14-22.   A 5.8-mH coil is in series with a 10-kΩ resistor. If the coil has a voltage of $25.8\underline{/36°}$ mV across it, what is the supply voltage? The frequency of the supply is 200 kHz.

14-23.   A 68-mH coil is in series with a 68-kΩ resistor. If the voltage across the resistance is $3.9\underline{/60°}$ V at a frequency of 15 kHz, what is the voltage across the entire circuit?

14-9. PROJECTS

14-1.   Measure the inductance of several coils by either measuring on an impedance bridge or supplying a 1-kHz signal and measuring the current. Then connect the coils in series and repeat the measurement. Do the measured results agree with the computed results?

14-2.   Repeat Project 14-1 with the coils connected in parallel.

14-3.   Connect an 8-H 50-mA coil in series with a 1-kΩ resistor and a signal generator. Keeping the voltage output by the generator constant, vary its frequency, taking measurements of the voltage across the inductor and the resistor. Do the results agree with the computed values?

14-4.   Connect three inductors in parallel, apply an ac voltage, and measure the branch currents and the total current. Do the results agree with the computed values?

15

CAPACITANCE

In Part I we studied the resistor, a device in which current and voltage are in phase. Then, in Chapters 13 and 14, we studied the inductor, a device in which the voltage leads the current by 90°. Would it now surprise you if we completed this happy triangle by studying a device in which the current leads the voltage by 90°? The capacitor is such a component and, in practically every way, complements the action of an inductor. We shall study this device in detail, then put several together and analyze the result. Finally, we shall examine the capacitor as it influences dc circuits. The sections include:

15-1. CAPACITANCE

The property of capacitance is present any time two conductors are separated by an insulator. Consider Fig. 15-1(a), in which two conductive plates of metal are separated separated by an insulator, also called a *dielectric*. The atoms that form this dielectric hold their electrons very tightly in orbit, preventing any free electrons from being generated. Thus, under neutral conditions, the electrons happily travel about their

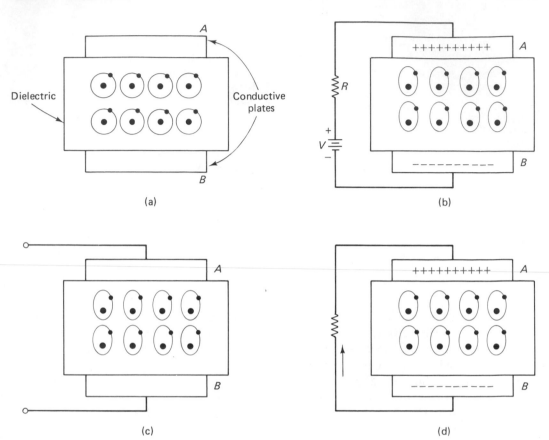

FIGURE 15-1: *Electron-orbit distortion: (a) initial conditions; (b) charging the capacitor; (c) charged capacitor; (d) discharging the capacitor.*

nucleus. However, when a positive charge is applied to plate *A* and a negative charge to *B* as in Fig. 15-1(b), a strange thing happens. Electrons flow from plate *A* through the resistor and the voltage source to plate *B*. Since the dielectric is an insulator, they cannot flow through it, so a negative charge builds up on plate *B* and a positive charge on plate *A*. This charge creates an electric field within the dielectric such that the orbiting electrons distort their orbits. This orbital distortion moves electrons closer to the positive plate and away from the negative plate, resulting in a polarizing of the dielectric. That is, one side of the dielectric is more negative than the other side. This is analogous to a ferromagnetic material when it is subjected to a magnetic field; it is magnetically polarized, whereas the dielectric is electrically polarized. (Some dielectrics also exhibit electrical hysteresis.)

This electron shift further enhances electron flow in the conductor, for the polarized dielectric repels additional electrons from conductor *B* toward conductor *A*.

Once the orbital distortion has been completed, no more electrons flow in the conductive paths and the capacitor is said to be *charged*. Thus, a voltage equal to V_s

exists across the capacitor. If the circuit were then to be opened as in Fig. 15-1(c), the charge would still be present and this voltage would continue to exist across the capacitor. Thus, we can see that the capacitor is capable of storing a charge after the source has been removed.

If we were now to short the capacitor through the resistor as shown in Fig. 15-1(d), the excess electrons on plate *B* would flow through the resistor to plate *A*, neutrallizing the conductor and causing the orbits to return to their neutral position. Once this has been accomplished, current would cease and the capacitor is said to be *discharged*.

A Mathematical Model

The unit for capacitance is the *farad* (F), named after Michael Faraday. A 1-F capacitor will displace a charge of 1 coulomb (C)[1] under an influence of 1 V. That is, if we were to apply 1 V across the plates of a 1-F capacitor, 1 C would flow in the external circuit. Mathematically,

$$C = \frac{Q}{V}$$
(15-1)

where *C* is the capacitance in farads, *Q* the charge in coulombs, and *V* the voltage in volts. Note that, with a 1-F capacitor, 1 V applied to the device will displace 1 C. Similarly, 5 V will displace 5 C and 10 V will displace 10 C. Thus, the greater the voltage, the greater the charge displacement. This is to be expected since the greater the applied voltage, the greater the orbital shift, resulting in a greater number of electrons moving closer to plate *A*, which in turn pushes more electrons through the conductor to plate *B*.

Let us now play a bit with Eq. (15-1). Assume that our 1-F capacitor has displaced 1 C by a source of 1 V. What happens if we increase the voltage to 2 V? The circuit will, of course, conduct one more coulomb of charge. But what if we increased the voltage steadily at a rate of 1 V/s? We would expect an increase of 1 C/s. Examine this mathematically by transposing the equation:

$$C = \frac{Q}{V}$$

$$Q = CV$$

From this, if we were to increase *V* by *dV* in *dt* seconds, we would have an increase of *dQ* in this same time:

$$\frac{dQ}{dt} = C\frac{dV}{dt}$$

---

[1] In the SI system of units, *C* represents capacitance measured in farads (F) and *Q* represents charge in coulombs (C). Thus, the letter C is doing double duty. In summary, whenever C is used in an equation, it represents capacitance, and whenever it is used after a number (10.65 C, for example), it represents coulombs.

Thus, an increase of 3 V in 1 s would result in Q increasing:

$$\frac{dQ}{dt} = C\frac{dV}{dt}$$

$$= 1 \cdot \frac{3}{1} = 3 \text{ C/s}$$

Q would increase 3 C/s. But what does a charge flow 3 C/s represent? According to our definition of current, it represents 3 A. Therefore, equating $I = dQ/dt$:

$$I = C\frac{dV}{dt} \tag{15-2}$$

where dV/dt represents the rate of change of voltage per unit time. The d's can be interpreted as "a small change in." We can see that Eq. (15-2) is the fundamental definition of capacitance, for it relates voltage, current, and capacitance.

EXAMPLE 15-1: A certain 1-F capacitor has a supply voltage across it that increases 20 V in 5 s. What is the current in the wires feeding the capacitor?

SOLUTION: Because current flows in the wires connected to the capacitor, we often refer to current through a capacitor. According to Eq. (15-2), the current is

$$I = C\frac{dV}{dt} = 1 \times \frac{20}{5} = 4 \text{ A}$$

Thus, the current is 4 A dc when the voltage is increasing at a rate of 20 V in 5 s.

EXAMPLE 15-2: A voltage increase of 20 V/s across a capacitor results in a current of 1 μA. What is the value of the capacitor?

SOLUTION:

$$I = C\frac{dV}{dt}$$

$$10^{-6} = C \times \frac{20}{1}$$

$$C = 0.05 \ \mu\text{F}$$

We should take note at this point that common capacitors are in the 1-pF (a picofarad is 10^{-12} farad) to 5000-μF range. A value of 1 F is very impractical and we have used units of this range merely to explain the various equations. However, all the equations require that capacitance be expressed in farads, such as was done in Example 15-2.

EXAMPLE 15-3: What will the current through a 100-μF capacitor be if the voltage across it increases at a rate of 26 mV/s?

SOLUTION:

$$I = C\frac{dV}{dt} = (100 \times 10^{-6}) \times (0.026) = 2.60 \ \mu\text{A}$$

Note that C was express in the equation as 100×10^{-6} F.

Capacitors and Sine Waves

Since most of the work in ac involves sine waves, let us examine the current waveform of a sinusoidal voltage across a capacitor, Fig. 15-2. At point A, the voltage is increasing quite rapidly per unit time, resulting in a rapid transfer of charge from one plate to the other, and, therefore, a great current flow. (Remember that current represents charge flow per second and note that this flow is maximum at point A.) At point B, however, the voltage is not changing; therefore, charge remains fixed and, since no electrons are moving, current is zero. At point C, the voltage is moving in the negative direction, resulting in rapid charge movement and a large, negative current flow. At point D, the voltage is again unchanging and there is no charge movement, resulting in zero current flow. As the voltage returns to point E, it is changing rapidly in the positive direction, resulting in a large charge movement and, therefore, maximum current flow.

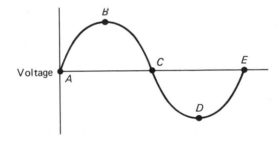

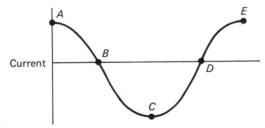

FIGURE 15-2: *Voltage and current waveforms of a capacitor.*

Having completed our graph, we should note that the current waveform is a sine wave and that it leads the voltage waveofre by 90°. Thus,

In an inductive circuit, the voltage leads the current by 90°, whereas in a capacitive circuit, the current leads the voltage by 90°.

A simple acronym will keep this straight in your mind: remember ELI the ICE man. ELI means voltage (*E* for emf) leads current (*I*) in an inductive (*L*) circuit. ICE means current (*I*) leads voltage (*E*) in a capacitive (*C*) circuit.

Symbol for Capacitance

The symbol for the capacitor is a straight line plus a curved line, as shown in Fig. 15-3. When the capacitance is adjustable, an arrow is drawn through the symbol as shown. The former symbol for a capacitor was that of two parallel lines and may be seen on many older drawings. However, this symbol now represents a set of relay contacts as shown in Fig. 15-4 and should not be used to represent a capacitor in new drawings.

| Fixed | Variable | Normally open | Normally closed |

FIGURE 15-3: *Symbols for capacitors.*   **FIGURE 15-4:** *Symbols for relay contacts.*

Having been introduced to capacitance, we shall now examine some of the factors affecting capacitance.

15-2. DIELECTRICS

Capacitance is greatly affected by the type of dielectric chosen. In some dielectrics, 1 V will displace more coulombs than in others. Thus, some materials have more orbital distortion and therefore more charge redistribution under the influence of one volt than others. In order to develop some feeling for this capability, let us examine some of the relationships that exist in a capacitor. We can express the displacement of charge on the capacitor plates as

$$D = \frac{Q}{A}$$

where *Q* is charge in coulombs, *A* the area of the plates in square meters, and *D* the charge displacement per unit area expressed in coulombs per square meters. Thus, according to this equation, a 1-C charge on a 1-m² plate has a displacement of 1 C/m².

EXAMPLE 15-4: A certain capacitor has a plate size of 3 × 64 cm and has a charge of 5 C. What is its charge displacement?

SOLUTION:

$$D = \frac{Q}{A} = \frac{5}{0.03 \times 0.64} = 260.4 \text{ C/m}^2$$

Let us now consider the voltage between the plates. We can express this in unit terms as

$$E = \frac{V}{s}$$

where V is the voltage between the plates, s the distance between the plates measured in meters, and E the electric field strength expressed in volts per meter. Thus, a voltage of 5 V across a dielectric 1-mm thick would be a field intensity of

$$E = \frac{V}{s} = \frac{5}{0.001} = 5 \text{ kV/m}$$

We now have two "unitized" quantities: one a cause, the other the effect. The effect is D, in which a unit charge is displaced over a unit area. The cause is E, in which a unit voltage stresses a unit distance between the plates. Next, we shall put these together.

Permittivity

With these two definitions, we have both charge displacement and field intensity expressed in per unit quantities. The ratio of these two is the permittivity of the dielectric:

$$\epsilon = \frac{D}{E}$$

where ϵ is the permittivity of the material in farads per meter, D the charge displacement in coulombs per square meter, and E the electric field intensity in volts per meter. Certain materials will displace a very large charge per unit area (D) under the influence of a field intensity of 1 V/m, whereas other materials will displace very little charge per unit area under this intensity. Therefore, permittivity is a measure of a dielectric's capability to displace charge, and a capacitor with a high permittivity dielectric will have a greater capacitance than one with a low permittivity of the same physical dimensions.

The units of permittivity expressed above may, at first glance (or even at second glance), seem confusing: farads per meter. However, with a bit of juggling, we can develop them. Since

$$\epsilon = \frac{D}{E} = \frac{\text{coulombs/meters}^2}{\text{volts/meter}} = \frac{\text{coulombs}}{\text{meters} \times \text{volts}}$$

However, since $Q = CV$, substituting for the numerator,

$$\epsilon = \frac{\text{farads} \times \text{volts}}{\text{meters} \times \text{volts}} = \frac{\text{farads}}{\text{meter}}$$

Thus, permittivity is expressed in farads per meter.

EXAMPLE 15-5: A certain glass material 1-mm thick is used as a dielectric in a capacitor. The total plate dimensions are 1.5 cm by 5 m and there is a charge of 75 nC between the plates at a voltage of 20 V. What is the permittivity of this glass?

SOLUTION: We must first find D and E. To find D:

$$D = \frac{Q}{A} = \frac{75 \times 10^{-9}}{0.015 \times 5} = 1 \ \mu\text{C/m}^2$$

To find E:

$$E = \frac{V}{s} = \frac{20}{0.001} = 20 \ \text{kV/m}$$

We can now find the permittivity of the material:

$$\epsilon = \frac{D}{E} = \frac{10^{-6}}{20,000} = 5.000 \times 10^{-11} \ \text{F/m}$$

Dielectric Constant

If we were to place two parallel plates within a vacuum and measure the permittivity, we would find that the permittivity of free space is 8.85×10^{-12} F/m. Solids have permittivities much greater than this, since they contain atoms within the dielectric capable of having orbital distortion. Thus, we can express a ratio of the permittivity of a material to that of free space as

$$\kappa = \frac{\epsilon}{\epsilon_v}$$

where κ is a ratio called the *dielectric constant* of a material (also called *relative permittivity*), ϵ_v the permittivity of free space, and ϵ the permittivity of the dielectric under consideration. This dielectric constant is, therefore, a figure of merit for the material when used in a capacitor. The higher the dielectric constant, the greater the capacitance, given fixed physical dimensions. Table 15-1 lists dielectric constants for several materials.

EXAMPLE 15-6: What is the dielectric constant of the glass in Example 15-5? Permittivity was computed as 5.000×10^{-11} F/m.

SOLUTION: The dielectric constant is the ratio of the permittivity of the glass to that of free space:

$$\kappa = \frac{\epsilon}{\epsilon_v} = \frac{5.000 \times 10^{-11}}{8.85 \times 10^{-12}} = 5.650$$

Therefore, the dielectric constant for this material is 5.650. Note that this is a dimensionless unit.

TABLE 15-1: *Dielectric properties of materials.*

Material	*Dielectric Constant*	*Dielectric Strength (V/mm)*
Air	1.0006	310
Aluminum oxide layer	10	
Bakelite	6.5	2000
Ceramic porcelain	6–10	1200–4700
Ceramic titanates	15–12,000	200–1200
Glass	4–10	1400–8000
Mica	6–9	8000
Niobium oxide layer	50	
Oil	4.2–4.7	1400
Paper, treated	2–6	7000
Polyethylene	2.3	4000
Polystyrene	2.5	2400
Quartz	4.3	1200
Water	80	

Dielectric Strength

Not only must these dielectrics be able to provide the capacitance needed, but they must be able to withstand the voltages encountered between the plates. Although we want to distort the orbits, we do not want to strip any electrons from their shells. Therefore, dielectrics have a specification called *dielectric strength*. It is defined as the voltage a 1-mm-thick dielectric can withstand without breaking down and stripping electrons from their orbits. These are also shown in Table 15-1.

EXAMPLE 15-7: What is the maximum voltage a 26-mm slab of glass can withstand if its dielectric strength is 2 kV/mm?

SOLUTION: Since a 1-mm-thick glass will withstand 2 kV, a 26-mm-thick glass will withstand 26 times as much. Thus,

$$V = d_s \times s = 2 \times 26 = 52 \text{ kV}$$

The term d_s represents the dielectric strength and s the thickness of the glass.

15-3. PLATE AREA AND SPACING

There are many shapes, sizes, and electrical values of capacitors on the market today. However, all of them can be described as a conductor separated by a dielectric. The capacity of a capacitor is dependent not only on the type of dielectric chosen, but upon several other factors, including plate area and distance between the plates.

Plate Area

Capacitance is directly proportional to plate area, Fig. 15-5. That is, if we double the area of the plates, we double capacitance. This can be expected, since doubling plate area will double the dielectric size and, in turn, double the number of orbits distorted within the dielectric.

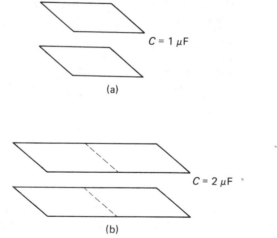

FIGURE 15-5: *Increasing capacitor plate area: (a) original capacitor; (b) plate area doubled.*

Distance Between Plates

Capacitance is inversely proportional to the square of the distance between the plates, Fig. 15-6. That is, if we double this distance, the capacitance is cut to one-fourth of its original value. This can best be understood by recognizing that orbital distortion is caused by the electric field that exists between the plates. When the plates are moved apart, the voltage stress from one side of a particular electron orbit to the other side is reduced, resulting in less orbital distortion.

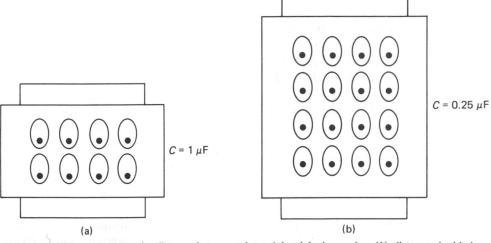

FIGURE 15-6: *Increasing the distance between plates: (a) original capacitor; (b) distance doubled.*

15-4. CAPACITOR TYPES

Capacitors are "typed" according to the dielectric used. The more common ones include paper, polyester/polystyrene, mica, ceramic, electrolytic, and air. We shall examine each of these types more closely in the following paragraphs.

Paper

Paper capacitors, Fig. 15-7(a), consist of lengths of wax- or oil-impregnated paper and metal foil rolled into a cylinder, Fig. 15-7(b). Electrodes are then connected to the plates and the whole assembly is encased in an insulating material, such as epoxy

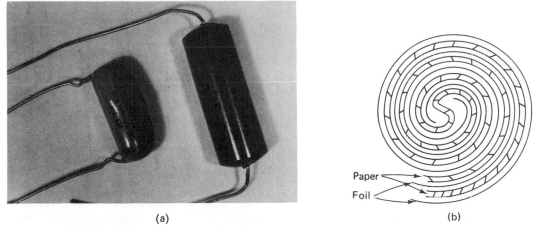

FIGURE 15-7: *Paper capacitor: (a) photograph (photo by Ruple); (b) construction.*

323

resin. Capacitances of 0.001 to 1 μF are available in dc working voltages (WV dc) of 75 to 400 V. The working voltage is the maximum dc voltage that can be placed across the plates without damaging the capacitor.

Paper capacitors have the advantage of being inexpensive, but the disadvantage of high dielectric hysteresis, making them unsuitable for higher frequencies. This hysteresis is caused by failure of the orbits to return completely to normal after the applied voltage is removed. Additional energy must be pumped into them to overcome this sluggishness, resulting in unwanted heat dissipation. Thus, this dielectric hysteresis is analogous to magnetic hysteresis.

Polyester and Polystyrene

Polyester and polystyrene capacitors, Fig. 15-8, are formed by rolling foil and polyester or foil and polystyrene into a cylinder. These dielectrics have much smaller dc leakage and reduced hysteresis compared with paper and are replacing the paper types. Values of 0.001 to 5.0 μF are available in working voltages from 50 to 2000 WV dc. Sometimes both paper and polyester are used as the dielectric.

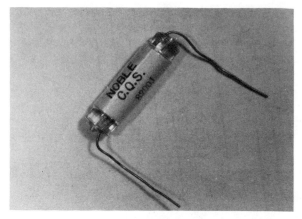

FIGURE 15-8: *Polystyrene capacitor (photo by Ruple).*

Mica

Mica capacitors have been with us for a long time and are formed by depositing thin metal films on both sides of a mica wafer, Fig. 15-9. They can also be made by inserting the substance between metal plates. Where larger values are needed, several of these wafers can be stacked. The whole assembly is then dipped in an insulating material. Mica is a very high quality dielectric, having very little leakage and hysteresis. However, because it is a solid, large capacitances are not possible, values of 3 pF to 0.047 μF being typical, with working voltages of 50 to 600 WV dc. Mica capacitors have fairly tight tolerances, $\pm 5\%$ being typical, and maintain their rated capacitance over a wide range of temperatures.

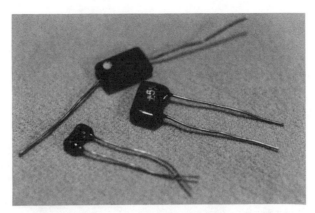

FIGURE 15-9: *Mica capacitors (photo by Ruple).*

Ceramic

Ceramic capacitors, Fig. 15-10, are formed into two different shapes: disk and cylindrical. They are fired with a conductive surface, then encapsulated in an insulator, such as epoxy. Ceramic dielectrics have the highest of all dielectric constants, making it possible to build them small physically, but large electrically. Values ranging from 1 pF to 2 μF and working voltages from 3 to 4000 WV dc are available, in 5% tolerance. In addition, they are very stable over a wide range of temperatures. Some ceramics are adjustable and are used by the technician to obtain a precise response. When so used, they are called a *trimmer capacitor*. Adjustment is performed by rotating a screw bringing the two metal plates closer together.

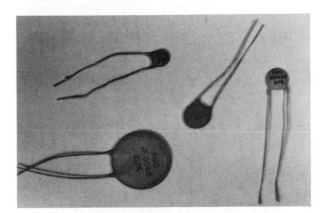

FIGURE 15-10: *Ceramic capacitors (photo by Ruple).*

Electrolytic

Electrolytic capacitors are formed by rolling up strips of metal that are separated by gauze which has been impregnated with a conducting electrolyte, Fig. 15-11. The whole assembly is then encased in a metal can. Aluminum and tantalum are the most

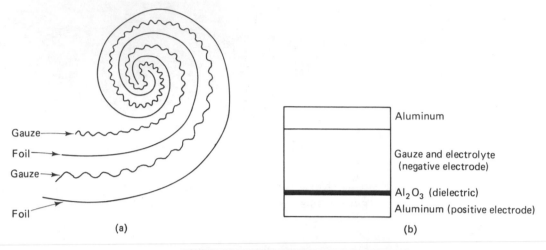

Gauze
Foil
Gauze

Foil

(a)

Aluminum

Gauze and electrolyte
(negative electrode)

Al_2O_3 (dielectric)
Aluminum (positive electrode)

(b)

(c)

FIGURE 15-11: *Electrolytic capacitor: (a) construction; (b) formation of dielectric; (c) photograph (photo by Ruple).*

common metals used for the foil, with tantalum favored for its ability to be made into thinner strips than aluminum. When a dc voltage is applied between the plates of the aluminum type, current flows, causing a buildup of an insulating coating of aluminum oxide (Al_2O_3) on the positive electrode. This process is known as *forming*, and soon the dc current reduces to a value near zero. Thus, the positive electrode forms one plate of the capacitor, the electrolyte the negative plate (connected to the outside world via the negative electrode), and the Al_2O_3 the dielectric. Because the dielectric is so thin, typically 10^{-5} in., values of up to 50,000 μF are attainable in electrolytics, ranging from 3 to 700 WV dc. However, the electrolytics are known for their high leakage currents.

A dc voltage must always be maintained across an electrolytic capacitor or it will lose its Al_2O_3 coating. Therefore, these devices are always used where dc is present, or where a dc voltage is present along with an ac voltage. Not only must a dc voltage be present, but it must be of correct polarity. Reversing the polarity of the leads of this type capacitor can cause it to conduct so heavily that it may heat up and explode.

Nonpolarized electrolytics are available and consist of two capacitors connected in series.

Air

Air is used as a dielectric in many adjustable capacitors, Fig. 15-12. The unit consists of two parts, a fixed set of plates and a movable set. The capacitance is increased by rotating a shaft that brings the plates of the movable part opposite those of the fixed part. Values of 1.5 to 75 pF are available. The primary use for these is in tuning circuits such as in a radio receiver, where the shaft is rotated to select the desired station.

FIGURE 15-12: *Air variable capacitor (photo by Ruple).*

15-5. CAPACITORS IN PARALLEL

When capacitors are connected in parallel, the total capacitance is the sum of the individual capacitances. This can be visualized by observing that the plate area has merely been increased. Thus,

$$C_t = C_1 + C_2 + C_3 + \ldots$$

EXAMPLE 15-8: Three capacitors, of 20, 50, and 80 μF, are connected in parallel. What is their total capacitance?

SOLUTION:

$$C_t = C_1 + C_2 + C_3$$
$$= 20 + 50 + 80$$
$$= 150 \ \mu\text{F}$$

Figure 15-13 illustrates the circuit.

EXAMPLE 15-9: Four capacitors are connected in parallel: 5000 pF, 0.015 μF, 0.002 μF, and 6000 pF. What is the total capacitance of the circuit?

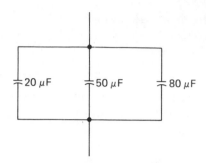

FIGURE 15-13: *Example 15-8, capacitors in parallel.*

SOLUTION: Compute in μF:

$$C_t = C_1 + C_2 + C_3 + C_4$$
$$= 0.005 + 0.015 + 0.002 + 0.006$$
$$= 0.028 \; \mu\text{F}$$

15-6. CAPACITORS IN SERIES

When capacitors are connected in series, they obey an inverse equation,

$$\frac{1}{C_t} = \frac{1}{C_1} + \frac{1}{C_2} + \frac{1}{C_3} + \cdots$$

where C_t is the total capacitance. Note that capacitors in series obey this inverse law, whereas resistors in parallel obey the inverse law. Conversely, capacitors in parallel are additive, whereas resistors in series are additive.

EXAMPLE 15-10: What is the total capacitance of a 1-, 3-, and 5-μF capacitor in series, Fig. 15-14?

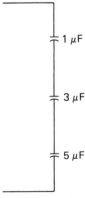

FIGURE 15-14: *Example 15-10, capacitors in series.*

SOLUTION: We can compute the whole problem in μF:

$$\frac{1}{C_t} = \frac{1}{C_1} + \frac{1}{C_2} + \frac{1}{C_3}$$

$$= \frac{1}{1} + \frac{1}{3} + \frac{1}{5}$$

$$C_t = 0.6522\ \mu F$$

EXAMPLE 15-11: Four capacitors are connected in series: 2000 pF, 5000 pF, 0.001 μF, and 0.0015 μF. What is the resultant capacitance?

SOLUTION:

$$\frac{1}{C_t} = \frac{1}{C_1} + \frac{1}{C_2} + \frac{1}{C_3} + \frac{1}{C_4}$$

$$= \frac{1}{0.002} + \frac{1}{0.005} + \frac{1}{0.001} + \frac{1}{0.0015}$$

$$C_t = 422.5\ \text{pF}$$

15-7. RC *TIME CONSTANT*

One of the most commonly used timing circuits in electronics is that of a resistor and a capacitor connected in series. It is used to determine frequency of many oscillators, the length of pulses, and to delay an event for a preset time. The basic circuit is that of Fig. 15-15. In Fig. 15-15(a) the resistor is effectively across C, resulting in there being 0 V across both R and C (after a length of time).

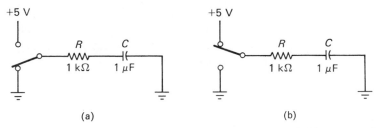

FIGURE 15-15: RC *time constant: (a) connected to ground; (b) connected to +5 V.*

Upon switching as shown in Fig. 15-15(b), the capacitor starts charging, causing current to flow through R. The value of this initial current is determined by the voltage across R. Since the circuit has 5 V across it and the capacitor 0 V, the resistor has 5 V across it initially. Thus, the initial current is

$$I = \frac{V}{R} = \frac{5}{1} = 5\ \text{mA}$$

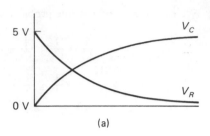

(a)

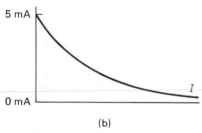

(b)

FIGURE 15-16: RC *charge curve:* (a) *voltages;* (b) *current.*

However, the capacitor soon develops a voltage across it, decreasing current flow as shown in Fig. 15-16. Finally, the capacitor has the full 5 V across it, current is 0 mA, and the voltage across the resistor is zero.

The instantaneous voltage across the charging capacitor is expressed by the equation

$$V_C = V_s(1 - \epsilon^{-t/RC})$$

where V_C is the instantaneous voltage at any time t, V_s is the voltage of the supply (in this case, 5 V), ϵ the constant 2.71828, R the value of the resistance in ohms, and C the value of the capacitor in μF. Note that at $t = 0$ (just after the switch has been closed), the equation becomes

$$V_C = V_s(1 - \epsilon^{-0/RC})$$
$$= V_s(1 - 1)$$
$$= 0$$

Note further that after a very large time:

$$V = V_s(1 - \epsilon^{-\text{large}/RC})$$
$$= V_s(1 - \epsilon^{-\text{large}})$$
$$= V_s(1 - 0)$$
$$= V_s$$

Thus, initially, the voltage across the capacitor is zero volts. However, when it is fully charged, this voltage is equal to the supply voltage. Using these equations we can find the voltage at any time, t:

EXAMPLE 15-12: A 10-kΩ resistor and a 0.02-μF capacitor are connected in series. If the capacitor has an initial charge of 0 V, what is its voltage 500 μs after 12 V has been applied to the circuit?

$$V_C = V_s(1 - \epsilon^{-t/RC})$$

$$= 12\left[1 - \exp\left(-\frac{500 \times 10^{-6}}{10^4 \times 0.02 \times 10^{-6}}\right)\right]$$

$$= 11.01 \text{ V}$$

EXAMPLE 15-13: A 5-MΩ resistor is in series with a 5-μF capacitor. If the capacitor has an initial charge of 0 V, what is the voltage across it 3.2 s after 25 V is applied to the circuit?

SOLUTION:

$$V_C = V_s(1 - \epsilon^{-t/RC})$$

$$= 25[1 - \epsilon^{-3.2/(5 \times 10^6 \times 5 \times 10^{-6})}]$$

$$= 3.004 \text{ V}$$

Not only can the equation be used for computing instantaneous voltage, but it can be used for selecting the proper resistor to obtain a particular voltage at a specified time.

EXAMPLE 15-14: What value of resistance must be placed in series with a 0.5-μF capacitor in order that it may charge to 12 V from a 15-V supply in 1 s?

SOLUTION:

$$V_C = V_s(1 - \epsilon^{-t/RC})$$

$$12 = 15[1 - \epsilon^{-1/(R \times 0.5 \times 10^{-6})}]$$

$$0.8 = 1 - \epsilon^{-1/(R \times 0.5 \times 10^{-6})}$$

$$\epsilon^{-1/(R \times 0.5 \times 10^{-6})} = 0.2$$

Take the natural log of both sides:

$$-\frac{1}{R \times 0.5 \times 10^{-6}} = \ln 0.2 = -1.609$$

$$R = 1.243 \text{ M}\Omega$$

EXAMPLE 15-15: A 1000-pF capacitor must be charged to 0.5 V in 25 μs from a 10-V supply. What resistance must be placed in series with it?

SOLUTION:

$$V_C = V_s(1 - \epsilon^{-t/RC})$$

$$0.5 = 10\left[1 - \exp\left(-\frac{25 \times 10^{-6}}{R \times 10^3 \times 10^{-12}}\right)\right]$$

$$\exp\left(-\frac{25 \times 10^{-6}}{R \times 10^{-9}}\right) = 0.95$$

Taking the ln of both sides:

$$-\frac{25 \times 10^{-6}}{R \times 10^{-9}} = \ln 0.95 = -0.05129$$

$$R = 487.4 \text{ k}\Omega$$

Discharge

A capacitor being discharged through a resistance as shown in Fig. 15-17 follows the discharge path shown in Fig. 15-18. The equation for the curve is

$$V_C = V_s\epsilon^{-t/RC}$$

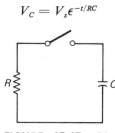

FIGURE 15-17: *Discharging a capacitor.*

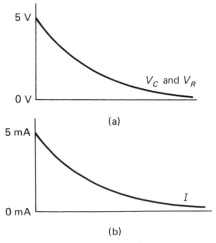

FIGURE 15-18: RC *discharge curve: (a) voltages; (b) current.*

where V_C is the voltage at any time t and V_s the voltage to which the capacitor was charged.

EXAMPLE 15-16: A 1-μF capacitor has been charged to 100 V. What will its voltage be 1 s after placing a 1-MΩ resistor across its terminals?

SOLUTION:

$$V_C = V_s\epsilon^{-t/RC}$$
$$= 100\epsilon^{-1/(10^6 \times 10^{-6})}$$
$$= 36.79 \text{ V}$$

EXAMPLE 15-17: A 10-μF capacitor has been charged to 20 V. What will a 400-kΩ voltmeter read 2 s after placing it across the capacitor terminals?

SOLUTION:

$$V_C = V_s\epsilon^{-t/RC}$$
$$= 20\epsilon^{-2/(400 \times 10^3 \times 10 \times 10^{-6})}$$
$$= 12.13 \text{ V}$$

Time Constant

The *time constant* (tc) of a circuit is defined as

$$\text{tc} = RC$$

where tc is the time constant in seconds, R the resistance, and C the capacitance. One time constant is defined as one RC, two time constants as $2RC$, three time constants as $3RC$. Thus, a 150-kΩ resistor in series with a 5-μF capacitor would have a time constant of

$$\text{tc} = RC = 150 \times 10^3 \times 5 \times 10^{-6} = 750 \text{ ms}$$

For this combination, three time constants would be (3×750) ms, or 2.25 s.

A capacitor under discharge will decrease to a fixed percent of its initial charge in one time constant:

$$V_C = V_s\epsilon^{-t/RC}$$
$$= V_s\epsilon^{-RC/RC}$$
$$= V_s(0.3639)$$

Thus, a capacitor that has been charged to 100 V will discharge to about 36 V in 1 tc. Furthermore, it will discharge to 36% of 36% in 2 tc's and $(36\%)^3$ in 3 tc's. This enables us to estimate the voltage-discharge curve. For instance, a 1-μF capacitor that has been charged to 50 V will discharge through a 1-MΩ resistor to 36% of 50 V, or 18 V in 1 s $(R \times C = 1)$, 36% of 18 V, or about 6.5 V in 1 more second, and 36% of 6.5 V, or about 2.3 V in 1 more second.

These same principles apply to the charging curve. At the end of 1 tc, the voltage reaches (100 — 36), or 64% of maximum, and at the end of 2 tc's, will increase 64% of the remainder to a total of 87% of maximum. Thus, the time constant is very useful in estimating the charge curve.

15-8. SUMMARY

Capacitance is present any time two conductors are separated by an insulator. The charge is stored in the dielectric of a capacitor. Current leads sinusoidal voltage by 90° in a capacitor. The permittivity of a dielectric determines how much charge it can store, whereas the dielectric constant is a measure of the permittivity of a material compared to that of free space. Dielectric strength indicates the voltage a dielectric can withstand without breaking down. Capacitance varies directly with the plate area and dielectric constant and inversely with the square of the distance between the plates.

Capacitors come in various dielectrics including paper, polyester, polystyrene, ceramic, mica, and electrolytic. The paper types have difficulty at high frequencies. The polyester and polystyrene devices have relatively low capacitances, although they are quite stable with humidity and age. The ceramic types provide relatively high capacitances, owing to the extremely high permittivities attainable. The mica types are very stable with temperature, but provide only very low capacitances. The electrolytic types have very high capacitances but relatively high leakage currents.

Capacitors connected in parallel are additive, whereas those in series have an inverse relationship. The *RC* time constant is a measure of the time a capacitive circuit will charge or discharge.

15-9. REVIEW QUESTIONS

15-1. Define capacitance.

15-2. Where is the charge stored in a capacitor?

15-3. Theoretically, what is the current through a capacitor if the voltage across it changes 0.1 V in 0 s?

15-4. Assuming that capacitance remains constant, what happens to charge if voltage across the capacitor is doubled?

15-5. What is the phase relationship between sinusoidal current and voltage waveforms within a capacitor?

15-6. What is the symbol for a capacitor?

15-7. What is meant by electric field strength? Charge displacement?

15-8. How is dielectric constant related to permittivity?

15-9. What is the difference between dielectric constant and dielectric strength?

15-10. If plate area is tripled, what happens to capacitance?

15-11. If the distance between the plates is tripled, what happens to the capacitance?

15-12. Why are paper capacitors ineffective at high frequencies?

15-13. Compare the characteristics of ceramic and mica capacitors.

15-14. What is meant by forming an electrolytic capacitor?

15-15. What is the advantage of tantalum over aluminum in electrolytics?

15-16. What will be the result of providing the wrong polarity across an electrolytic capacitor?

15-17. What is the equation for finding the total capacitance of capacitors connected in parallel? In series?

15-18. What is the initial current of a charging capacitor? The final current?

15-19. What is the initial current in a discharging capacitor? The final current?

15-20. What is the time constant of a series *RC* circuit and what does it represent?

15-10. PROBLEMS

15-1.   What is the voltage across a 1-μF capacitor that has a charge of 3 μC?

15-2.   A 10-μF capacitor has a supply voltage across it that increases at a rate of 2.0 V/s. What is the current through the capacitor?

15-3.   A 2000-pF capacitor has a supply voltage that increases at a rate of 25 V/s. What is the current through the capacitor?

15-4.   What is the charge displacement of a capacitor with a plate size of 2.5 $\times$ 50 cm and a charge of 3.6 C?

15-5.   What is the field strength of a capacitor's dielectric if the dielectric is 1-mm thick and has 75 V across it?

15-6.   A certain material 1.2-mm thick is used as a dielectric in a capacitor. With plate dimensions of 2.0 cm by 10 m, and a charge of 100 nC between the plates, at a voltage of 25 V, what is the permittivity of the material?

15-7.   Polystyrene has a dielectric constant of 2.5. What is its permittivity?

15-8.   What voltage will a 1-cm-thick slab of quartz withstand?

15-9.   Two capacitors, 1 μF and 4 μF, are connected in parallel. What is the total capacitance of the circuit?

15-10. Four capacitors are connected in parallel, 0.015 μF, 2000 pF, 0.002 μF, and 1500 pF. What is the total capacitance?

15-11. Four capacitors are connected in parallel, 1.5 μF, 0.05 μF, 2000 pF, and 25 pF. What is the total capacitance of the circuit?

15-12. Repeat Prob. 15-9 with the capacitors wired in series.

15-13. Repeat Prob. 15-10 with the capacitors wired in series.

15-14. Repeat Prob. 15-11 with the capacitors wired in series.

15-15. A 1-MΩ resistor and a 0.1-μF capacitor are connected in series. If the capacitor has an initial charge of 0 V across it, what is its voltage 30 ms after 25 V has been applied to the circuit?

15-16. A 25-kΩ resistor and a 0.5-μF capacitor are in series. If the capacitor has an initial charge of 0 V across it, what will its voltage be 8 ms after 8.6 V is applied to the circuit?

15-17. What value of resistor must be placed in series with a 0.1-μF capacitor in order that it may charge to 25 V from a 30-V supply in 2.5 ms?

15-18. What value resistor must be placed in series with a 250-pF capacitor in order that it may charge to 1 mV from a 5-mV supply in 25 ns?

15-19. A 0.015-μF capacitor has been charged to 25 V. What will its voltage be 25 ms after placing a 2-MΩ resistor across its terminals?

15-20. A 500-μF capacitor has been charged to 50 V. If an 11-MΩ electronic voltmeter is placed across its terminals, how long will it be before the meter reads 30 V? Assume a perfect capacitor.

15-11. PROJECTS

15-1. Experimentally compare the phase relationship of voltage and current in a capacitor using an oscilloscope.

15-2. What is meant by a sintered capacitor and how does it affect capacitance?

15-3. A roll of metal foil contains inductance. How can this inductance that exists within a capacitor be reduced?

15-4. Compare the voltage ratings and the capacitances of electrolytics with their physical size. Develop a mathematical relationship among these three factors.

15-5. What is the chemical composition of the dielectric of a tantalum capacitor?

15-6. Measure the capacitance of three capacitors. Connect them in parallel and measure the total capacitance. Then, connect them in series and repeat the measurement. How do your measured results compare with your computed results?

15-7. Connect a meter across a charged, 10-μF capacitor and plot its discharge with time. How do your results agree with the equation for discharge.

15-8. Charge a capacitor through a voltmeter and compare your measured results with those of the charge equation.

16

CAPACITIVE REACTANCE

In Chapter 15 we considered the property of capacitance and the problems associated with capacitors. In this chapter we shall put our blinders on and consider the capacitor only as it relates to sinusoidal waveforms. We shall consider the reactance (ac resistance) of a capacitor, then put these fellows in series, then parallel. Finally, we shall throw resistors into the circuit, then voltages and currents. The sections include:

16-1. Capacitive reactance

16-2. Reactances in series

16-3. Reactances in parallel

16-4. Series RC circuits

16-5. Ohm's law in series capacitive reactive circuits

16-1. CAPACITIVE REACTANCE

Capacitive reactance can be defined as the opposition a capacitor offers to a sinusoidal voltage waveform. Note that we are now concerned only with sinusoidal waveforms. This reactance is computed using the equation

$$X_C = \frac{1}{2\pi f C}$$

where X_C is the capacitive reactance in ohms, f the frequency in hertz, and C the capacitance in farads. We can observe that, according to this equation, increasing capacitance (C) will decrease X_C. This is to be expected, for if we double the plate area of a capacitor, more charges will flow, thus more current, implying a smaller reactance to the applied voltage. The equation also indicates that if we increase frequency, reactance will decrease. This can be reasoned out by observing that a 1-V/s increase across a capacitor will result in more charge per second (current) than a 0.1-V/s signal. At higher frequencies we are changing the voltage more rapidly, and the current must increase, implying that the reactance will decrease. Figure 16-1 is a plot of X_C as frequency increases across a 1-μF capacitor.

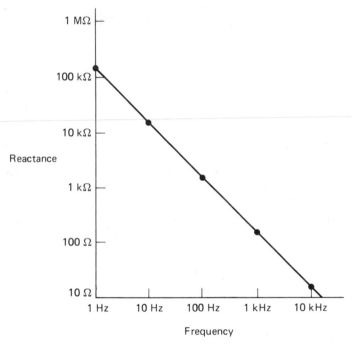

FIGURE 16-1:   *Capacitive reactance versus frequency for a 1-μF capacitor.*

EXAMPLE 16-1: Compute the capacitive reactance of a 100-pF capacitor for frequencies of 1 Hz, 1 kHz, and 1 MHz.

SOLUTION: For 1 Hz,

$$X_C = \frac{1}{2\pi fC} = \frac{1}{2 \times \pi \times 1 \times 100 \times 10^{-12}} = 1.592 \text{ G}\Omega$$

For 1 kHz,

$$X_C = \frac{1}{2\pi fC} = \frac{1}{2 \times \pi \times 10^3 \times 100 \times 10^{-12}} = 1.592 \text{ M}\Omega$$

For 1 MHz,

$$X_C = \frac{1}{2\pi fC} = \frac{1}{2 \times \pi \times 10^6 \times 100 \times 10^{-12}} = 1.592 \text{ k}\Omega$$

Note that, as frequency increases, X_C decreases.

Current and Voltage

Using Ohm's law, we can easily compute voltages and currents within a capacitive circuit.

EXAMPLE 16-2: A 5-μF capacitor is connected across a 5-V 1-kHz supply. What is the current within the circuit, Fig. 16-2?

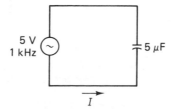

FIGURE 16-2: *Example 16-2.*

SOLUTION: We must first find the capacitive reactance:

$$X_C - \frac{1}{2\pi fC} = \frac{1}{2 \times \pi \times 10^3 \times 5 \times 10^{-6}} = 31.83 \ \Omega$$

Knowing this, we can now compute current:

$$I = \frac{V}{X_C} = \frac{5}{31.83} = 157.1 \ \text{mA}$$

The circuit has 157.1 mA flowing through it.

EXAMPLE 16-3: A capacitor of 0.015 μF has a current of 2.5 mA through it at a frequency of 7.5 kHz. What is the voltage across the capacitor?

SOLUTION: We must first find X_C:

$$X_C = \frac{1}{2\pi fC} = \frac{1}{2 \times \pi \times 7500 \times 0.015 \times 10^{-6}} = 1.415 \ \text{k}\Omega$$

We can now solve for the voltage using Ohm's law:

$$V = IX_C = 2.5 \times 1.415 = 3.537 \ \text{V}$$

Phasors

Capacitive reactance is always considered at an angle of $-90°$ compared to resistance, Fig. 16-3(a). Note that this puts inductive reactance and capacitive reactance $180°$ apart. Current always leads the voltage by $90°$ in a purely capacitive circuit, and may be shown vectorally as in Fig. 16-3(b).

339

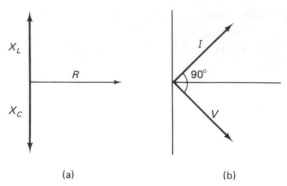

(a)   (b)

FIGURE 16-3: *Phasors in a capacitive circuit:* (a) *reactance and resistance;* (b) *voltage and current.*

EXAMPLE 16-4: A voltage of $30\underline{/-20°}$ V is impressed across a 2500-pF capacitor at a frequency of 500 kHz. What is the current through the capacitor, Fig. 16-4?

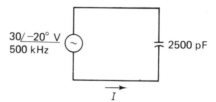

FIGURE 16-4: *Example 16-4.*

SOLUTION: We must first compute X_C:

$$X_C = \frac{1}{2\pi fC} = \frac{1}{2 \times \pi \times ,500 \times 10^3 \times 2500 \times 10^{-12}} = 127.3\underline{/-90°}\ \Omega$$

Note that, since this is capacitive reactance, we assigned it an angle of $-90°$. Computing current:

$$I = \frac{V}{X_C} = \frac{30\underline{/-20°}}{127.3\underline{/-90°}} = 236\underline{/70°}\ \text{mA}$$

We can, therefore, apply any of the techniques used in dc, as long as we use phasors.

16-2. REACTANCES IN SERIES

As with inductive reactances, capacitive reactances add when placed in series, Fig. 16-5. Thus:

$$X_{Ct} = X_{C1} + X_{C2} + X_{C3} + \dots$$

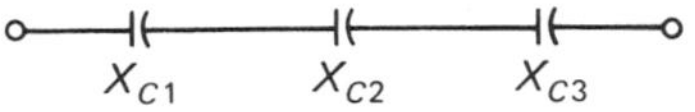

$$X_{Ct} = X_{C1} + X_{C2} + X_{C3}$$

FIGURE 16-5: *Capacitive reactances in series.*

Note that capacitors in series use the inverse equation, but capacitive reactances in series are summed.

EXAMPLE 16-5: Three capacitors, of 2, 5, and 10 μF, are connected in series. What is the total reactance at 400 Hz?

SOLUTION: We will solve the problem in two ways.
Method 1: Compute total capacitance, then total reactance. The total capacitance for capacitors in series is

$$\frac{1}{C_t} = \frac{1}{C_1} + \frac{1}{C_2} + \frac{1}{C_3}$$

$$= \frac{1}{2} + \frac{1}{5} + \frac{1}{10}$$

$$C_t = 1.25 \ \mu F$$

The total reactance is

$$X_C = \frac{1}{2\pi fC} = \frac{1}{2 \times \pi \times 400 \times 1.25 \times 10^{-6}} = 318.3 \underline{/-90°} \ \Omega$$

Method 2: Find the individual reactances for each capacitor, then sum them. The individual reactances are

$$X_{C1} = \frac{1}{2\pi fC} = \frac{1}{2 \times \pi \times 400 \times 2 \times 10^{-6}} = 198.94 \underline{/-90°} \ \Omega$$

$$X_{C2} = \frac{1}{2\pi fC} = \frac{1}{2 \times \pi \times 400 \times 5 \times 10^{-6}} = 79.58 \underline{/-90°} \ \Omega$$

$$X_{C3} = \frac{1}{2\pi fC} = \frac{1}{2 \times \pi \times 400 \times 10 \times 10^{-6}} = 39.79 \underline{/-90°} \ \Omega$$

The total reactance is the sum

$$X_{Ct} = X_{C1} + X_{C2} + X_{C3}$$

$$= 198.94 + 79.58 + 39.79 = 318.3 \underline{/-90°} \ \Omega$$

EXAMPLE 16-6: Three capacitors, of 2000 pF, 4000 pF, and 0.005 μF, are connected in series. What is the total reactance of the circuit at 250 kHz?

SOLUTION:

Method 1:

$$\frac{1}{C_t} = \frac{1}{C_1} + \frac{1}{C_2} + \frac{1}{C_3}$$

$$= \frac{1}{2000} + \frac{1}{4000} + \frac{1}{5000}$$

$$C_t = 1052.6 \text{ pF}$$

$$X_{Ct} = \frac{1}{2\pi f C_t} = \frac{1}{2 \times \pi \times 250 \times 10^3 \times 1052.6 \times 10^{-12}} = 604.8\underline{/-90°}\ \Omega$$

Method 2:

$$X_{C1} = \frac{1}{2\pi f C_1} = \frac{1}{2 \times \pi \times 250{,}000 \times 2000 \times 10^{-12}} = 318.31\underline{/-90°}\ \Omega$$

$$X_{C2} = \frac{1}{2\pi f C_2} = \frac{1}{2 \times \pi \times 250{,}000 \times 4000 \times 10^{-12}} = 159.15\underline{/-90°}\ \Omega$$

$$X_{C3} = \frac{1}{2\pi f C_3} = \frac{1}{2 \times \pi \times 250{,}000 \times 0.005 \times 10^{-6}} = 127.32\underline{/-90°}\ \Omega$$

$$X_t = X_{C1} + X_{C2} + X_{C3}$$

$$= 318.31 + 159.15 + 127.32 = 604.8\underline{/-90°}\ \Omega$$

16-3. REACTANCES IN PARALLEL

As with inductive reactances, capacitive reactances in parallel require the inverse equation, Fig. 16-6. Thus,

$$\frac{1}{X_{Ct}} = \frac{1}{X_{C1}} + \frac{1}{X_{C2}} + \frac{1}{X_{C3}} + \cdots$$

Note that capacitors in parallel add, but capacitive reactances require the inverse equation.

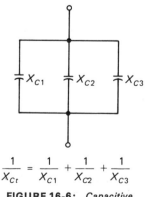

$$\frac{1}{X_{Ct}} = \frac{1}{X_{C1}} + \frac{1}{X_{C2}} + \frac{1}{X_{C3}}$$

FIGURE 16-6: *Capacitive reactances in parallel.*

EXAMPLE 16-7: Three capacitive reactances, of 250, 100, and 300 kΩ, are connected in parallel. Compute the total reactance.

SOLUTION:

$$\frac{1}{X_{C_t}} = \frac{1}{X_{C1}} + \frac{1}{X_{C2}} + \frac{1}{X_{C3}}$$

$$= \frac{1}{250} + \frac{1}{100} + \frac{1}{300}$$

$$X_{C_t} = 57.69\underline{/-90°}\text{ kΩ}$$

EXAMPLE 16-8: Three capacitors, one each of 0.05 μF and 0.022 μF, and one with a reactance of 150 Ω are connected in parallel at a frequency of 75 kHz. What is the total reactance of the circuit?

SOLUTION: We can combine the two capacitances, then compute the reactance of this combination. Finally, we can combine this reactance with the 150 Ω. For capacitors in parallel:

$$C_{EQ} = C_1 + C_2 = 0.05 + 0.022 = 0.072\ \mu\text{F}$$

$$X_{C_{EQ}} = \frac{1}{2\pi f C_{EQ}} = \frac{1}{2 \times \pi \times 75{,}000 \times 0.072 \times 10^{-6}} = 29.473\underline{/-90°}\ \Omega$$

$$\frac{1}{X_{C_t}} = \frac{1}{X_{C_{EQ}}} + \frac{1}{X_{C3}}$$

$$= \frac{1}{29.473} + \frac{1}{150}$$

$$X_{C_t} = 24.63\underline{/-90°}\ \Omega$$

16-4. SERIES RC CIRCUITS

In our study of inductance we learned that impedance is the phasor sum of resistance and reactance. This also applies to capacitive reactance; to find impedance of a resistance and a capacitive reactance connected in series, we phasor-add the two.

EXAMPLE 16-9: A resistance of 1 kΩ and a capacitor of 2500 pF are connected in series, Fig. 16-7(a). What is the total impedance of the circuit at 40 kHz?

SOLUTION: We must first find the capacitive reactance:

$$X_C = \frac{1}{2\pi f C} = \frac{1}{2 \times \pi \times 40 \times 10^3 \times 2500 \times 10^{-12}} = 1592\underline{/-90°}\ \Omega$$

The $\underline{/-90°}$ was assigned because it was capacitive reactance. Find the impedance:

$$Z = R - jX_C = 1000 - j1592 = 1880\underline{/-57.86°}\ \Omega$$

The minus sign occurs because −90° is the same as −*j* in phasor notation. The phasor diagram is shown in Fig. 16-7.

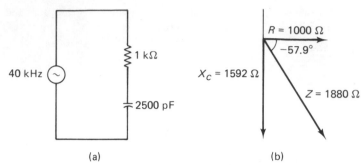

FIGURE 16-7: *Example 16-9, resistance and capacitance in series:*
(a) circuit; (b) impedance phasor diagram.

EXAMPLE 16-10: Two resistors, 1 kΩ and 3.3 kΩ, are connected in series with two capacitors, 0.001 μF and 0.002 μF. What is the total impedance of the circuit at 25 kHz?

SOLUTION: We can first find the equivalent capacitance for the two series capacitors, then the reactance:

$$\frac{1}{C_t} = \frac{1}{C_2} + \frac{1}{C_3}$$

$$= \frac{1}{0.001} + \frac{1}{0.002}$$

$$C_t = 666.7 \text{ pF}$$

The reactance is

$$X_C = \frac{1}{2\pi f C_t} = \frac{1}{2 \times \pi \times 25 \times 10^3 \times 666.7 \times 10^{-12}} = 9549\underline{/-90°} \ \Omega$$

We can now find total impedance:

$$Z = R_1 + R_2 - jX_C = 1000 + 3300 - j9549 = 10.47\underline{/-65.76°} \text{ k}\Omega$$

16-5. OHM'S LAW IN SERIES CAPACITIVE REACTIVE CIRCUITS

By applying Ohm's law to ac circuits, we can compute voltage drops, currents, and impedances. The only requirement is that phasors be used for all electrical quantities.

EXAMPLE 16-11: A capacitor of 0.1 μF is in series with a 500-Ω resistor across a 50-V 800-Hz source, Fig. 16-8(a). Compute the voltages across the capacitor and the resistor.

SOLUTION: We must first find the capacitive reactance:

$$X_C = \frac{1}{2\pi f C} = \frac{1}{2 \times \pi \times 800 \times 0.1 \times 10^{-6}} = 1989\underline{/-90°} \ \Omega$$

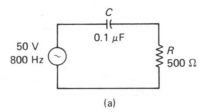

(a)

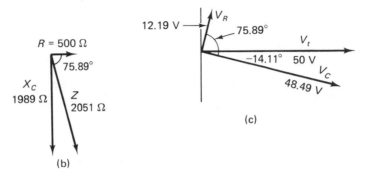

(b)

(c)

FIGURE 16-8: *Example 16-11, voltages in a series* RC *circuit: (a) circuit; (b) impedance phasors; (c) voltage phasors.*

Knowing this, we can compute impedance:

$$Z = R - jX_C = 500 - j1989 = 2051\underline{/-75.89°} \ \Omega$$

Knowing voltage and impedance, we can use Ohm's law to compute current:

$$I = \frac{V}{Z} = \frac{50\underline{/0°}}{2051\underline{/-75.89°}} = 24.37\underline{/75.89°} \ \text{mA}$$

Knowing current, we can compute the voltages across the components:

$$V_C = IX_C = (24.37\underline{/75.89°})(1.989\underline{/-90°}) = 48.49\underline{/-14.11°} \ \text{V}$$

$$V_R = IR = (24.37\underline{/75.89°})(0.5\underline{/0°}) = 12.19\underline{/75.89°} \ \text{V}$$

Just as a check, let us see if the applied voltage equals the sum of the voltage drops across the components:

$$V_t = V_C + V_R = 48.49\underline{/-14.11°} + 12.19\underline{/75.89°}$$

$$= 47.03 - j11.82 + 2.97 + j11.82$$

$$= 50.00 + j0 \quad \text{or} \quad 50\underline{/0°} \ \text{V}$$

Thus, Kirchhoff's law has not been repealed. The phasor diagrams are shown in Fig. 16-8.

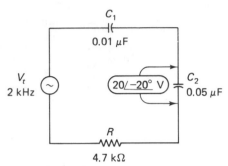

FIGURE 16-9: *Example 16-12.*

EXAMPLE 16-12: Find V_t of Fig. 16-9.

SOLUTION: We must first find the reactances of the capacitors:

$$X_{C1} = \frac{1}{2\pi fC} = \frac{1}{2 \times \pi \times 2000 \times 0.01 \times 10^{-6}} = 7.958\underline{/-90°}\ k\Omega$$

$$X_{C2} = \frac{1}{2\pi fC} = \frac{1}{2 \times \pi \times 2000 \times 0.05 \times 10^{-6}} = 1.592\underline{/-90°}\ k\Omega$$

Knowing X_{C2} and the voltage across C_2, we can calculate its current:

$$I = \frac{V}{X_{C2}} = \frac{20\underline{/-20°}}{1.592\underline{/-90°}} = 12.56\underline{/70°}\ mA$$

Next, we must calculate the total impedance:

$$Z = R - jX_{C1} - jX_{C2} = 4.7 - j7.958 - j1.592 = 10.64\underline{/-63.80°}\ k\Omega$$

We can now calculate the total voltage:

$$V_t = IZ = (12.56\underline{/70°})(10.64\underline{/-63.80°}) = 133.6\underline{/6.20°}\ V$$

16-6. SUMMARY

Capacitive reactance is the opposition a capacitor offers to sinusoidal voltages. Expressed in ohms at a phase of $-90°$ to resistance, it is computed using the equation $X_C = 1/(2\pi fC)$. Reactances in series are additive, whereas those in parallel require the inverse equation. All the dc analysis techniques can be used for ac as long as phasors are used for the electrical quantities.

16-7. REVIEW QUESTIONS

16-1. Define capacitive reactance.

16-2. What is the difference between capacitance and capacitive reactance?

16-3.   What is the equation for finding capacitive reactance, given capacitance and frequency?

16-4.   As frequency increases, what happens to capacitive reactance? Inductive reactance?

16-5.   What is the phase angle of capacitive reactance compared to resistance? Compared to inductive reactance?

16-6.   What is the equation for total capacitive reactance for several in series? How does this compare with that for inductive reactance?

16-7.   What is the equation for finding total reactance of several capacitive reactances in parallel? How does that compare with inductive reactance?

16-8. PROBLEMS

16-1.   Compute the capacitive reactance for a 0.047-μF capacitor at frequencies of (a) 1 Hz, (b) 10 Hz, (c) 100 Hz, and (d) 1 kHz.

16-2.   Repeat Prob. 16-1 for a 47-μF capacitor.

16-3.   A 0.022-μF capacitor is connected across a 2-V 10-kHz source. What is its reactance and current?

16-4.   A 5-pF capacitor is connected across a 10-V 1-kHz source. What is its reactance and current?

16-5.   A 2-μF capacitor has a current of 20 mA through it at a frequency of 7.5 kHz. What is its voltage?

16-6.   A 0.015-μF capacitor has a current of 1 mA through it at a frequency of 60 kHz. What is its voltage?

16-7.   What is the frequency of a source if it provides 10 V at 20 mA to a 0.47-μF capacitor?

16-8.   What is the frequency of a source if it provides 20 V at 2.5 mA to a 200-μF capacitor?

16-9.   Three capacitors, of 0.47, 0.15, and 0.22 μF, are connected in series. What is the total reactance of the circuit at 5 kHz?

16-10.   A 1.5-μF capacitor is in series with capacitive reactances of 1.5 kΩ and 890 Ω. What is the total reactance of the circuit at 120 Hz?

16-11.   Three capacitors are connected in series, one each of 2000 pF and 2500 pF, and one with a reactance of 1 kΩ at 250 kHz. What is the total reactance of the circuit?

16-12.   Three capacitors, of 0.5, 0.8, and 1.2 μF, are connected in parallel. What is the total reactance of the circuit at 4.5 kHz?

16-13.   Four capacitors, of 350, 400, 2000, and 2500 pF, are connected in parallel. What is the total reactance of the circuit at 45 MHz?

16-14.   A 12,000-pF capacitor is connected in parallel with capacitive reactances of 10 kΩ and 45 kΩ at a frequency of 1 kHz. What is the total reactance of the circuit?

16-15.   A resistance of 1 kΩ and a capacitor of 1000 pF are connected in series at 30 kHz. What is the impedance of the circuit?

16-16.   Repeat Prob. 16-15 for a frequency of 30 MHz.

16-17. Repeat Prob. 16-15 for a frequency of 30 Hz.

16-18. Resistances of 4.7 kΩ and 8.6 kΩ are connected in series with capacitors of 20 μF and 5 μF at a frequency of 10 Hz. Compute the impedance.

16-19. A capacitor of 0.15 μF is in series with a 1-kΩ resistor across a 75-V 1-kHz source. What are the voltages across the components?

16-20. A capacitor of 0.47 μF is in series with a 10-kΩ resistor across a 23-V 750-Hz supply. What are the voltages across the components?

16-21. Compute the voltage across C_1 in Fig. 16-10.

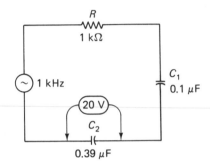

FIGURE 16-10: *Problem 16-21.*

16-22. In Fig. 16-8(a), at what frequency will the voltage across R be equal in magnitude to the voltage across C?

16-9. PROJECTS

16-1. Connect an ac source across a 1-μF capacitor. Vary the frequency of the source, keeping the voltage constant and measure the current through the capacitor. Draw a graph of X_C versus frequency. How do your computed results compare with the calculated results?

16-2. Connect two 1-μF capacitors in series across an ac supply. Measure the current through the circuit. Next, short out one of the capacitors and again measure the current (keeping supply voltage the same). What does this illustrate about reactances in series?

16-3. After completing Project 16-2, connect the capacitors in parallel and note the current (again holding voltage constant). Does this verify the equations for reactances in parallel?

16-4. Devise a circuit to illustrate that phasor addition of voltages across the components in a series RC circuit will equal the supply voltage.

17

RLC CIRCUITS

Thus far we have examined both inductive and capacitive circuits. In this chapter we shall combine these with resistive elements in various circuits. We shall first examine power relationships within ac circuits. Then, parallel *RLC* circuits and series *RLC* circuits will be considered. Finally, we shall examine circuits that must be solved using the network theorems presented in Chapter 8. The sections include:

17-1. Power in ac circuits

17-2. Series circuits

17-3. Parallel circuits

17-4. Series–parallel circuits

17-5. Network theorems

17-1. *POWER IN ac CIRCUITS*

We have so far quite conveniently ignored the problem of power in ac circuits. However, lest you be lulled into a false sense of security, ac power is a reality that must be considered. The ac power relationships are shown in Fig. 17-1 and, as might be expected, are related vectorially. These components consist of real power, reactive power, and apparent power.

Real Power

The I^2R of a resistive element is called *real power* (P) and is measured in watts (W). It represents actual heat given off by the element and has a phase angle of $0°$, since the voltage and current are in phase in a resistor. Any of the following equations

349

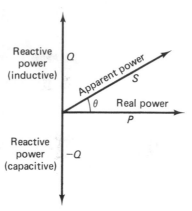

FIGURE 17-1: *Power phasors.*

can be used for computing power, as long as V is the voltage across only the resistive element, I the current through that element, and R the resistance of the element.

$$P = VI = I^2R = \frac{V^2}{R}$$

EXAMPLE 17-1: A 5-Ω resistor has 7 V ac across it. What is its power?

SOLUTION:

$$P = \frac{V^2}{R} = \frac{7^2}{5} = 9.800 \text{ W}$$

Reactive Power

Multiplying voltage across a reactive element by the current through that element results in reactive power (Q) measured in a unit called a *var* (volt-ampere reactive). This quantity differs from real power in that it does not represent a heat loss. Thus, a pure capacitance conducting 15 A at a voltage of 50 V will be very cool to the touch. For this reason, *reactive power* is also called *imaginary power*. The phase angle of reactive power is considered to be $+90°$ for inductive and $-90°$ for capacitive loads, corresponding to the phase of current and voltage in these elements. All the conventional power equations apply to reactive power as long as the voltage represents only that which is across the reactive element, and the current that which flows through the element. Thus,

$$Q = VI = I^2X = \frac{V^2}{X}$$

EXAMPLE 17-2: A capacitor has a current of 50 mA and a reactance of 30 Ω. What is its reactive power?

SOLUTION: $Q = I^2X = (0.05)^2(30) = 75$ mvar

This reactive power provides a problem to electric power utilities, for we are only charged for real power used within our homes. If, however, we have a high reactive component, resulting in many var being required, the power company receives very little revenue but must supply large wires to carry this high reactive current and large transformers for voltage conversion. Thus, they have a large investment but little return. We shall see later that capacitors can be used to cancel inductive reactive power and inductors can be used to cancel capacitive reactive power.

Apparent Power

The total voltage across a complex ac circuit multiplied by the current results in *apparent power* (*S*), measured in volt-amperes (VA). Since it is a measurement of the entire circuit, it is the phasor sum of real power and reactive power:

$$S = P + Q$$

Thus, it has a reactive component and a real component. Again, all the conventional power equations apply, as long as total voltage, total current, and total impedance are used:

$$S = VI = I^2Z = \frac{V^2}{Z}$$

EXAMPLE 17-3: A circuit with an impedance of 70 Ω has 56 V impressed across it. What is its apparent power?

SOLUTION:

$$S = \frac{V^2}{Z} = \frac{56^2}{70} = 44.80 \text{ VA}$$

Power Factor

The *power factor* (PF) is the cosine of the phase angle between real power and apparent power. As such, it represents a figure of merit to the power company: the nearer it is to 1.0, the larger the real power component of apparent power.

EXAMPLE 17-4: A circuit has apparent power of 50 VA and real power of 40 W. What is its power factor?

SOLUTION: The power factor is the cosine of the phase angle between real and apparent power. But, from Fig. 17-1, this cosine is

$$PF = \cos \theta = \frac{\text{real power}}{\text{apparent power}} = \frac{40}{50} = 0.8$$

Therefore, the power factor of this circuit is 0.8.

Power Calculations

Let us put all these definitions together and consider some ac problems.

EXAMPLE 17-5: An inductive reactance of 1 kΩ is in series with a 500-Ω resistor across a 50-V source. Compute the real power, apparent power, reactive power, and power factor.

SOLUTION: We must first find the total impedance:

$$Z = R + jX_L = 500 + j1000 = 1118\underline{/63.43°}\ \Omega$$

We can now find the current:

$$I = \frac{V}{Z} = \frac{50\underline{/0°}}{1118\underline{/63.43°}} = 44.72\underline{/-63.43°}\ \text{mA}$$

We are now in a position to compute the power components:

$$P = I^2R = (0.04472)^2(500) = 1.000\ \text{W across the resistor}$$
$$Q = I^2X = (0.04472)^2(1000) = 2.000\ \text{var across the inductor}$$
$$S = I^2Z = (0.04472)^2(1118) = 2.236\ \text{VA across the total circuit}$$

The phase angle is easier to compute separately, knowing that P is at 0° and Q at +90°. From Fig. 17-1,

$$\theta = \tan^{-1}\left(\frac{Q}{P}\right) = \tan^{-1}\left(\frac{2}{1}\right) = 63.43°$$

But, we also know that $S = P + Q$. Just as a check, let us use this equation to find the apparent power and power factor:

$$S = P + Q = 1.000 + j2.000 = 2.236\underline{/63.43°}\ \text{VA}$$

This checks with our previous calculations. The power factor is the cosine of the phase angle:

$$\text{PF} = \cos\theta = \cos 63.43° = 0.4473$$

The phasor diagram for these electrical quantities is shown in Fig. 17-2.

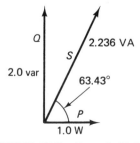

FIGURE 17-2: *Example 17-5.*

EXAMPLE 17-6: A capacitive reactance of 40 Ω is in series with a 100-Ω resistor and has 0.5 A flowing through it. Compute the circuit's power components and power factor.

SOLUTION:

$$P = I^2R = 0.5^2 \times 100 = 25 \text{ W}$$

$$Q = I^2X = 0.5^2 \times 40 = 10 \text{ var}$$

Note that this Q has a phase angle of $-90°$ because it is capacitive. Solve for the apparent power:

$$S = P + Q = 25 - j10 = 26.93\underline{/-21.80°} \text{ VA}$$

Solve for PF:

$$\text{PF} = \cos\theta = \cos(-21.80°) = 0.9285$$

Even though the cosine of $-21.80°$ is positive, the PF is said to be -0.9285 to show that the circuit is capacitive rather than inductive. The phasor diagram is shown in Fig. 17-3.

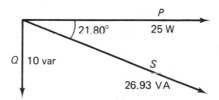

FIGURE 17-3: *Example 17-6.*

EXAMPLE 17-7: A circuit has a power factor of 0.85 and dissipates 200 W. What are its reactive and apparent power?

SOLUTION: The positive sign of 0.85 indicates an inductive circuit. Solve for the phase angle:

$$\text{PF} = \cos\theta$$

$$0.85 = \cos\theta$$

$$\theta = 31.79°$$

We now know the adjacent side and an angle, Fig. 17-4. Compute the other two sides:

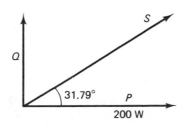

FIGURE 17-4: *Example 17-7.*

$$S = \frac{P}{\cos\theta} = \frac{200}{0.85} = 235.3\underline{/31.79°} \text{ VA}$$

$$Q = P \tan\theta = 200 \tan 31.79° = 124.0 \text{ var}$$

Power Measurement

Real power in ac circuits is measured using a wattmeter, Fig. 17-5. The unit consists of two types of coils: fixed-series coils to measure current and a movable coil in parallel with the load to measure voltage. Thus, it measures *VI*, power.

Reactive power is measured using a similar meter in which the resistor of Fig. 17-5(b) is replaced with a resistor and coil to change the phase in the movable coil by 90°.

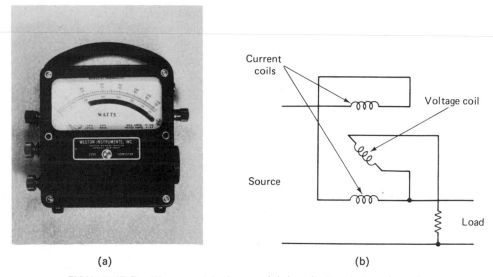

(a)   (b)

FIGURE 17-5:   *Wattmeter: (a) photograph (photo by Ruple); (b) schematic.*

17-2. SERIES CIRCUITS

Inductive reactance is considered at +90°, resistance at 0°, and capacitive reactance at −90° in series circuits. This means that inductive and capacitive reactance will cancel, since they are 180° apart. We can find the total impedance of a series *RLC* circuit by summing the phasors:

$$Z = R + jX_L - jX_C$$

EXAMPLE 17-8: An inductive reactance of 100 Ω is in series with a capacitive reactance of 120 Ω and a resistance of 150 Ω. Compute the total impedance, Fig. 17-6.

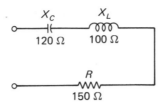

FIGURE 17-6: *Example 17-8,*
RLC *series circuit.*

SOLUTION:

$$Z = R + jX_L - jX_C$$
$$= 150 + j100 - j120$$
$$= 150 - j20$$
$$= 151.33\underline{/-7.595°}\ \Omega$$

EXAMPLE 17-9: A resistance of 1 kΩ is in series with a 0.33-μF capacitor and a 200-mH coil at a frequency of 500 Hz. Compute the total impedance.

SOLUTION: We must first calculate the reactances:

$$X_L = 2\pi fL = 2 \times \pi \times 500 \times 0.2 = 628.3\ \Omega$$
$$X_C = \frac{1}{2\pi fC} = \frac{1}{2 \times \pi \times 500 \times 0.33 \times 10^{-6}} = = 964.6\ \Omega$$

We can now calculate the impedance:

$$Z = R + jX_L - jX_C$$
$$= 1000 + j628.3 - j964.6$$
$$= 1000 - j336.3$$
$$= 1055\underline{/-18.59°}\ \Omega$$

Note the canceling effect of X_L and X_C.

Voltage and Current

The voltages and current within a series *RLC* circuit can easily (there's that word again) be computed using phasor algebra. One very interesting quality of such a circuit is the large voltages that can occur across the reactive elements. Study the following example.

EXAMPLE 17-10: Compute the current and voltages across each of the elements in the series *RLC* circuit of Fig. 17-7(a). Also compute the power phasors.

SOLUTION: We must first find the total impedance:

$$Z = R + jX_L - jX_C$$
$$= 1 + j2.0 - j2.2$$
$$= 1 - j0.2$$
$$= 1.020\underline{/-11.31°}\ \text{k}\Omega$$

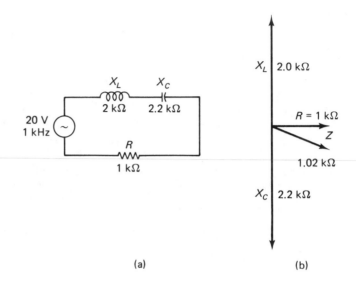

(a)   (b)

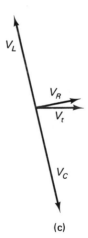

(c)

FIGURE 17-7: *Example 17-10: (a) schematic; (b) impedance phasors; (c) voltage phasors.*

We can now calculate the current:

$$I = \frac{V}{Z} = \frac{20\underline{/0°}}{1.020\underline{/-11.31°}} = 19.61\underline{/11.31°}\ \text{mA}$$

Knowing this, we can calculate the voltage across each of the elements:

$$V_R = IR = (19.61\underline{/11.31°})(1\underline{/0°}) = 19.61\underline{/11.31°}\ \text{V}$$
$$V_L = IX_L = (19.61\underline{/11.31°})(2\underline{/90°}) = 39.22\underline{/101.31°}\ \text{V}$$
$$V_C = IX_C = (19.61\underline{/11.31°})(2.2\underline{/-90°}) = 43.15\underline{/-78.69°}\ \text{V}$$

Note that the voltages across the reactive elements are larger than the source voltage. However, the phasor sum of the voltage drops is

$$\begin{aligned}
V_t &= V_L + V_C + V_R \\
&= 39.22\underline{/101.31°} + 43.15\underline{/-78.69°} + 19.61\underline{/11.31°} \\
&= (-7.69 + j38.46) + (8.46 - j42.31) + (19.23 + j3.85) \\
&= 20.00 + j0.00 \\
&= 20.00\underline{/0°}\ \text{V}
\end{aligned}$$

Thus, the sum of the voltage drops is still equal to the supply voltage. The phasor diagram is shown in Fig. 17-7(c). The power phasors can be calculated as follows:

$$P = V_R I = 19.61 \times 0.01961 = 384.6\ \text{mW}$$

Reactive power is the phasor sum of the inductive and capacitive components:

$$\begin{aligned}
Q &= V_L I - V_C I = I(V_L - V_C) \\
&= (0.019612)(39.223 - 43.15) = -76.93\ \text{mvar}
\end{aligned}$$

The apparent power can be found by multiplying the total current by total voltage or by adding the real power phasor to the reactive power phasor. We shall use the latter method.

$$S = P - jQ_C = 384.6 - j76.93 = 392.2\underline{/-11.31°}\ \text{mVA}$$

EXAMPLE 17-11: A 0.39-μF capacitor is in series with a 150-mH inductor and an 890-Ω resistor. If the supply voltage is 10 V at 2.5 kHz, compute the voltage across each of the elements and the power phasors.

SOLUTION: Find the reactances:

$$X_L = 2\pi f L = 2 \times \pi \times 2500 \times 0.15 = 2356\ \Omega$$
$$X_C = \frac{1}{2\pi f C} = \frac{1}{2 \times \pi \times 2500 \times 0.39 \times 10^{-6}} = 163.24\ \Omega$$

The total impedance is

$$Z = R + jX_L - jX_C = 890 + j2356 - j163.24 = 2367\underline{/67.91°}\ \Omega$$

The current is

$$I = \frac{V}{Z} = \frac{10\underline{/0°}}{2367\underline{/67.91°}} = 4.225\underline{/-67.91°}\ \text{mA}$$

The voltages are

$$V_R = IR = (4.225\underline{/-67.91°})(0.89\underline{/0°}) = 3.760\underline{/-67.91°}\text{ V}$$

$$V_L = IX_L = (4.225\underline{/-67.91°})(2.356\underline{/90°}) = 9.954\underline{/22.09°}\text{ V}$$

$$V_C = IX_C = (4.225\underline{/-67.91°})(0.1632\underline{/-90°}) = 689.5\underline{/-157.9°}\text{ mV}$$

The power phasors are

$$P = I^2R = (0.004225)^2 \times (890) = 15.89\text{ mW}$$

$$Q = I^2(X_L - X_C) = (0.004225)^2 \times (2356 - 163) = 39.15\text{ mvar}$$

$$S = I^2Z = (0.004225)^2 \times (2367) = 42.25\underline{/67.91°}\text{ mVA}$$

The phase angle of the power phasors is always the same as the phase angle of the impedance phasors.

17-3.   PARALLEL CIRCUITS

Whether a capacitive reactance is in parallel with a resistor or an inductive reactance is in parallel with a resistor, the approach to the problem solution is the same. Each element is, of course, expressed as a phasor.

RL *Impedance*

When an inductor is connected in parallel with a resistor, the resultant impedance is found using an inverse equation,

$$\frac{1}{Z} = \frac{1}{R} + \frac{1}{jX_L}$$

where Z is the total impedance, R the value of the resistance, and X_L the value of the inductive reactance, all measured in ohms.

> **EXAMPLE 17-12:** An inductive reactance of 1 kΩ is connected in parallel with a 1-kΩ resistance, Fig. 17-8. What is the total impedance?

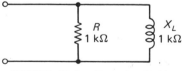

FIGURE 17-8: *Example 17-12, the* RL *parallel circuit.*

SOLUTION:

$$\frac{1}{Z} = \frac{1}{R} + \frac{1}{jX_L}$$

$$= \frac{1}{1\underline{/0°}} + \frac{1}{1\underline{/90°}}$$

$$Z = 0.707\underline{/45°} \text{ k}\Omega$$

RC *Impedance*

In a similar manner, a capacitive reactance in parallel with a resistance has the following impedance relationship:

$$\frac{1}{Z} = \frac{1}{R} + \frac{1}{-jX_C}$$

EXAMPLE 17-13: A capacitive reactance of 50 kΩ is in parallel with a resistance of 33 kΩ, Fig. 17-9. What is the impedance of the circuit?

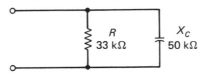

FIGURE 17-9: *Example 17-13, the RC parallel circuit.*

SOLUTION:

$$\frac{1}{Z} = \frac{1}{R} + \frac{1}{-jX_C}$$

$$= \frac{1}{33\underline{/0°}} + \frac{1}{50\underline{/-90°}}$$

$$Z = 27.54\underline{/-33.42°} \text{ k}\Omega$$

Inverse ac Quantities

These equations afford an opportunity to introduce some new terms. When we studied dc, we defined the inverse of resistance as conductance. In a similar manner, we can use the definitions of ac inverse quantities, as shown in Table 17-1, to compute circuit elements. Therefore, the equation for an inductive reactance in parallel with a resistance becomes

$$\frac{1}{Z} = \frac{1}{R} + \frac{1}{jX_L}$$

$$Y = G + \frac{B_L}{j}$$

$$Y = G - jB_L \tag{17-1}$$

Note that moving the j from the denominator to the numerator resulted in its sign changing. We can save a bit of j fiddling by using this form of the parallel equation rather than. the inverse impedance form. When using this method, admittance is first found, then converted into impedance for the final answer.

TABLE 17-1: *Definitions of ac inverse quantities.*

Fraction	Symbol	Name	Unit of Measure
$\dfrac{1}{Z}$	Y	Admittance	Siemens (S)
$\dfrac{1}{R}$	G	Conductance	Siemens (S)
$\dfrac{1}{X}$	B	Susceptance	Siemens (S)

EXAMPLE 17-14: A resistance of 10 kΩ is in parallel with an inductive reactance of 7.5 kΩ. Compute the impedance.

SOLUTION:

$$Y = G - jB_L$$

$$= \frac{1}{10} - j\frac{1}{7.5}$$

$$= 0.1 - j0.13333$$

$$= 0.16667\underline{/-53.13°} \text{ mS}$$

Since, by definition, impedance is the inverse of admittance,

$$Z = \frac{1}{Y} = \frac{1}{0.16667\underline{/-53.13°}} = 6.000\underline{/53.13°} \text{ k}\Omega$$

In a similar manner, the equation for a capacitive reactance in parallel with a resistance becomes

$$\frac{1}{Z} = \frac{1}{R} + \frac{1}{-jX_C}$$

$$Y = G + \frac{1}{-j} \cdot \frac{1}{X_C}$$

$$Y = G + jB_C \qquad (17\text{-}2)$$

where B_C is the inverse of capacitive reactance.

EXAMPLE 17-15: A 5-kΩ resistor is in parallel with a capacitive reactance of 6 kΩ. Find the total impedance.

SOLUTION:

$$Y = G + jB_C$$

$$= \frac{1}{5} + j\frac{1}{6}$$

$$= 0.2 + j0.16667$$

$$Y = 0.2603\underline{/39.81°}\ \text{mS}$$

But

$$Z = \frac{1}{Y} = \frac{1}{0.2603\underline{/39.81°}} = 3.841\underline{/-39.81°}\ \text{k}\Omega$$

Using this method, inductors are considered $-jB$'s and capacitors $+jB$'s—just the opposite of their usual consideration.

RLC *Impedance*

As long as all electrical quantities are expressed as phasors, any combinations of reactances and resistances in parallel can easily be computed. Note, however, the effect of equal inductive and capacitive reactances in parallel; they cancel each other.

EXAMPLE 17-16: A 1-kΩ resistor is connected in parallel with an inductive reactance of 1.2 kΩ and a capacitive reactance of 1.2 kΩ. Compute total impedance.

SOLUTION: We can combine Eqs. (17-1) and (17-2):

$$Y = G - jB_L + jB_C$$

$$= \frac{1}{R} - j\left(\frac{1}{X_L}\right) + j\left(\frac{1}{X_C}\right)$$

$$= \frac{1}{1} - j\left(\frac{1}{1.2}\right) + j\left(\frac{1}{1.2}\right)$$

$$= 1.0\ \text{mS}$$

$$Z = \frac{1}{Y} = \frac{1}{1.0} = 1.0\ \text{k}\Omega$$

Note that equal capacitive and inductive reactances in parallel cancel.

EXAMPLE 17-17: The following six elements are in parallel: resistances of 4 kΩ and 7 kΩ, inductive reactances of 5 kΩ and 6 kΩ, and capacitive reactances of 3 kΩ and 4 kΩ. Compute the impedance.

SOLUTION:

$$Y = G_1 + G_2 - jB_{L1} - jB_{L2} + jB_{C1} + jB_{C2}$$

$$= \frac{1}{4} + \frac{1}{7} - j\frac{1}{5} - j\frac{1}{6} + j\frac{1}{3} + j\frac{1}{4}$$

$$= 0.3929 + j0.2167$$

$$= 0.4486\underline{/28.88°}\ \text{mS}$$

$$Z = \frac{1}{0.4486\underline{/28.88°}} = 2.229\underline{/-28.88°}\ \text{k}\Omega$$

Currents and Voltages

The voltage across each of the elements of a parallel circuit is, of course, the same. In addition, the currents within each branch, when summed, equal the supply current.

EXAMPLE 17-18: Compute total impedance, branch currents, and total current for the circuit shown in Fig. 17-10.

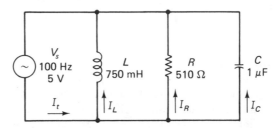

FIGURE 17-10: *Example 17-18.*

SOLUTION: We must first compute the reactances:

$$X_L = 2\pi f L = 2 \times \pi \times 100 \times 0.75 = 471.24 \ \Omega$$

$$X_C = \frac{1}{2\pi f C} = \frac{1}{2 \times \pi \times 100 \times 10^{-6}} = 1591.5 \ \Omega$$

Thus, the total impedance is

$$Y = G - jB_L + jB_C$$

$$= \frac{1}{510} - j\frac{1}{471.24} + j\frac{1}{1591.5}$$

$$= 2.4649 / -37.30° \ \text{mS}$$

$$Z = \frac{1}{Y} = \frac{1}{2.4649 \times 10^{-3} / -37.30°} = 405.7 / 37.30° \ \Omega$$

The total current is

$$I_t = \frac{V}{Z} = \frac{5 / 0°}{405.7 / 37.30°} = 12.32 / -37.30° \ \text{mA}$$

The branch currents are

$$I_L = \frac{V}{X_L} = \frac{5 / 0°}{471.24 / 90°} = 10.610 / -90° \ \text{mA}$$

$$I_C = \frac{V}{X_C} = \frac{5 / 0°}{1591.5 / -90°} = 3.142 / 90° \ \text{mA}$$

$$I_R = \frac{V}{R} = \frac{5 / 0°}{510 / 0°} = 9.804 / 0° \ \text{mA}$$

362

Just as a check, total current should equal the sum of the branch currents. Note how capacitive reactive current cancels inductive reactive current.

$$I_t = I_R + I_L + I_C$$
$$= 9.804\underline{/0°} + 10.610\underline{/-90°} + 3.142\underline{/90°}$$
$$= 9.804 - j\,10.610 + j\,3.142$$
$$= 9.804 - j\,7.468$$
$$= 12.32\underline{/-37.30°}\text{ mA}$$

This checks with our previous calculation.

As can be seen from this example, inductive and capacitive current are 180° apart and, therefore, cancel.

EXAMPLE 17-19: Compute the total current, voltage, and impedance for the circuit shown in Fig. 17-11.

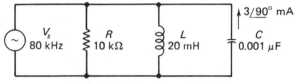

FIGURE 17-11: *Example 17-19.*

SOLUTION: We must first calculate reactances:

$$X_L = 2\pi f L = 2 \times \pi \times 80,000 \times 0.02 = 10.053 \text{ k}\Omega$$

$$X_C = \frac{1}{2\pi f C} = \frac{1}{2 \times \pi \times 80,000 \times 0.001 \times 10^{-6}} = = 1.989 \text{ k}\Omega$$

The impedance is

$$Y = G - jB_L + jB_C$$

$$= \frac{1}{10} - j\,\frac{1}{10.053} + j\,\frac{1}{1.989}$$

$$= 0.4155\underline{/76.07°}\text{ mS}$$

$$Z = \frac{1}{Y} = \frac{1}{0.4155\underline{/76.07°}} = 2.407\underline{/-76.07°}\text{ k}\Omega$$

The voltage across the circuit can be found by

$$V_s = I_C X_C = (3\underline{/90°})(1.989\underline{/-90°}) = 5.968\underline{/0°}\text{ V}$$

The total current is, therefore,

$$I_t = \frac{V_s}{Z} = \frac{5.968\underline{/0°}}{2.407\underline{/-76.07°}} = 2.479\underline{/76.07°}\text{ mA}$$

17-4. SERIES–PARALLEL CIRCUITS

The analysis techniques used in dc series–parallel circuits apply to ac circuits, if phasors are used.

EXAMPLE 17-20: Compute total impedance for the circuit shown in Fig. 17-12.

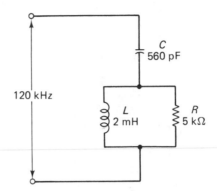

FIGURE 17-12: *Example 17-20, series–parallel ac circuit.*

SOLUTION: We must first calculate all the reactances:

$$X_C = \frac{1}{2\pi f C} = \frac{1}{2 \times \pi \times 120{,}000 \times 560 \times 10^{-12}} = 2368.4 \ \Omega$$

$$X_L = 2\pi f L = 2 \times \pi \times 120{,}000 \times 0.002 = 1508.0 \ \Omega$$

Computing the impedance of the parallel network:

$$Y_{EQ} = G - jB_L$$

$$= \frac{1}{5} - j\frac{1}{1.5080} = 0.2 - j0.6631 = 0.6926\underline{/-73.217^\circ} \ \text{mS}$$

$$Z_{EQ} = \frac{1}{Y_{EQ}} = \frac{1}{0.6926\underline{/-73.217^\circ}} = 1.444\underline{/73.22^\circ} \ \text{k}\Omega$$

Total impedance is the phasor sum of the capacitive reactance and Z_{EQ}:

$$Z_t = -jX_C + Z_{EQ}$$

$$= 2.3684\underline{/-90^\circ} + 1.444\underline{/73.22^\circ}$$

$$= -j2.3684 + 0.4169 + j1.3822$$

$$= 0.4169 - j0.9861$$

$$= 1.071\underline{/-67.08^\circ} \ \text{k}\Omega$$

Circuit elements expressed as impedances in parallel can be computed by noting that the total admittance is the sum of the individual admittances:

$$Y_t = Y_1 + Y_2 + \dots \tag{17-3}$$

Remember, this is applicable only to parallel impedances, not to those in series.

EXAMPLE 17-21: Compute the total impedance for the circuit shown in Fig. 17-13.

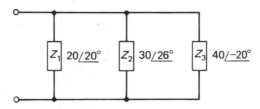

FIGURE 17-13: *Example 17-21, parallel impedances.*

SOLUTION: We can use Eq. 17-3 to find total admittance:

$$Y_t = Y_1 + Y_2 + Y_3$$

$$= \frac{1}{20\underline{/20°}} + \frac{1}{30\underline{/26°}} + \frac{1}{40\underline{/-20°}} = 0.10307\underline{/-12.987°}\ \text{mS}$$

$$Z_t = \frac{1}{Y_t} = \frac{1}{0.10307\underline{/-12.987°}} = 9.702\underline{/12.99°}\ \text{k}\Omega$$

Current and Voltage

Using the law made famous by Georg Simon Ohm, we can calculate currents, voltages, voltage drops, and the power phasors of any ac circuit. The techniques are merely an extension of those used in dc.

EXAMPLE 17-22: Compute the currents through C and L_1 and the voltage drop across L_1 in Fig. 17-14.

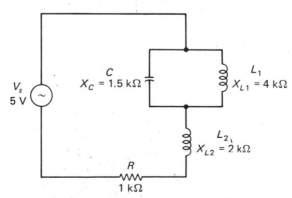

FIGURE 17-14: *Example 17-22.*

SOLUTION: We must first find total impedance and total current. To do this, we must calculate the equivalent impedance for C and L_1:

$$Y_{EQ} = -jB_{L1} + jB_C = -j\frac{1}{4} + j\frac{1}{1.5} = j0.4167 \text{ mS}$$

$$Z_{EQ} = \frac{1}{Y_{EQ}} = \frac{1}{j0.4167} = -j2.400 \text{ k}\Omega$$

The total impedance is

$$\begin{aligned} Z_t &= Z_{EQ} + jX_{L2} + R \\ &= -j2.4 + j2 + 1 \\ &= 1 - j0.4 \\ &= 1.077\underline{/-21.80°} \text{ k}\Omega \end{aligned}$$

The total current is

$$I_t = \frac{V}{Z_t} = \frac{5\underline{/0°}}{1.077\underline{/-21.80°}} = 4.642\underline{/21.80°} \text{ mA}$$

The voltage drop across the parallel combination of C and L_1 can be found by

$$V_{L1} = I_t Z_{EQ} = (4.642\underline{/21.80°})(2.4\underline{/-90°}) = 11.14\underline{/-68.20°} \text{ V}$$

Note that this exceeds the supply voltage. This occurs because the voltage across L_2 is out of phase with V_{L1}. Thus, total voltage is still equal to the sum of the individual voltages and Kirchhoff can rest in peace. Solve for current through L_1:

$$I_{L1} = \frac{V_{L1}}{X_{L1}} = \frac{11.14\underline{/-68.20°}}{4\underline{/90°}} = 2.785\underline{/-158.2°} \text{ mA}$$

Solve for current through C:

$$I_C = \frac{V_C}{X_C} = \frac{11.14\underline{/-68.20°}}{1.5\underline{/-90°}} = 7.428\underline{/21.80°} \text{ mA}$$

EXAMPLE 17-23: Compute V_s and I_t of Fig. 17-15.

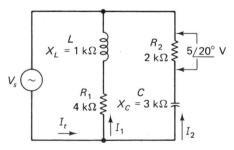

FIGURE 17-15: *Example 17-23.*

SOLUTION: Knowing the voltage across R_2, we can find the current:

$$I_2 = \frac{V_{R2}}{R_2} = \frac{5\underline{/20°}}{2\underline{/0°}} = 2.5\underline{/20°} \text{ mA}$$

Compute V_s:

$$V_s = I_2(R_2 - jX_C) = (2.5\underline{/20°})(2 - j3) = 9.014\underline{/-36.31°} \text{ V}$$

Compute I_1:

$$I_1 = \frac{V_s}{R_1 + jX_L} = \frac{9.014\underline{/-36.31°}}{4 + j1} = 2.186\underline{/-50.35°} \text{ mA}$$

Solve for the total current:

$$
\begin{aligned}
I_t &= I_1 + I_2 \\
&= 2.186\underline{/-50.35°} + 2.5\underline{/20°} \\
&= (1.395 - j1.683) + (2.349 + j0.855) \\
&= 3.744 - j0.828 \\
&= 3.835\underline{/-12.47°} \text{ mA}
\end{aligned}
$$

17-5. NETWORK THEOREMS

Although the math at times becomes somewhat sticky, all the dc network theorems are applicable to ac, as long as phasor algebra is used. Thus, we can apply loops, nodes, Thévenin's theorem, Norton's theorem, superposition, and wye–delta transformation to solve ac problems. The following examples illustrate the techniques required for application of these methods.

EXAMPLE 17-24: Compute the voltage drops across each of the elements shown in Fig. 17-16.

SOLUTION: We shall use the loop method of analysis. First, analyze M_1:

Ⓐ $V_1 - V_R - V_{L1} - V_{XC} = 0$
Ⓐ $100 - 100I_1 - j200I_1 + j300(I_1 - I_2) = 0$
Ⓐ $(100 - j100)I_1 + j300I_2 = 100$
Ⓐ $(1 - j)I_1 + j3I_2 = 1$

Then, analyze M_2:

Ⓑ $V_2 - V_{XC} - V_{XL2} = 0$
Ⓑ $100 - (-j300)[-(I_1 - I_2)] - j200I_2 = 0$
Ⓑ $j300I_1 - j100I_2 = 100$
Ⓑ $j3I_1 - jI_2 = 1$

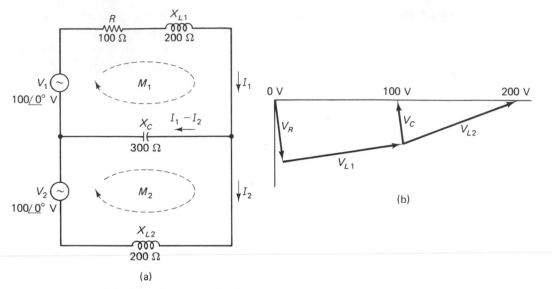

FIGURE 17-16: *Example 17-24, the loop method in ac: (a) circuit; (b) phasor diagram.*

Solve for I_1 and I_2:

$$I_1 = \frac{\begin{vmatrix} 1 & j3 \\ 1 & -j1 \end{vmatrix}}{\begin{vmatrix} 1-j & j3 \\ j3 & -j1 \end{vmatrix}} = \frac{-j1-j3}{-j1+j^2 1 - j^2 9} = \frac{-j4}{8-j1}$$

$$I_1 = \frac{4\underline{/-90°}}{8.062\underline{/-7.13°}} = 0.4962\underline{/-82.87°} \text{ A}$$

$$I_2 = \frac{\begin{vmatrix} 1-j & 1 \\ j3 & 1 \end{vmatrix}}{8.062\underline{/-7.13°}} = \frac{4.123\underline{/-75.96°}}{8.062\underline{/-7.13°}} = 0.5114\underline{/-68.84°} \text{ A}$$

Finally, solve for the voltage drops across each of the components:

$$V_R = I_1 R = (0.4962\underline{/-82.87°})(100\underline{/0°}) = 49.62\underline{/-82.87°} \text{ V}$$
$$V_{L1} = I_1 X_{L1} = (0.4962\underline{/-82.87°})(200\underline{/90°}) = 99.24\underline{/7.13°} \text{ V}$$
$$V_{L2} = I_2 X_{L2} = (0.5114\underline{/-68.84°})(200\underline{/90°}) = 102.28\underline{/21.16°} \text{ V}$$
$$V_C = (I_1 - I_2)(X_C)$$
$$= (0.4962\underline{/-82.87°} - 0.5114\underline{/-68.84°})(300\underline{/-90°})$$
$$= 37.22\underline{/97.13°} \text{ V}$$

The complete head-to-tail voltage phasor diagram is shown in Fig. 17-16(b).

EXAMPLE 17-25: Compute V in Fig. 17-17.

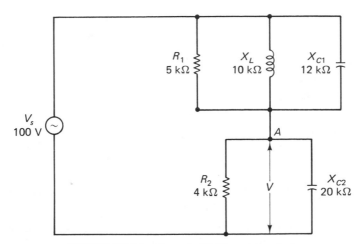

FIGURE 17-17: *Example 17-25, nodal analysis.*

SOLUTION: We shall use nodal analysis; use node A:

$$I_{R1} + I_{XL} + I_{XC1} = I_{R2} + I_{XC2}$$

$$\frac{100 - V}{5} + \frac{100 - V}{j10} + \frac{100 - V}{-j12} = \frac{V}{4} + \frac{V}{-j20}$$

$$0.2(100 - V) - j0.1(100 - V) + j0.08333(100 - V) = 0.25V + j0.05V$$

$$0.45V + j0.0333V = 20 - j1.6667$$

$$V = \frac{20 - j1.6667}{0.45 + j0.03333} = 44.48\underline{/-9.000°}\ \text{V}$$

17-6. SUMMARY

There are three types of power within an ac circuit: (1) real power, the power developed across a resistor, (2) reactive power, the power developed across a pure reactance, and (3) apparent power, the power developed across an impedance. Inductive reactance and capacitive reactance in series tend to cancel to zero ohms; those in parallel tend to cancel to infinite ohms. All the network-analysis techniques used in dc are applicable to ac if phasor algebra is used.

17-7. REVIEW QUESTIONS

17-1. Give three equations by which real power can be found.

17-2. Give three equations by which reactive power can be found.

17-3. Give three equations by which apparent power can be found.

17-4. What is the unit of measure for real power? For reactive power? For apparent power?

17-5. What is the symbol for real power? For reactive power? For apparent power?

17-6. How are real power, reactive power, and apparent power related in phase?

17-7. How can the power factor be found?

17-8. What is the usual phasor designation for inductive reactance? For capacitive react-ance? For resistance?

17-9. Does Kirchhoff's law apply to series *RLC* circuits?

17-10. How are inductive current and capacitive current in a parallel circuit related in phase?

17-11. How are inductive voltage and capacitive voltage in a series circuit related in phase?

17-12. What is admittance? Conductance? Susceptance? What are the symbols and units of measure of each of these?

17-13. Give an equation for finding total admittance of both inductive and capacitive sus-ceptances in parallel with a conductance.

17-14. Can Norton's theorem be applied to ac circuits?

17-8. PROBLEMS

17-1. A 2-kΩ resistor has 6.4 V ac across it. What is its power?

17-2. A 450-Ω resistor has 23 mA flowing through it. Compute its power.

17-3. A capacitive reactance of 250 Ω has 25 A flowing through it. Compute its reactive power.

17-4. An inductive reactance of 2.5 kΩ has 45 V across it. Compute its reactive power.

17-5. An impedance of $4.5\underline{/31°}$ kΩ has 250 mA flowing through it. What is the circuit's apparent power, real power, and power factor?

17-6. An impedance of $250\underline{/67°}$ Ω has 25 V across it. What is the circuit's apparent power, reactive power, real power, and power factor?

17-7. A circuit has a power factor of 0.90 (inductive) and has an apparent power of 450 VA. What is the circuit's real power and reactive power?

17-8. Compute the impedance of a series circuit consisting of an inductive reactance of 2.5 kΩ, a capacitive reactance of 3.5 kΩ, and a resistance of 8 kΩ.

17-9. Compute the impedance of a series circuit consisting of an inductive reactance of 45 kΩ, a capacitive reactance of 26 kΩ, and a resistance of 20 kΩ.

17-10. A resistance of 2.5 kΩ is in series with a 2.5-H coil and a 0.5-μF capacitor at a frequency of 100 Hz. Compute the total impedance.

17-11. A resistance of 4.7 kΩ is in series with a 5-H coil and a 1 μF capacitor across a 60-Hz source. Compute the total impedance.

17-12. A resistance of 1.8 kΩ is in series with a 1.5-H coil and a 0.47-μF capacitor across a 25-V 60-Hz source. Compute the current and the voltages across the elements.

17-13. A 100-mH coil, a 12-kΩ resistor, and an 800-pF capacitor are in series with a 20-kHz 8-V supply. Compute the current and the voltages across the elements.

17-14. A 10-mH coil, a 5-kΩ resistor, and a 300-pF capacitor are in series with a 200-kHz supply. If the capacitor has 6 V across it, what is the supply voltage?

17-15. A 50-mH inductor is in parallel with a 7.5-kΩ resistor. (a) What is the impedance of this combination at 150 kHz? (b) At 15 kHz?

17-16. A 0.001-μF capacitor is in parallel with a 25-kΩ resistor. (a) What is the impedance of this circuit at 5 kHz? (b) At 50 kHz?

17-17. A 3.3-kΩ resistor is in parallel with an inductive reactance of 5.6 kΩ and a capacitive reactance of 3.6 kΩ. Compute the impedance.

17-18. A parallel circuit consists of a 500-mH coil, a 750-Ω resistor, a 0.8-μF capacitor, and a 250-Hz 10-V supply. Compute the branch currents and the total current.

17-19. A parallel circuit consists of a 10-kΩ resistor, a 50-mH coil, and a 0.005-μF capacitor. Compute the branch currents and the total current assuming a 16-V 8-kHz supply.

17-20. Compute the impedance of the circuit shown in Fig. 17-18 at 1 kHz.

17-21. If V_s in Fig. 17-18 is 3.5 V at 2.5 kHz, what are the currents through C_1 and C_2?

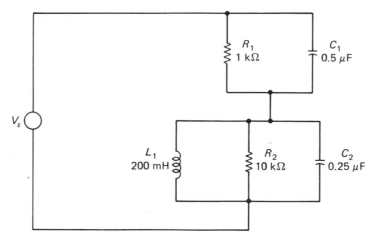

FIGURE 17-18: *Problems 17-20, 17-21, and 17-22.*

17-22. Using nodes, compute the voltage across L_1 in Fig. 17-18 if the supply voltage is 35 V at 1.2 kHz.

17-23. Compute the total impedance seen by the 1-kHz supply in Fig. 17-19.

17-24. If V_s in Fig. 17-19 is 20 V at 2 kHz, compute the voltage across C_3 using nodes.

17-25. Compute the currents through R_1 and R_2 of Fig. 17-20.

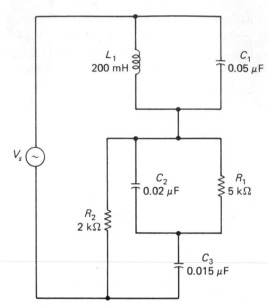

FIGURE 17-19: *Problems 17-23 and 17-24.*

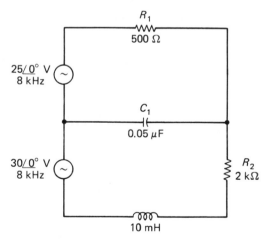

FIGURE 17-20: *Problem 17-25.*

17-9.   PROJECTS

17-1.   Connect a 1-μF capacitor, 8-H inductor, and a 180-Ω resistor in series with a 60-Hz supply. Measure the voltages across the elements. Do these results agree with the theoretical results?

17-2.   Connect the elements of Project 17-1 in parallel across a 60-Hz supply. Measure the branch currents and the total current. How do the measured results agree with the computed results?

17-3. Connect the series circuit of Project 17-1 in series with the parallel circuit of Project 17-2. Measure the voltage across each element and the current through each element. Do the measured results agree with the computed results?

17-4. Measure the dc resistance of an 8-H coil. Next, measure the inductance of the coil at 60 Hz using an ammeter–voltmeter method or an impedance bridge. The inductor can be considered a pure inductance in series with a pure resistance. Assuming this fact, connect the inductor in series with the 1-μF capacitor and measure the voltages and the current at 60 Hz. Theoretically, a capacitor has no effective series resistance. However, there is actually some present. Using your measured results, calculate this effective series resistance.

18

RESONANCE

Resonance can be defined as that frequency at which the capacitive reactance and the inductive reactance of a circuit are equal. Resonant circuits are used extensively in communications electronics for selecting a particular channel or station. In this chapter we shall examine both series and parallel resonant circuits: their currents, voltages, impedances, and selectivity. The sections include:

18-1. Series resonance

18-2. Frequency of series resonance

18-3. The quality factor (Q) of series resonant circuits

18-4. Summary of series resonance

18-5. Parallel resonance

18-6. Frequency of parallel resonance

18-7. The quality factor (Q) of parallel resonant circuits

18-8. Summary of parallel resonance

18-9. Tuned coupling circuits

18-1. SERIES RESONANCE

A series resonant circuit is a series RLC connection in which X_L equals X_C. Such a circuit is shown in Fig. 18-1, in which X_L and X_C are both 1 kΩ at a frequency of 100 Hz. The impedance of this circuit is

$$Z = R + jX_L - jX_C$$
$$= 100 + j1000 - j1000$$
$$= 100\underline{/0^\circ}\ \Omega$$

Thus, the impedance of a series resonant circuit is determined only by the resistance at the resonant frequency. If R were $0\ \Omega$, the impedance of the circuit would be $0\ \Omega$, representing a direct short circuit across the voltage source, resulting in an infinite current. As a practical matter, though, the inductor and the capacitor have a finite resistance, limiting current flow.

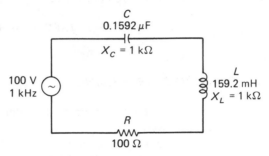

FIGURE 18-1: *Series resonant circuit.*

Let us look more closely at Fig. 18-1 and examine the currents and voltages. The current can be found by

$$I = \frac{V}{Z} = \frac{100\underline{/0^\circ}}{100\underline{/0^\circ}} = 1\underline{/0^\circ}\ \text{A}$$

The voltage across the capacitor is

$$V_C = IX_C = (1\underline{/0^\circ})(1000\underline{/-90^\circ}) = 1000\underline{/-90^\circ}\ \text{V}$$

Note that this represents 10 times the source voltage. Similarly, the voltage across the inductor is

$$V_L = IX_L = (1\underline{/0^\circ})(1000\underline{/90^\circ}) = 1000\underline{/90^\circ}\ \text{V}$$

Observe that, from this analysis, the voltage drops across the inductor and the capacitor will always be equal in magnitude and 180° apart in phase at resonant frequency. Also, note that their voltages may be many times the magnitude of the supply voltage. By the definition of resonance the magnitudes are

$$X_L = X_C$$
$$IX_L = IX_C$$
$$V_L = V_C$$

On the other hand, the resistor drops:

$$V_R = IR = (1\underline{/0°})(100\underline{/0°}) = 100\underline{/0°} \text{ V}$$

Note that, at the resonant frequency, the voltage across the resistor will always be equal to the supply voltage. This can be shown by

$$V_s = IZ = IR$$

since R is the total impedance. Note further that the sum of the voltages is equal to the supply voltage, so Kirchhoff's laws remain intact:

$$V_s = V_R + V_C + V_L = 100 - j1000 + j1000 = 100\underline{/0°} \text{ V}$$

18-2. *FREQUENCY OF SERIES RESONANCE*

Resonance occurs at only one frequency for a particular *RLC* circuit. In Fig. 18-1, for example, if the frequency were to increase above 1 kHz, X_L would increase and X_C decrease; thus, they would no longer be equal in magnitude and impedance must increase. Similarly, a decrease below 1 kHz would result in a decrease of X_L and an increase in X_C, increasing impedance. We can develop an equation for the resonant frequency of a circuit based upon the fact that X_L and X_C must be equal:

$$X_L = X_C$$

$$2\pi f L = \frac{1}{2\pi f C}$$

Solve for f:

$$f_r = \frac{1}{2\pi\sqrt{LC}} \qquad\qquad (18\text{-}1)$$

EXAMPLE 18-1: What is the resonant frequency of a series circuit consisting of a 100-pF capacitor, a 30-μH coil, and a 500-Ω resistor? Also, compute the reactance of the capacitor and the inductor at resonance.

SOLUTION: The resonant frequency is

$$f_r = \frac{1}{2\pi\sqrt{LC}} = \frac{1}{2 \times \pi \times \sqrt{30 \times 10^{-6} \times 100 \times 10^{-12}}}$$

$$= 2.9058 \text{ MHz}$$

The inductive reactance at this frequency is

$$X_L = 2\pi f L = 2 \times \pi \times 2.9058 \times 10^6 \times 30 \times 10^{-6}$$

$$= 547.7 \ \Omega$$

The capacitive reactance at this frequency is

$$X_C = \frac{1}{2\pi fC} = \frac{1}{2 \times \pi \times 2.9058 \times 10^6 \times 100 \times 10^{-12}}$$
$$= 547.7 \ \Omega$$

Note that the capacitive reactance is equal in magnitude to the inductive reactance at resonance.

EXAMPLE 18-2: A capacitor must be selected for a resonance of 10 kHz using an inductor of 50 mH. Compute the value required.

SOLUTION:

$$f_r = \frac{1}{2\pi\sqrt{LC}}$$

Solve for C:

$$C = \frac{1}{4\pi^2 f_r^2 L}$$
$$= \frac{1}{4 \times \pi^2 \times (10^4)^2 \times 50 \times 10^{-3}}$$
$$= 5066 \text{ pF}$$

The Series Resonance Curve

Consider the circuit of Fig. 18-2(a). If V_s is maintained at 1 V and its frequency varied from 100 Hz to 2 kHz, the values of X_L and X_C would be as shown in Fig. 18-2(b). Thus, total impedance will vary, as the table indicates. Note that there is only one frequency at which X_L and X_C are equal, the resonant frequency of 1 kHz. If we now plot the current against the frequency, a bell-shaped curve results, Fig. 18-2(c), in which current is maximum at resonance. If frequency is plotted logarithmically

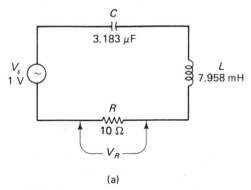

(a)

FIGURE 18-2: *Frequency response of a series resonant circuit: (a) circuit; (b) table of values; (c) graph.*

Frequency (Hz)	X_L (Ω)	X_C (Ω)	Z (Ω)	I (mA)	V_R (V)
100	5.0	500.0	495	2	0.02
200	10.0	250.0	240	4	0.04
300	15.0	166.7	152	7	0.07
400	20.0	125.0	105	9	0.09
500	25.0	100.0	76	13	0.13
600	30.0	83.3	54	18	0.18
700	35.0	71.4	38	26	0.26
800	40.0	62.5	25	41	0.41
900	45.0	55.6	15	69	0.69
1000	50.0	50.0	10	100	1.00
1100	55.0	45.5	14	72	0.72
1200	60.0	41.7	21	48	0.48
1300	65.0	38.5	28	35	0.35
1400	70.0	35.7	36	28	0.28
1500	75.0	33.3	43	23	0.23
1600	80.0	31.3	50	20	0.20
1700	85.0	29.4	56	18	0.18
1800	90.0	27.8	63	16	0.16
1900	95.0	26.3	69	14	0.14
2000	100.0	25.0	76	13	0.13
10000	500.0	5.0	495	2	0.02

Figure 18-2(b): *Table of values.*

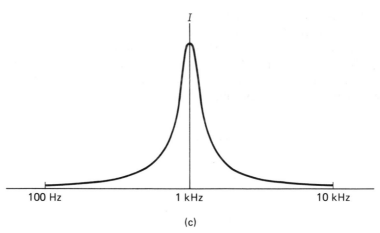

(c)

FIGURE 18-2: *Cont.*

along the base of the graph instead of linearly, the bell curve is perfectly symmetrical. By examining this curve, we can draw two very important conclusions about series resonant circuits:

1. Minimum impedance occurs at resonance.

2. Maximum current occurs at resonance.

According to Eq. 18-1, the frequency of resonance is determined by the product of inductance and capacitance. That is, with an LC product of 2.533×10^{-8}, only one resonant frequency can be obtained, in this case 1 kHz. This product could be obtained using an L of 1 H and a C of 0.02533 μF, an L or 100 mH and a C of 0.2533 μF, an L of 10 mH and a C of 2.533 μF, or any of a myriad of combinations. Each will result in a bell-shaped curve with a resonant frequency of 1 kHz. However, different combinations of L and C will make the bell wider or narrower, allowing more or less frequencies to have higher currents. A broad curve is said to be less selective than one which permits only a very narrow band of frequencies to have the high current.

18-3. THE QUALITY FACTOR (Q) OF SERIES RESONANT CIRCUITS

It seems that every time we turn around we find a different meaning for the symbol Q—charge, reactive power, and now a quality factor for resonant circuits. Q in this sense is a measure of the narrowness of the resonance curve. A very narrow curve will have a high Q, a broad curve a low Q, Fig. 18-3. For this purpose, Q can be defined as

$$Q = \frac{f_r}{f_{bw}}$$

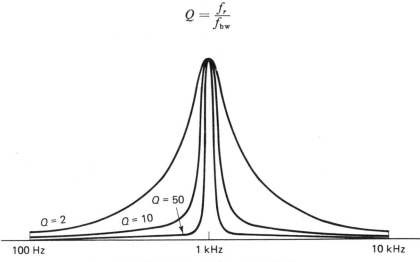

FIGURE 18-3: *Q of resonant circuits.*

where f_r is the resonant frequency and f_{bw} the bandwidth, defined as the range of frequencies with amplitudes above 0.707 times the maximum rms amplitude, Fig. 18-4. Thus, Q compares the bandwidth with the resonant frequency.

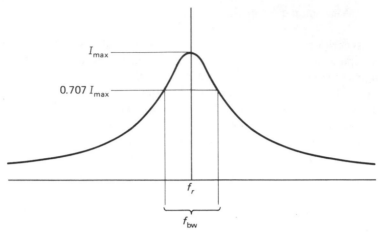

FIGURE 18-4: *Bandwidth.*

EXAMPLE 18-3: A series circuit has a resonance of 5 kHz and a Q of 20. Compute the bandwidth.

SOLUTION:

$$Q = \frac{f_r}{f_{bw}}$$

$$f_{bw} = \frac{f_r}{Q} = \frac{5000}{20} = 250 \text{ Hz}$$

Thus, there is a band of 250 Hz about the resonant frequency that will exceed 0.707 times the amplitude at resonance.

Q can also be found by the following definition:

$$Q = \frac{X}{R}$$

where X is the reactance at resonance and R the series resistance. From this equation we can see that bandwidth can be controlled by controlling Q, and Q is determined by ratio of X to R. Thus, our choice of L and C is determined by two factors:

1. The Q desired determines the specific value of X and, thus, L or C:

$$Q = \frac{X}{R} = \frac{2\pi f L}{R} = \frac{1}{2\pi f C R}$$

2. The resonant frequency determines the LC product:

$$f_r = \frac{1}{2\pi\sqrt{LC}}$$

EXAMPLE 18-4: A circuit must resonate at 10 kHz and have a bandwidth of 2 kHz. Assuming that the circuit has a resistance of 500 Ω, compute the values of L and C.

SOLUTION: We must first determine Q:

$$Q = \frac{f_r}{f_{bw}} = \frac{10,000}{2000} = 5$$

We can next determine X at resonance:

$$Q = \frac{X}{R}$$

Therefore,

$$X = QR = 5 \times 500 = 2.5 \text{ k}\Omega$$

Find the proper value of inductance:

$$X_L = 2\pi f L$$

$$L = \frac{X_L}{2\pi f} = \frac{2500}{2 \times \pi \times 10,000} = 39.79 \text{ mH}$$

We can next use either the resonance equation or the capacitive reactance equation to solve for C. We shall use the resonance equation:

$$f_r = \frac{1}{2\pi\sqrt{LC}}$$

Solve for C:

$$C = \frac{1}{4\pi^2 f^2 L}$$

$$= \frac{1}{4 \times \pi^2 \times (10^4)^2 \times 39.79 \times 10^{-3}}$$

$$= 6366 \text{ pF}$$

Check this answer with the capacitive reactance equation:

$$X_C = \frac{1}{2\pi f C}$$

$$C = \frac{1}{2\pi f X_C} = \frac{1}{2 \times \pi \times 10^4 \times 2500} = 6366 \text{ pF}$$

Note that when the constraint of Q is placed upon the resonant frequency equation, unique values of L and C must be found.

EXAMPLE 18-5: Compute resonant frequency, Q, and bandwidth of a series circuit having a 1-μF capacitor, 1-mH coil, and 10-Ω resistor.

SOLUTION: We must first compute the resonant frequency:

$$f_r = \frac{1}{2\pi\sqrt{LC}} = \frac{1}{2 \times \pi \times \sqrt{10^{-6} \times 10^{-3}}} = 5.033 \text{ kHz}$$

We must now calculate X (note that X_L equals X_C at resonance).

$$X_L = 2\pi f L = 2 \times \pi \times 5033 \times 10^{-3} = 31.62 \ \Omega$$

We can now calculate the Q of the circuit:

$$Q = \frac{X}{R} = \frac{31.62}{10} = 3.162$$

The bandwidth is

$$Q = \frac{f_r}{f_{bw}}$$

$$f_{bw} = \frac{f_r}{Q} = \frac{5033}{3.162} = 1.592 \text{ kHz}$$

Thus, resonance is at 5.033 kHz, Q is 3.162, giving a bandwidth of 1.592 kHz. Figure 18-5 illustrates the resulting curve.

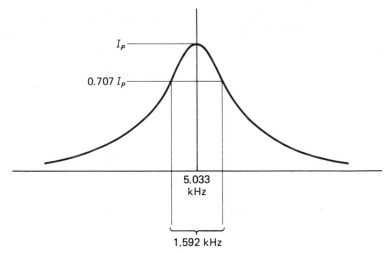

FIGURE 18-5: *Example 18-5.*

Q of Coils

An ideal inductor has no dc resistance, for it is a pure reactor. However, this is impossible to achieve for the coil, being wire, has resistance. The ratio of the reactance and the resistance of a coil is its Q.

EXAMPLE 18-6: A 20-mH coil has a resistance of 10 Ω. What will be its Q at 4 kHz?

SOLUTION:

$$Q = \frac{X_L}{R} = \frac{2\pi fL}{R} = \frac{2 \times \pi \times 4 \times 10^3 \times 0.02}{10} = 50.27$$

This Q would appear from the equations to increase linearly with frequency. However, in actual practice, as frequency in increased, Q reaches a maximum, then decreases. At higher frequencies, hysteresis and eddy currents add to the loss and are, in effect, an additional resistance in series with the coil Moreover, the capacitance between adjacent windings reduces the effective inductance of the device. Thus, the maximum Q attainable for most coils is around 100 and coils are rated at this maximum Q, specifying the frequency at which this Q is realized.

EXAMPLE 18-7: The specifications for an 820-μH coil state that it has a Q of 67 at a frequency of 790 kHz. What is its effective resistance?

SOLUTION:

$$Q = \frac{X_L}{R}$$

$$R = \frac{X_L}{Q} = \frac{2\pi fL}{Q} = \frac{2 \times \pi \times 790 \times 10^3 \times 820 \times 10^{-6}}{67} = 60.75 \ \Omega$$

18-4. SUMMARY OF SERIES RESONANCE

We can, therefore, summarize the characteristics of a series resonant circuit as follows:

1. Minimum impedance occurs at resonance.

2. Maximum line current occurs at resonance.

3. Inductive reactance and capacitive reactance are equal and, therefore, cancel, leaving zero ohms.

4. The total impedance of the circuit is merely the resistance of that circuit.

5. The voltage drops across the inductor and capacitor are equal in magnitude and 180° apart in phase and, therefore, cancel.

6. The voltage across the resistor is equal to the source voltage.

7. The product LC determines the resonance of the circuit, but the ratio of X to R determines the Q.

18-5. PARALLEL RESONANCE

Not only can resonance occur in series RLC circuits, but it can also occur in parallel circuits, Fig. 18-6. Let us assume that an inductive reactance of 2000.001 Ω is in

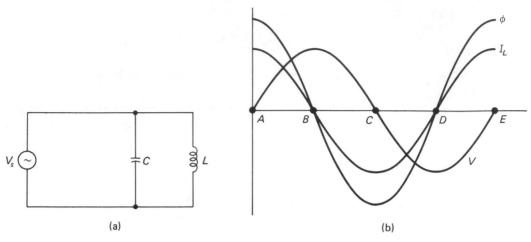

FIGURE 18-6: *Parallel resonance: (a) circuit; (b) waveforms.*

parallel with a capacitive reactance of 2000.000 Ω. The admittance of the parallel circuit is

$$Y = -jB_L + jB_C$$

$$= -j\frac{1}{2000.001} + j\frac{1}{2000.000}$$

$$= -j(2.500 \times 10^{-10})\,\text{S}$$

But

$$Z = \frac{1}{Y} + \frac{1}{-j(2.500 + 10^{-10})} = 4\underline{/90°}\,\text{G}\Omega$$

Thus, the impedance of the circuit is very large, 4 GΩ in this case. From this example it can be seen that pure inductive reactance in parallel with an equal capacitive reactance results in an infinite impedance. Thus, whereas the series resonant circuit has minimum impedance at resonance, the parallel resonant circuit has maximum impedance at resonance.

Some very fascinating electronic gymnastics occur within a parallel resonant circuit. Energy is continually being transferred from the magnetic field of the coil to the dielectric of the capacitor, Fig. 18-6(b). We pick up on the action as the current is maximum, resulting in a large static magnetic field, point A. The capacitor has been completely discharged to 0 V. At this point the magnetic field starts to collapse, tending to maintain the diminishing current by generating a counter emf. This, in turn, charges the capacitor until, at point B, it is charged to its maximum voltage. At point B, the field of the coil is entirely collapsed and current starts to flow from the capacitor to the coil, discharging the capacitor and building up the coil's magnetic field until, at point C, the capacitor has been completely discharged. At this point the field again collapses, producing an emf in such a direction as to tend to maintain the current flow. This current, in turn, charges the capacitor until, at point D, it is again fully charged. From point D to point E the capacitor again discharges its energy into the field of the coil.

In circuits such as these, huge currents can flow between the capacitor and the inductor, restricted only by the resistance of the coil and the effective resistance of the capacitor. Scientists have conducted experiments in a supercooled environment using parallel resonant circuits and have shown that, if it were not for the resistance, these *oscillations*, as they are called, would continue forever. At these extremely low temperatures, resistance is virtually nil and the circuits oscillate for very long periods of time.

We have, in the parallel resonant circuit, a configuration that extracts very little energy from the supply, for the impedance is very high, but produces a large current flow. It is very much like a child's swing, in which just a very little push every cycle results in a very large weight being transported from one excursion of the cycle to another, Fig. 18-7. We need supply only enough energy to overcome the resistance of the system. Let's examine these oscillating currents a little closer in the following example.

FIGURE 18-7: *A child's swing resona-nance (drawing by Kirishian).*

EXAMPLE 18-8: Compute the line current, I_t, and the branch currents for the circuit shown in Fig. 18-8.

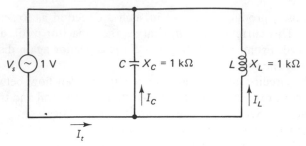

FIGURE 18-8: *Example 18-8.*

SOLUTION:

$$I_L = \frac{V}{X_L} = \frac{1}{1\underline{/90°}} = 1\underline{/-90°} \text{ mA}$$

$$I_C = \frac{V}{X_C} = \frac{1}{1\underline{/-90°}} = 1\underline{/90°} \text{ mA}$$

$$I_t = I_L + I_C = 1\underline{/-90°} + 1\underline{/90°} = 0.0 \text{ mA}$$

Note that, even though no line current flows, there is current in each of the branches.

When a resistor is placed in parallel with a pure resonant circuit, the total line current equals the resistive current. That is, the capacitive and inductive currents cancel as far as line current is concerned. Additionally, instead of impedance being infinite, it is equal to the resistance itself.

EXAMPLE 18-9: A 1-kΩ capacitive reactance, 1-kΩ inductive reactance, and a 10-kΩ resistor are in parallel across a 5-V source. Compute total impedance, branch currents, and total current.

SOLUTION: Total impedance can be found by finding admittance:

$$Y = G + jB_C - jB_L$$

$$= \frac{1}{10} + j\left(\frac{1}{1}\right) - j\left(\frac{1}{1}\right)$$

$$= 0.1 + j0.0 \text{ S}$$

$$Z = \frac{1}{Y} = \frac{1}{0.1 + j0.0} = 10\underline{/0°} \text{ k}\Omega$$

Note that the total impedance is merely the value of the parallel resistance. Compute the branch currents:

$$I_R = \frac{V}{R} = \frac{5\underline{/0°}}{10\underline{/0°}} = 0.5\underline{/0°} \text{ mA}$$

$$I_L = \frac{V_s}{X_L} = \frac{5\underline{/0°}}{1\underline{/0°}} = 5\underline{/-90°} \text{ mA}$$

$$I_C = \frac{V_s}{X_C} = \frac{5\underline{/0°}}{1\underline{/-90°}} = 5\underline{/90°} \text{ mA}$$

The total current is

$$I_t = I_R + I_C + I_L$$
$$= 0.5 + j5 - j5 = 0.5\underline{/0°}\text{ mA}$$

Note that the total line current is only due to that drawn by the parallel resistance.

18-6. FREQUENCY OF PARALLEL RESONANCE

In an *RLC* parallel circuit, the resonant frequency is determined by the equation

$$f_r = \frac{1}{2\pi\sqrt{LC}}$$

EXAMPLE 18-10: A 0.33-μF capacitor, a 120-mH inductor, and a 3-kΩ resistor are connected in parallel. What is the resonant frequency of the circuit?

SOLUTION:

$$f_r = \frac{1}{2\pi\sqrt{LC}}$$

$$= \frac{1}{2 \times \pi \times \sqrt{0.12 \times 0.33 \times 10^{-6}}}$$

$$= 799.8 \text{ Hz}$$

Thus, the resonant frequency is 799.8 Hz.

The equation above assumes that the resistor is in parallel with the inductor and capacitor. The equation is also valid for circuits in which there are equal resistances in the parallel branches, Fig. 18-9(a). As a practical matter, however, the inductor has

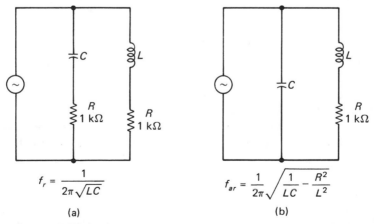

$$f_r = \frac{1}{2\pi\sqrt{LC}}$$

(a)

$$f_{ar} = \frac{1}{2\pi}\sqrt{\frac{1}{LC} - \frac{R^2}{L^2}}$$

(b)

FIGURE 18-9: *Branch resistors in parallel resonant circuits: (a) equal branch resistors; (b) unequal branch resistors.*

dc resistance, unbalancing the branches of the circuit as shown in Fig. 18-9(b). In such a circuit, resonance can be defined in three different ways, each yielding a different frequency:

1. The frequency at which the circuit acts as a pure resistance, also called the *antiresonant frequency.*
2. The frequency at which X_L equals X_C.
3. The frequency at which the line current is minimum.

In order to demonstrate that each of these definitions yields a different frequency, imagine the resistor, R, of Fig. 18-9(b) to be of infinite value, making the right branch an open circuit. Under definition 1, resonance occurs when the circuit acts as a pure resistance. Obviously, our modified circuit is purely capacitive for all frequencies, so there is no frequency satisfying this definition. On the other hand, X_L and X_C can, indeed, be equal at some frequency determined by Eq. (18-1), so definition 2 can easily be computed. According to the third definition, resonance occurs when line current is minimum. For our modified circuit, this will occur when the frequency is 0 Hz, representing direct current. Thus, when R is infinite, resonance occurs at the following frequencies:

1. Definition 1: impossible.
2. Definition 2: a nonzero frequency.
3. Definition 3: 0 Hz, or dc.

Therefore, the three definitions yield three different frequencies.

The equation for determining resonance at which the circuit acts as a pure resistance is

$$f_{ar} = \frac{1}{2\pi}\sqrt{\frac{1}{LC} - \frac{R^2}{L^2}} \tag{18-2}$$

This is the frequency that is of most importance in the treatment of parallel circuits. Note that if R is zero (there is no resistance in series with the inductor), the equation becomes the same as Eq. (18-1). Note further that any resistance in series with the inductor will lower the resonant frequency from that at which X_L equals X_C. If, for example, X_L and X_C were equal at 1000 Hz, adding a resistor in series with the inductor would cause the circuit to act as a pure resistance at a frequency less than 1000 Hz.

In order to establish resonance according to the second definition, the following equation is used:

$$f_r = \frac{1}{2\pi\sqrt{LC}} \tag{18-1}$$

This is, of course, the same equation as is used in series resonant circuits.

The equation for resonance according to the third definition is of no practical importance.

It should be noted that when R_L is less than one-tenth of X_L, Eq. (18-1) can be used and will result in an error of less than 0.51 %.

EXAMPLE 18-11: Compute the antiresonant frequency for the circuit in Fig. 18-10. Also compute the frequency at which X_L equals X_C.

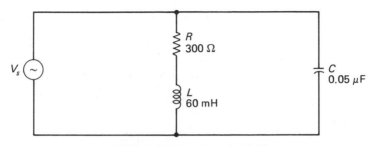

FIGURE 18-10: *Example 18-11.*

SOLUTION: To find the antiresonant frequency:

$$f_{ar} = \frac{1}{2\pi}\sqrt{\frac{1}{LC} - \frac{R^2}{L^2}}$$

$$= \frac{1}{2 \times \pi}\sqrt{\frac{1}{0.06 \times 0.05 \times 10^{-6}} - \frac{(300)^2}{(0.06)^2}}$$

$$= 2.795 \text{ kHz}$$

To find the frequency at which X_L equals X_C:

$$f_r = \frac{1}{2\pi\sqrt{LC}}$$

$$= \frac{1}{2 \times \pi\sqrt{0.06 \times 0.05 \times 10^{-6}}}$$

$$= 2.906 \text{ kHz}$$

Note that the two differ.

The Parallel Resonance Curve

As frequency is varied across a parallel *RLC* circuit, impedance becomes maximum and current minimum at resonance, as shown in Fig. 18-11. Note that this is just the opposite of the series resonant circuit.

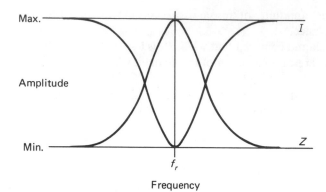

FIGURE 18-11: *Parallel resonance curves.*

18-7. *THE QUALITY FACTOR* (Q)
OF PARALLEL RESONANT CIRCUITS

The Q for a parallel resonant circuit can be found according to the following equation:

$$Q = \frac{R_p}{X} = \frac{f_r}{f_{bw}}$$

where R_p is the parallel resistance, X the reactance, f_r the resonant frequency, and f_{bw} the bandwidth. In this case the bandwidth can be defined as that range of frequencies with line currents below:

$$I_{max} - 0.707(I_{max} - I_{min})$$

This ends up being the 0.707 point on the resonance curve.

EXAMPLE 18-12: A circuit has a 1-mH coil, a 0.005-μF capacitor, and an 8-kΩ resistor in parallel. Compute the resonant frequency, Q, and bandwidth.

SOLUTION: The resonant frequency can be found by

$$f_r = \frac{1}{2\pi\sqrt{LC}} = \frac{1}{2 \times \pi \times \sqrt{0.001 \times 0.005 \times 10^{-6}}} = 71.18 \text{ kHz}$$

At that frequency the reactance is

$$X_L = 2\pi fL = 2 \times \pi \times 71{,}180 \times 0.001 = 447.2 \ \Omega$$

Q can be found by

$$Q = \frac{R}{X} = \frac{8000}{447.2} = 17.89$$

The bandwidth can be found by

$$Q = \frac{f_r}{f_{bw}}$$

$$f_{bw} = \frac{f_r}{Q} = \frac{71.18}{17.89} = 3.979 \text{ kHz}$$

EXAMPLE 18-13: A resistance of 10 kΩ must be placed in parallel with a capacitor and a parallel inductor such that the circuit has a resonance of 100 kHz and a bandwidth of 10 kHz. Compute L and C.

SOLUTION: We can find the required Q by

$$Q = \frac{f_r}{f_{bw}} = \frac{100}{10} = 10$$

Knowing this, we can find the inductance:

$$Q = \frac{R}{X_L} = \frac{R}{2\pi f L}$$

$$L = \frac{R}{2\pi f Q} = \frac{10,000}{2 \times \pi \times 10^5 \times 10} = 1.592 \text{ mH}$$

We can now find the capacitor:

$$f_r = \frac{1}{2\pi\sqrt{LC}}$$

$$C = \frac{1}{2^2\pi^2 f^2 L} = \frac{1}{4 \times \pi^2 \times (10^5)^2 \times 1.592 \times 10^{-3}} = 1592 \text{ pF}$$

18-8.   SUMMARY OF PARALLEL RESONANCE

We can summarize the characteristics of a parallel resonant circuit as follows:

1. Maximum impedance occurs at resonance.
2. Inductive reactance and capacitive reactance are equal and, therefore, cancel leaving infinite ohms.
3. The total impedance of the circuit is merely the parallel resistance of that circuit.
4. The currents through the inductor and capacitor are equal and 180° out of phase resulting in zero line current due to the reactances.
5. The line current is equal to the current flowing through the parallel resistance.

18-9. TUNED COUPLING CIRCUITS

Parallel resonant circuits are used very frequently to transfer energy of one band of frequencies to another circuit. In Fig. 18-12 a radio antenna receives all radio frequencies and induces them into the parallel resonant circuit. However, the tuned circuit provides a low impedance path to ground for all frequencies except those around resonance. Thus, the next stage sees the relative voltages shown in Fig. 18-12(b).

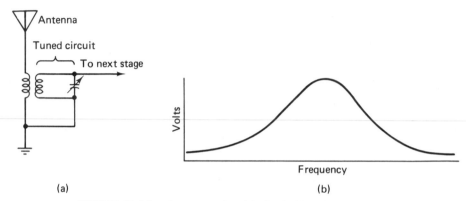

(a)

(b)

FIGURE 18-12: *Antenna tuning: (a) circuit; (b) voltage to next stage.*

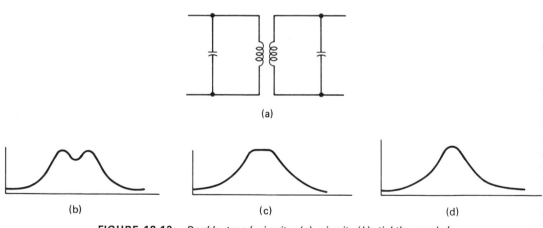

FIGURE 18-13: *Doubly tuned circuits: (a) circuit; (b) tightly coupled (k > 0.05); (c) critically coupled (k = 0.05); (d) loosely coupled (k = 0.005).*

If one tuned circuit is good, two tuned circuits are twice as good, Fig. 18-13. Here, both the primary and secondary are tuned. If they are tuned to precisely the same frequency and the coefficient of coupling is greater than 0.05 (closely coupled), the frequency response is that shown in Fig. 18-3(b). The double humps are caused by the primary loading the seondary, effectively putting a resistor in series with the secondary coil, shifting the resonant frequency. In addition, the secondary also loads the primary, effectively putting a resistor in series with the primary inductor, shifting its frequency response. One shifts up, the other down, resulting in the double humps.

With coefficients of coupling less than 0.05, the frequency-response shift is not so pronounced, and with a k of 0.005 (loosely coupled), no shift is noticeable.

18-10. SUMMARY

Table 18-1 compares the characteristics of both series and parallel resonant circuits. Note that in practically every case, the parallel and series circuits are opposites. If a resistor is placed in one of the branches of a parallel resonant circuits, the frequency at which X_L equals X_C is not the same as the frequency at which the circuit acts as a pure resistance. The Q of a resonant circuit represents the narrowness of the bell curve—the higher the Q, the narrower the curve. Bandwidth is the range of frequencies about resonance with values exceeding 0.707 of the resonant value, and is determined by the Q of the circuit. The ratio of resistance and reactance determines Q. Coils are rated by the highest Q the coil can attain. They suffer from excessive eddy current, hysteresis, and capacitive losses at high frequencies, lowering the Q. A tuned coupling circuit is one in which the inductor is part of a coupling transformer. Such a circuit, when tightly coupled, will produce a double-humped frequency-response curve.

TABLE 18-1. *Series and parallel resonance.*

Series	Parallel
Impedance minimum	Impedance maximum
Current maximum	Current minimum
Source voltage equals drop across series resistor	Source current equals current through parallel resistance
X_L cancels X_C, yielding zero ohms	X_L cancels X_C, yielding infinite ohms
$Z = R_s$	$Z = R_p$
$Q = X/R_s$	$Q = R_p/X$

18-11. REVIEW QUESTIONS

18-1.   What is the impedance of a series resonant circuit in terms of X_L, X_C, and R?

18-2.   What is the phase between the voltage across the inductor and the voltage across the capacitor in a series resonant circuit?

18-3.   What happens to the impedance of a series resonant circuit as frequency is increased above resonance?

18-4.   Under what conditions is the resonance curve symmetrical?

18-5.   Since resonance is determined only by the LC product, what is the effect of choosing a larger L and smaller C over a smaller L and larger C?

18-6.   What is meant by the bandwidth of a circuit?

18-7.   What does Q have to be in order to provide a very small bandwidth?

18-8.   Since Q is determined by X_L, why can't we increase Q to 1000 by increasing the frequency across the coil?

18-9.   Energy is constantly being transferred between what destinations in a parallel resonant circuit?

18-10.   What is the effect of inserting resistance in series with the inductor in a parallel resonant circuit?

18-11.   What is the impedance of a perfect parallel resonant circuit?

18-12.   What is the shape of the frequency-response curve in a doubly tuned close-coupled coupling circuit?

18-12.   PROBLEMS

18-1.   What is the resonant frequency of a 0.11-μF capacitor and a 30-mH coil?

18-2.   What is the resonant frequency of a 30-pF capacitor and a 20-μH coil?

18-3.   What capacitor must be connected in series with a 50-mH coil to cause resonance at 25 kHz?

18-4.   What size of coil must be connected in series with a 0.015-μF capacitor to cause resonance at 250 kHz?

18-5.   A series circuit has a resonance of 12 kHz and a Q of 25. Compute the bandwidth.

18-6.   A circuit must have a bandwidth of 25 kHz and a resonant frequency of 1.2 MHz. What must its Q be?

18-7.   A series circuit must resonate at 18 kHz and have a bandwidth of 3 kHz. Assuming that the circuit has a resistance of 800 Ω, compute the values of L and C.

18-8.   A series circuit must resonate at 2 MHz and have a bandwidth of 280 kHz. Compute L and C for a resistance of 1 kΩ.

18-9.   Compute resonant frequency, Q, and bandwidth of a series circuit having a 0.022-μF capacitor, a 40-mH coil, and a resistance of 75 Ω.

18-10.   A 50-mH coil has a resistance of 80 Ω. What will its Q be at 20 kHz?

18-11.   A 68-μH coil has a Q of 40 at a frequency of 2.5 MHz. What is its resistance?

18-12.   A parallel resonant circuit has a 3.5-μF capacitor, a 580-mH coil, and a 4.7-kΩ resistor in parallel. Compute branch and line currents for a supply voltage of 10 V.

18-13.   A 0.150-μF capacitor, a 75-mH coil, and a 10-kΩ resistor are connected in parallel. Compute branch currents, line current, and impedance at resonance for a supply voltage of 1 V.

18-14.   Compute the resonant frequency for a 25-μF capacitor and a 25-H coil.

18-15.   Compute the antiresonant frequency for a 25-mH coil with a resistance of 20 Ω and a 3.5-μF capacitor. Also compute the frequency at which X_L equals X_C.

18-16. Compute the antiresonant frequency for a circuit consisting of a 2600-pF capacitor in parallel with a series circuit of an 800-μH coil and a 400-Ω resistor.

18-17. A parallel resonant circuit has a Q of 64, R_p of 15 kΩ, and an f_r of 25 kHz. Compute L and C.

18-18. What is the resonant frequency, Q, and bandwidth of a circuit with a 5-mH coil, 0.010-μF capacitor, and 10-kΩ resistor in parallel?

18-13. PROJECTS

18-1. Connect a 150-mH coil in series with a 0.5-μF capacitor across a signal generator. Determine the resonant frequency by varying the frequency of the generator and observing the voltage across the capacitor. Vary the frequency on both sides of resonance and plot the frequency versus inductor voltage, the frequency versus capacitor voltage, and the frequency versus current.

18-2. Measure the dc resistance of the coil in Project 18-1. How much of the resistance of the resonant circuit in Project 18-1 was due to this dc resistance?

18-3. Plot the impedance of a 150-mH coil versus frequency up to at least 2 MHz. At what frequency is the impedance maximum?

18-4. Repeat Project 18-1 with a 1-kΩ resistor in series with the circuit.

18-5. Connect a 150-mH coil in parallel with a 0.5-μF capacitor. Run a frequency-response curve on the parallel combination and plot the line current, capacitor current, and inductor current against frequency. Compute the impedance at each point and plot it against frequency.

18-6. Repeat Project 18-5 with a 4-kΩ resistor connected in parallel with the circuit.

18-7. Repeat Project 18-5 with a 390-Ω resistor connected in series with the inductor.

19

FILTERS

A *filter* is an electronic circuit used to selectively admit or prevent a range of frequencies from being received or transmitted. It is used extensively in communications equipment to enable reception or transmission of signals. In this chapter we shall first examine the single capacitor and single inductor filter, then those using both elements. We shall then examine filters containing resonant sections and finally will investigate a rather new entry into the field, the active filter. The sections include:

19-1. Filters

19-2. The decibel

19-3. Single element filters

19-4. Resonant filters

19-5. Constant-*K* filters

19-6. *m*-Derived filters

19-7. Active filters

19-1. FILTERS

A *filter* is a device that is frequency-selective such that it will admit or restrict (attenuate) certain frequencies. A block diagram of a filter is shown in Fig. 19-1(a), in which all frequencies may appear at the input, but only certain frequencies will appear at the output. Filters fall into four broad classifications: high pass, low pass, band pass, and band eliminate. The *high-pass* filters only admit higher frequencies; the *low-pass* filters permit only lower frequencies to pass through the filter. The

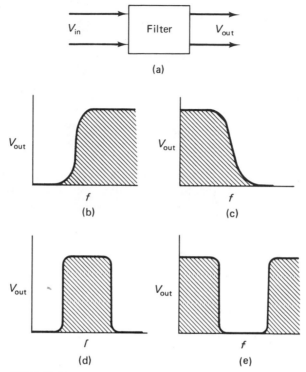

FIGURE 19-1: *Filters: (a) block diagram; (b) high pass; (c) low pass; (d) band pass; (e) band eliminate. Shaded areas are passed frequencies.*

band-pass device attenuates both high and low frequencies permitting only midrange ones to pass. The *band eliminate* allows all to pass except a prescribed range of frequencies.

19-2. THE DECIBEL

Before progressing further, we must examine the unit commonly used for expressing the ratio of the output to the input, the decibel. This unit has three definitions, depending upon whether we are comparing voltages, currents, or powers (wattages):

$$dB = 20 \log \frac{V_{out}}{V_{ref}} \qquad dB = 20 \log \frac{I_{out}}{I_{ref}} \qquad dB = 10 \log \frac{P_{out}}{P_{ref}}$$

where ref is the reference unit. Thus, a filter with an input of 1 V that provides 0.6 V of output will have a gain of

$$dB = 20 \log \frac{V_{out}}{V_{ref}} = 20 \log \frac{0.6}{1.0} = -4.437 \, dB$$

The minus sign indicates the filter has a loss—its output is lower than its input. A filter that has an input of 5 V and an output of 5 V will have a gain of

$$dB = 20 \log \frac{V_{out}}{V_{ref}} = 20 \log \frac{5.0}{5.0} = 0.0 \text{ dB}$$

An amplifier that has an input of 1 mW and an output of 26 W has a gain of

$$dB = 10 \log \frac{P_{out}}{P_{ref}} = 10 \log \frac{26}{0.001} = 44.15 \text{ dB}$$

Note that in every case the denominator is the reference signal—the 0-db point.

The unit can also compare the relative strengths of two output signals. If the output of a 1-kHz filter is 0.7 V and the output of the filter with a 20-kHz signal at its input is 0.015 V, the 20-kHz signal is

$$dB = 20 \log \frac{V}{V_{ref}} = 20 \log \frac{0.015}{0.7} = -33.38 \text{ dB}$$

Thus, the 20-kHz signal is 33.38 dB below the 1-kHz signal.

EXAMPLE 19-1: Two frequencies appear at the input to a filter, each 700 mV: a 1-kHz signal and a 5-kHz signal. The filter provides outputs of 0.4 V for the 5-kHz signal and 1.5 mV for the 1-kHz signal. What is the loss for each signal through the filter and what is the output level of the 5-kHz signal referenced to the output level of the 1-kHz signal?

SOLUTION: For the 5-kHz signal:

$$dB_5 = 20 \log \frac{V_{out}}{V_{ref}} = 20 \log \frac{0.4}{0.7} = -4.861 \text{ dB}$$

The output level of the 1-kHz signal is

$$dB_1 = 20 \log \frac{V_{out}}{V_{ref}} = 20 \log \frac{0.0015}{0.7} = -53.38 \text{ dB}$$

Thus, the 5 kHz has a loss of 4.861 dB through the filter and the 1 kHz has a loss of 53.38 dB through the filter. The 5-kHz output referenced to the 1-kHz output is

$$dB_r = 20 \log \frac{V}{V_{ref}} = 20 \log \frac{0.0015}{0.4} = -48.52 \text{ dB}$$

The 1-kHz signal is, therefore, 48.52 dB below the 5-kHz signal.

One of the very useful properties of decibels is its ability to find differences in output levels merely by subtraction. In the previous example the 5-kHz signal is -4.861 dB and the 1-kHz signal -53.38 dB. Thus, the 1-kHz signal is

$$-4.86 - (-53.38) = 48.52 \text{ dB}$$

below the 5-kHz signal, which checks with the computation of dB_r.

EXAMPLE 19-2: A certain filter provides an input of 1 V and an output of 0.09 V at 20 MHz and an input of 1 V and an output of 21 μV at 400 MHz. What is the loss through the filter at each frequency and what is the difference in losses between the two signals?

SOLUTION: For the 20-MHz signal,

$$dB_{20} = 20 \log \frac{V_{\text{out}}}{V_{\text{ref}}} = 20 \log \frac{0.09}{1.0} = -20.92 \text{ dB}$$

For the 400-MHz signal,

$$dB_{400} = 20 \log \frac{V_{\text{out}}}{V_{\text{ref}}} = 20 \log \frac{21 \times 10^{-6}}{1.0} = -93.56 \text{ dB}$$

The difference between the two signals can be found by subtraction:

$$dB_{\text{diff}} = 93.56 - 20.92 = 72.64 \text{ dB}$$

19-3. SINGLE-ELEMENT FILTERS

A capacitor is, by definition, a filter, for it will not pass a frequency of 0 Hz, a dc voltage. Such a filter is shown in Fig. 19-2(a), and its frequency response in Fig. 19-2(b). If the frequency is plotted on a log scale and decibels are plotted on the vertical axis, the frequency response can be approximated by the straight lines shown in dashes in the figure. These two lines intersect at a frequency called the cutoff frequency (f_c) and the slope of the slanted line is 20 dB/decade. This means that the voltage across the load decreases by 20 dB as the frequency is cut by one-tenth.

This cutoff frequency can be found by setting X_C equal to the sum of R_s and R_L:

$$X_C = R_s + R_L$$

Using this fact, we can select capacitors to give a specific cutoff frequency.

EXAMPLE 19-3: Select a capacitor for a cutoff frequency of 100 Hz if the load resistor is 12 kΩ and the source resistance is 5 kΩ.

SOLUTION:

$$X_C = R_s + R_L = 5 + 12 = 17 \text{ k}\Omega$$

$$X_C = \frac{1}{2\pi f C}$$

$$C = \frac{1}{2\pi f X_C} = \frac{1}{2 \times \pi \times 100 \times 17,000} = 0.0936 \ \mu\text{F}$$

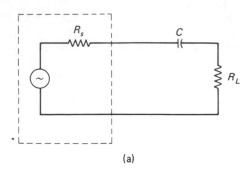

(a)

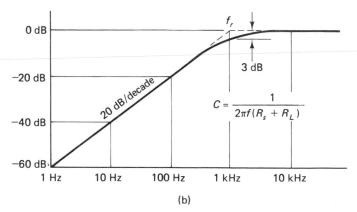

(b)

FIGURE 19-2: *Capacitor as a high-pass filter: (a) circuit; (b) frequency response.*

Its roll-off would be 20 dB/decade. Thus, if the output were 0 dB at a high frequency, it would be -20 dB at 10 Hz, -40 dB at 1 Hz, and -60 dB at 0.1 Hz.

Although the slope of the straight-line segments of Fig. 19-2(b) is 20 dB/decade, the actual response follows the solid, curved line. Thus, at the cutoff frequency, the response is actually 3 dB down from what it is at a high frequency. Thus can be shown by recognizing that, at f_c, the total impedance of the circuit is

$$Z_c = (R_s + R_L) - jX_C$$

But $R_s + R_L = X_C$; therefore,

$$Z_c = (R_s + R_L) - j(R_s + R_L) = (1.414\underline{/-45°})(R_s + R_L)$$

At a very high frequency X_C is effectively zero and the impedance is

$$Z_h = (R_s + R_L)\underline{/0°}$$

Comparing these two impedances, note that $Z_c = 1.414Z_h$. Thus, a voltage at f_c will produce $1/1.414$, or 0.707 times the current that same voltage would produce at a high frequency (f_h), resulting in 0.707 times the voltage drop across R_L at f_c as at f_h, a -3-dB point on the curve.

A single capacitor can also be used as a low-pass filter as in Fig. 19-3(a). As in all single-element circuits, the cutoff frequency is determined by setting X_C equal to whatever Thévenized resistance the capacitor sees, in this case R_s in parallel with R_L.

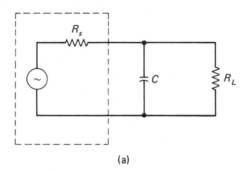

(a)

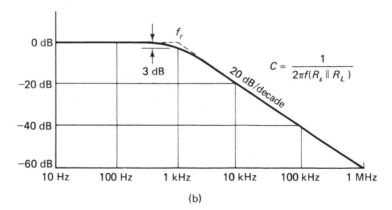

(b)

FIGURE 19-3: *Capacitor as a low-pass filter: (a) circuit; (b) frequency response.*

EXAMPLE 19-4: Compute C in Fig. 19-3(a) for an f_c of 20 kHz if R_s is 600 Ω and R_L, 500 Ω.

SOLUTION: Thévenizing the resistances seen by C:

$$R_{\text{th}} = R_s \| R_L$$

$$= \frac{600 \times 500}{600 + 500} = 272.7 \ \Omega$$

Setting X_C equal to this value,

$$C = \frac{1}{2\pi f X_C} = \frac{1}{2 \times \pi \times 20{,}000 \times 272.7} = 0.02918 \ \mu\text{F}$$

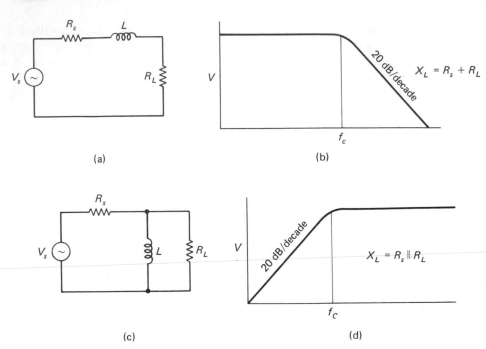

FIGURE 19-4: *Inductive filters: (a) low-pass filter; (b) frequency-response of the low-pass filter; (c) high-pass filter; (d) frequency-response of the high-pass filter.*

Inductors can also be used as high-pass and low-pass filters as shown in Fig. 19-4. In each case f_c is selected by setting X_L equal to the Thévenized resistance seen by the inductor.

EXAMPLE 19-5: Design a low-pass filter for an f_c of 10 Hz using an inductor if R_s is 50 Ω and R_L is 100 Ω.

SOLUTION: The circuit of Fig. 19-4(a) must be used. Calculate L:

$$X_L = R_s + R_L = 50 + 100 = 150 \ \Omega$$
$$= 2\pi f L$$
$$L = \frac{X_L}{2\pi f} = \frac{150}{2 \times \pi \times 10} = 2.387 \ \text{H}$$

The output will be -3 dB at 10 Hz, -20 dB at 100 Hz, -40 dB at 1000 Hz, and -60 dB at 10 kHz.

EXAMPLE 19-6: Design a simple high-pass filter with an f_c of 1 kHz if R_s is 1 kΩ and R_L is 1.2 kΩ.

SOLUTION: We can use the circuit shown in Fig. 19-4(c). Calculate L:

$$X_L = R_s R_L = 1 \,\|\, 1.2 = 545.5 \ \Omega$$

$$L = \frac{X_L}{2\pi f} = \frac{545.5}{2 \times \pi \times 10^3} = 86.81 \ \text{mH}$$

The output will be -3 dB at 1 kHz, -20 dB at 100 Hz, -40 dB at 10 Hz, and -60 dB at 1 Hz.

19-4. RESONANT FILTERS

A band-pass or band-eliminate filter can be made from either a series resonant or a parallel resonant circuit, Fig. 19-5. In each case the cutoff frequency is calculated using the equation

$$f_c = \frac{1}{2\pi\sqrt{LC}}$$

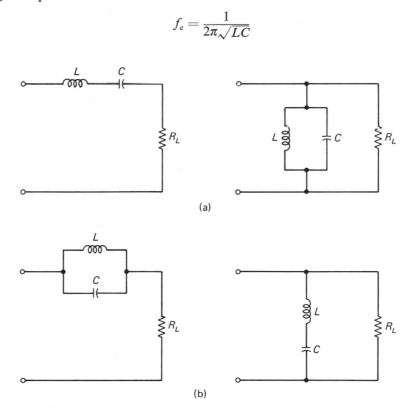

(a)

(b)

FIGURE 19-5: *Resonant filters: (a) band pass; (b) band eliminate.*

The band-eliminate filter is especially useful in eliminating an unwanted, interfering frequency from a television receiver.

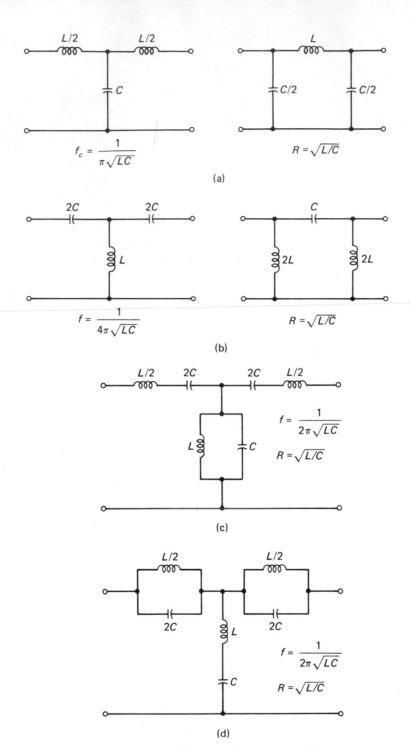

FIGURE 19-6: *Constant-*K *filter sections: (a) low pass; (b) high pass; (c) band pass; (d) band eliminate.*

19-5. CONSTANT-K FILTERS

Figure 19-6 illustrates the constant-K filter. Quite easy to design, the unit has the advantage of maintaining a relatively constant input and output impedance over a wide range of frequencies. This impedance is designated R in the figure. Let us examine this R a little closer, in Fig. 19-7, where a filter is shown being driven by a source and

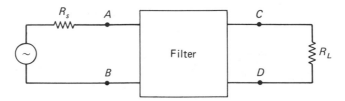

FIGURE 19-7: *Characteristic impedance.*

terminated in a load. Let us assume that $R_s = R_L = R$. This value, R, is called the filter's characteristic impedance and it satisfies the following conditions:

1. The impedance looking into the filter from terminals A and B with R_L connected is R.

2. The impedance looking back into the filter from terminals C and D with R_s connected is R.

Thus, the value of R used in the filter equations should be selected to match the load, R_L, and source, R_s.

The constant-K filter is named because the product of X_L and X_C is independent of frequency:

$$K = X_L X_C = 2\pi f L \times \frac{1}{2\pi f C} = \frac{L}{C}$$

Note that frequency dropped out of the expression and K is dependent only upon L and C.

EXAMPLE 19-7: Design a constant-K low-pass filter with a cutoff frequency of 1 kHz and an impedance of 20 kΩ.

SOLUTION: We must first massage the design equations. Let us solve for L in the second one:

$$R = \sqrt{\frac{L}{C}} \qquad R^2 = \frac{L}{C} \qquad L = R^2 C$$

Plug this into the first equation of the figure:

$$f_c = \frac{1}{\pi\sqrt{LC}} = \frac{1}{\pi\sqrt{R^2 C^2}} = \frac{1}{\pi R C}$$

405

Solve for C:

$$C = \frac{1}{\pi R f_c} = \frac{1}{\pi \times 20{,}000 \times 1000} = 0.015915 \ \mu\text{F}$$

We can now find L:

$$L = R^2 C = 20{,}000^2 \times 0.015915 \times 10^{-6} = 6.366 \ \text{H}$$

In actual construction, we can use the T filter or the Π filter of Fig. 19-6(a). For the T filter, the inductors would be 6.366/2 H or 3.183 H. For the Π type, the capacitors would be 0.015915/2 μF, or 7958 pF.

19-6. m-DERIVED FILTERS

The constant-K filter has two serious disadvantages:

1. Its input and output impedance vary too much with frequency for some critical applications, especially around the resonant frequency.

2. Its roll-off is not sufficiently steep for many applications.

These two disadvantages are overcome by the *m*-derived filter, Fig. 19-8. The circuit contains a resonant section designed to sharply increase roll-off. Thus, in the low-pass section, it can cut off extremely sharp just beyond f_c. However, the *m*-derived filter does a poorer job than the constant K at frequencies higher than f_c, Fig. 19-8(c). This rise in frequency response is called "pop up," and depends upon the value of m.

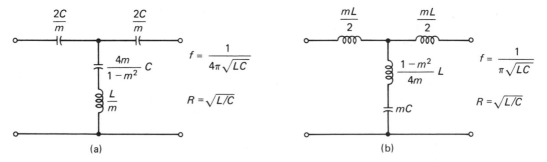

(a)   (b)

FIGURE 19-8: m-*Derived filter:* (a) *high pass;* (b) *low pass;* (c) *frequency-response of the low-pass filters.*

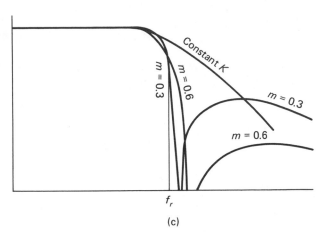

FIGURE 19-8: (*Cont.*)

The value of m can vary between 0 and 1.0 and is selected based upon a trade-off of pop-up versus sharp roll-off (also called *skirts*). A value of 0.2 provides very sharp skirts, but has substantial pop-up whereas a value of 0.6 has much poorer roll-off but much less pop-up.

For critical applications, the m-derived filter is connected in tandem with a constant K, providing the advantage of sharp skirts from the m-derived section and very little pop-up from the constant K.

EXAMPLE 19-8: Design a two-section high-pass filter consisting of a constant-K section in tandem with an m-derived section. The filter is to have a cutoff frequency of 2500 Hz and an impedance of 500 Ω.

SOLUTION: We shall use two T-type filters. For the constant-K filter, the values of L and C can be computed by messaging the design equations:

$$R = \sqrt{\frac{L}{C}}$$

Therefore,

$$L = CR^2$$

Substitute for L in the following equation:

$$f_c = \frac{1}{4\pi\sqrt{LC}} = \frac{1}{4\pi\sqrt{C^2 R^2}} = \frac{1}{4\pi CR}$$

Solve for C:

$$C = \frac{1}{4\pi f R} = \frac{1}{4 \times \pi \times 2500 \times 500} = 0.06366 \ \mu F$$

Compute L:

$$L = CR^2 = 0.06366 \times 10^{-6} \times 500^2 = 15.92 \text{ mH}$$

This constant-K section is shown in Fig. 19-9(a). Note that Fig. 19-6(b) requires that the capacitors be doubled in a T section.

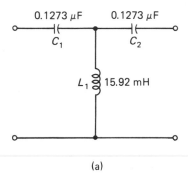

(a)

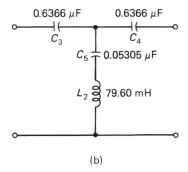

(b)

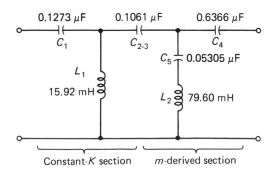

(c)

FIGURE 19-9: *Example 19-8:* (a) *constant-K section;* (b) *m-derived section;* (c) *finished filter.*

The *m*-derived section uses precisely the same design equations, except that the elements are modified by the *m* terms. We shall use an *m* of 0.2 to give sharp roll-off. Computing the elements in Fig. 19-9(b) using the equations in Fig. 19-8(a):

$$C_3 = \frac{2C}{m} = \frac{2 \times 0.06366}{0.2} = 0.6366 \ \mu\text{F}$$

$$C_4 = C_3 = 0.6366 \ \mu\text{F}$$

$$C_5 = \frac{4m}{1-m^2} C = \frac{4 \times 0.2}{1-0.2^2} \times 0.06366 = 0.05305 \ \mu\text{F}$$

$$L_2 = \frac{L}{m} = \frac{15.92}{0.2} = 79.60 \ \text{mH}$$

Next, the two sections are connected in tandem, Fig. 19-9(c). One final refinement: C_2 can be combined with C_3, yielding a single, 0.1061-μF capacitor referred to as C_{2-3}.

19-7. ACTIVE FILTERS

An *active filter* is formed from simple one-capacitor filters and an isolating amplifier, Fig. 19-10. By connecting two capacitors in a single amplifier circuit, roll-offs of 40 dB/decade are easily attainable. Such a device is called a *Butterworth filter*. A band-pass filter may be constructed by connecting a high-pass Butterworth filter in series with a low-pass Butterworth filter. For example, if the high-pass filter passes all frequencies above 1 kHz and the low-pass filter passes all frequencies below 10 kHz, the resultant circuit will pass all frequencies between 1 kHz and 10 kHz with 40 dB/decade roll-off on both ends of the pass band, Fig. 19-10(b).

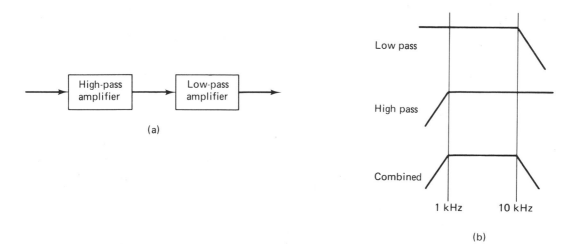

FIGURE 19-10: *Active bandpass filter; (a) block diagram; (b) response curves.*

A filter is a circuit that is frequency-selective and falls into one of four categories: high pass, low pass, band pass, and band eliminate. The decibel is a logarithmic unit for comparing two electrical quantities. A simple capacitor or inductor will provide a rolloff of 20 dB/decade and has a cutoff frequency at the -3-dB point. The constant-K and m-derived filters provide both impedance matching and frequency selection, with the m-derived device capable of extremely sharp skirts. The active filter uses a simple capacitive circuit and an amplifier to give rolloffs of 40 dB/decade. Two active filters can be cascaded to provide a band-pass filter.

19-9. *REVIEW QUESTIONS*

19-1. What are the four basic types of filters?

19-2. What is a decibel?

19-3. What is the roll-off of a single capacitor filter per decade?

19-4. How can a single capacitor be used as a low-pass filter?

19-5. What is the decibel level of a single capacitor filter at the cutoff frequency?

19-6. What is an application of the band-pass resonant filter?

19-7. What is the advantage of a constant-K filter over a simple capacitor filter?

19-8. What is the meaning of the name "constant K?"

19-9. What is the advantage of the m-derived filter over the constant-K filter. Its disadvantage?

19-10. What is the range of m?

19-11. What is the advantage and disadvantage of choosing a high value of m over a low value?

19-12. What is a Butterworth filter? What is its roll-off per decade?

19-13. How may a band-pass filter be made from two Butterworth filters?

19-10. *PROBLEMS*

19-1. A filter has an input of 750 mV and an output of 65 mV. What is its loss in decibels?

19-2. A filter provides an output of 45 mV with an input of 600 mV at 1 kHz; it provides an output of 12 mV when a 600-mV 10-kHz signal is received. What are the losses at each frequency in decibels, and what are the relative strengths of the two frequencies at the output in decibels?

19-3. Select a series capacitor for a cutoff frequency of 7.6 kHz if the load resistor is 12 kΩ and the source resistance is 11 kΩ.

19-4. Select a series capacitor for a cutoff frequency of 25 kHz if the load resistor is 10 Ω and the source resistance is 8 Ω.

19-5. Design a low-pass filter using a simple capacitor that has a cutoff frequency of 16 kHz if the source impedance is 1.5 kΩ and the load 1.2 kΩ.

19-6. Design a low-pass filter using a simple inductor that has a cutoff frequency of 100 Hz if the source impedance is 5 Ω and the load, 10 Ω.

19-7. Design a high-pass filter using a simple inductor that has a cutoff frequency of 120 Hz, a source impedance of 1 kΩ, and a load impedance of 750 Ω.

19-8. Design a constant-*K* T-section low-pass filter with a cutoff frequency of 2.5 kHz and an impedance of 8 kΩ.

19-9. Design a constant-*K* T-section high-pass filter with a cutoff frequency of 25 kHz and an impedance of 600 Ω.

19-10. Design a high-pass T-section *m*-derived filter that cuts off very sharply, has a cutoff frequency of 6.8 kHz, and has an impedance of 2 kΩ.

19-11. Design a low-pass T-section *m*-derived filter that cuts off very sharply, has a cutoff frequency of 20 kHz, and has an impedance of 600 Ω.

19-12. Design a two-section high-pass filter consisting of a constant-*K* T section in tandem with an *m*-derived T section. The filter is to have a cutoff frequency of 10 kHz and an impedance of 1 kΩ.

19-11. PROJECTS

19-1. Connect a 0.5-μF capacitor in series with a 180-Ω source resistance and a 330-Ω load resistor. Take a frequency response of the voltage across the load. Do your results agree with your computed results?

19-2. Design and build a constant-*K* filter high-pass section having a cutoff frequency of 5 kHz and a resistance of 1 kΩ. Using a 1-kΩ load resistor, take a frequency response on the filter.

19-3. Design and build the filter in Prob. 19-12. Take a frequency response of the filter across a 1-kΩ load.

PART III

ACTIVE ELEMENTS

The circuits and elements we have studied so far have been passive; that is, they provide less output energy than was supplied at the input. In the chapter that follows we shall investigate circuits in which a small input voltage or current provides a large voltage or current at the output. Aha! Perpetual motion! No, for the power that is provided at the output comes from a battery or power supply. Thus, the input merely controls what the power supply outputs much as a person controls a dimmer switch on a light.

The chapter within this part is:

20. Active elements

20

ACTIVE ELEMENTS

The study of electronics begins with the study of amplifiers, for they form the essential difference between electrical science and electronics. These amplifiers are formed from active devices such as vacuum tubes and transistors. In this chapter we shall first investigate the amplifier as a black box. Following this, we shall examine the semiconductors and vacuum tubes from which these boxes are formed. The sections include:

20-1. The amplifier

20-2. Operational amplifiers

20-3. Digital circuits

20-4. Semiconductors

20-5. Vacuum tubes

20-1. THE AMPLIFIER

An *amplifier* is a circuit or device that provides an electrical output that is proportional to and greater than the input. If, for example, an input of 1 mA causes an output of 1 A and an input of 5 mA causes an output of 5 A, the device is said to be a *current amplifier*, Fig. 20-1(a). We can plot input current versus output current of such an amplifier, Fig. 20-1(b); such a plot is called a *transfer curve*.

The amount of amplification a device supplies is called its gain and the current gain of an amplifier is found by the equation

$$A_i = \frac{I_o}{I_i} \tag{20-1}$$

415

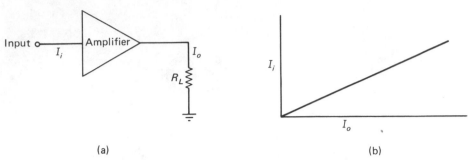

FIGURE 20-1   *Current amplifier: (a) block diagram; (b) transfer curve.*

where A_i is the current gain, I_o the output current, and I_i the input current. Note that gain has no units of measure, for we are dividing a current by a current.

EXAMPLE 20-1: Compute the current gain for an amplifier in which a 1-mA input causes a 1-A output.

SOLUTION: The gain can be computed using Eq. (20-1):

$$A_i = \frac{I_o}{I_i} = \frac{1000}{1} = 1000$$

Gain can also be computed in decibels. In this example,

$$dB = 20 \log \frac{I_o}{I_i} = 20 \log \frac{1000}{1} = 60 \text{ dB}$$

Thus, the amplifier has a current gain of 1000, or 60 dB.

EXAMPLE 20-2: A current amplifier with a gain of 400 has an input current of 20 μA. What is its output current?

SOLUTION:

$$A_i = \frac{I_o}{I_i} \qquad I_o = A_i I_i$$

$$= 400 \times 20$$

$$= 8 \text{ mA}$$

EXAMPLE 20-3: What input current must be supplied a current amplifier with a gain of 45 dB to supply 500 μA at its output?

SOLUTION:

$$dB = 20 \log \frac{I_o}{I_i}$$

$$45 = 20 \log \frac{500}{I_i}$$

$$2.250 = \log \frac{500}{I_i}$$

Convert both sides to exponential format:

$$177.83 = \frac{500}{I_i}$$

$$I_i = 2.812 \ \mu A$$

We can also analyze the response of an amplifier to input voltages, Fig. 20-2. If an input of 10 mV causes an output of 1 V, and an input of 20 mV causes an output of 2 V, the device is said to have a *voltage gain*. We can plot the amplifier's response on a voltage transfer curve, Fig. 20-2(b). Furthermore, this gain can be computed using the equation

$$A_v = \frac{V_o}{V_i} \tag{20-2}$$

where A_v is the voltage gain, V_o the output voltage, and V_i the input voltage.

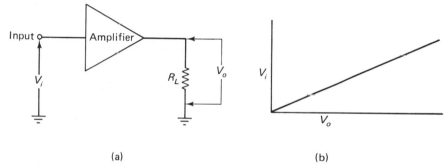

(a)   (b)

FIGURE 20-2:   *Voltage amplifier: (a) block diagram; (b) transfer curve.*

EXAMPLE 20-4: A voltage amplifier with a gain of 800 must supply a 20-V signal to its load. What must its input signal be?

SOLUTION:

$$A_v = \frac{V_o}{V_i}$$

$$V_i = \frac{V_o}{A_v} = \frac{20}{800} = 25 \ mV$$

EXAMPLE 20-5: A voltage amplifier with a gain of 24 dB has an input signal of 20 mV. What is its output voltage?

SOLUTION:

$$dB = 20 \log \frac{V_o}{V_i}$$

$$24 = 20 \log \frac{V_o}{0.02}$$

$$V_o = 317 \ mV$$

Not only can an amplifier have current gain and voltage gain, but it can also have a power gain. This power gain is the product of voltage and current gain:

$$A_p = A_i A_v \qquad (20\text{-}3)$$

Note that this is also

$$A_p = A_i A_v$$
$$= \frac{V_o}{V_i} \times \frac{I_o}{I_i}$$
$$= \frac{P_o}{P_i} \qquad (20\text{-}4)$$

where P_i is the input power and P_o the output power.

EXAMPLE 20-6: A power amplifier has input and output signals as shown in Fig. 20-3. What is its power gain?

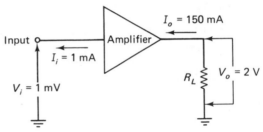

FIGURE 20-3: *Power amplifier.*

SOLUTION: Compute the voltage gain:

$$A_v = \frac{V_o}{V_i} = \frac{2}{0.001} = 2000$$

Compute the current gain:

$$A_i = \frac{I_o}{I_i} = \frac{150}{1} = 150$$

The power gain is

$$A_p = A_v A_i = 2000 \times 150 = 300{,}000$$

Some amplifiers have a voltage gain of 1 (called *unity gain*) and can still have substantial power gain, for they have a large current gain. Others have a unity current gain and substantial voltage gain, resulting in a power gain.

20-2. OPERATIONAL AMPLIFIERS

The *operational amplifier* is an almost ideal integrated-circuit voltage amplifier. As such, its assumed specifications include:

1. Infinite voltage gain.

2. Infinite input impedance.

3. Zero output impedance.

4. The input voltage is virtually ground.

Its symbol is shown in Fig. 20-4 with a series resistor, R_s, of 10 kΩ and a feedback resistor, R_f, of 100 kΩ. Assume that the device is an inverter and $+1$ V is applied to the input. Next, analyze the current at point A in the diagram. Since point A is virtually 0 V, the current through R_s is

$$I = \frac{V}{R} = \frac{1}{10} = 0.1 \text{ mA}$$

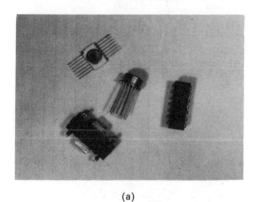

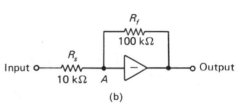

(a)   (b)

FIGURE 20-4: *Operational amplifier: (a) photograph (photo by Ruple); (b) circuit.*

But there is no current flow into or out of the $(-)$ input of the op amp. Therefore, this 0.1 mA must be coming from R_f. But if R_f is 100 kΩ, its voltage drop is

$$V = IR = 0.1 \times 100 = 10 \text{ V}$$

Since point A is a virtual ground, the output voltage must be -10 V and the gain of the amplifier 10.

Next, let us develop a general equation. Assume an input voltage V_i. The current through R_s is

$$I = \frac{V_i}{R_s}$$

Since this is the same current that flows through R_f, the voltage drop across R_f is

$$V_o = IR_f = -\frac{V_i}{R_s}R_f$$

The minus sign indicates that the current through R_f is in the opposite direction from the current through R_s.

Next, let us develop a formula for voltage gain. Solve for V_o/V_i:

$$A_v = \frac{V_o}{V_i} = -\frac{R_f}{R_s}$$

Note that the voltage gain of an operational amplifier is dependent only upon the values of the external resistors.

> **EXAMPLE 20-7:** Compute the gain for the op amp shown in Fig. 20-5. What is the output voltage if V_i is 0.8 V?

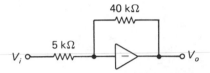

FIGURE 20-5: *Example 20-7.*

SOLUTION: Since $A_v = -R_f/R_s$:

$$A_v = -\frac{R_f}{R_s} = -\frac{40}{5} = -8$$

The voltage gain is, therefore, 8 and the amplifier is an inverter. If the input voltage is 0.8 V:

$$V_o = A_v V_i = -8 \times 0.8 = -6.4 \text{ V}$$

Not only can the op amp be used as a dc amplifier, but it will amplify ac signals. Thus, providing a sine wave on its input results in an amplified sine wave at its output, Fig. 20-6. Since the op amp inverts, the output will be 180° out of phase with the input.

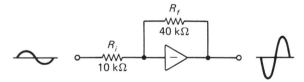

FIGURE 20-6: *ac amplifier.*

The operational amplifier can be used in many different applications, including voltage addition, current addition, pulse generation, integration, and many non-inverting circuit applications. Thus, it is truly a versatile device, being incorporated into many applications hitherto reserved for very complex circuits.

20-3. DIGITAL CIRCUITS

Digital circuits are used extensively in computers, calculators, and digital readout instruments. Like the op amp, they can be treated as black boxes, for most are now integrated circuits, Fig. 20-7. Digital circuits differ from linear circuits such as the op amp in that each input or output is permitted to have only one of two voltage levels, a high or a low. A low voltage, usually around zero volts, is considered a *logical 0*, whereas a high voltage, usually between $+3$ and $+5$ V, is called a *logical 1*. Thus, each point in a digital system is either a logical 1 or a logical 0.

FIGURE 20-7: *Digital integrated circuits (photo by Ruple).*

There are four types of building blocks used within a digital system: the AND gate, the OR gate, the inverter, and the flip-flop (FF), Fig. 20-8.

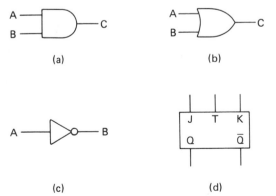

FIGURE 20-8: *Digital building blocks: (a) AND gate; (b) OR gate; (c) inverter; (d) flip-flop.*

The AND Gate

The *AND gate* will output a logical 1 only when all its inputs are logical 1s. Under any other condition, the AND gate outputs a logical 0. We can express this mathematically using a system called *Boolean algebra* as

$$C = A \cdot B$$

meaning that C equals A AND B. Note that the dot signifies the AND operation. This equation is most easily understood by realizing that A and B are input wires to the AND gate and C is its output wire. Thus, if the A wire is 1 and the B wire is 1, the C wire will be a 1. If, however, the *A* wire is 0 and the B wire is 1, the output wire C will be a 0.

> **EXAMPLE 20-8:** An AND gate has three inputs, an X lead, a Y lead, and a Z lead, and one output, R, Fig. 20-9. If X and Z are 1 and Y is 0, what will R be?

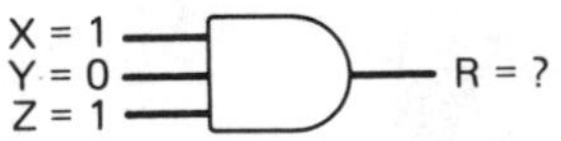

$$
\begin{array}{l}
X = 1 \\
Y = 0 \\
Z = 1
\end{array}
\!\!\!\!\!\!\Rightarrow\!\!\!- R = ?
$$

FIGURE 20-9: *Example 20-8.*

> *SOLUTION:* According to the definition of an AND gate, all inputs must be 1 in order for the output to be 1. Since all inputs are not 1, the output is 0.

The OR Gate

The *OR gate*, Fig. 20-10, will output a logical 1 if any of its input leads are 1s. If, however, all its input leads are 0, it will output a 0. This relationship can be expressed in Boolean algebra as

$$C = A + B$$

meaning that the C output lead equals the A input lead ORed with the B input lead. Note that the + sign represents the OR function. Remember that each letter of the equation represents a wire, and each can take on values of only 1 and 0. If, for example, the A wire were 1 and the B wire 0, the OR gate would respond by providing a 1 on the C wire.

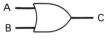

$$A + B = C$$

FIGURE 20-10: *OR gate.*

> **EXAMPLE 20-9:** An OR gate has four inputs, A, B, C, and D, and one output, M, Fig. 20-11. If A and B are 1s and C and D are 0s, what is M?

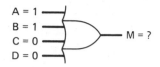

FIGURE 20-11: *Example 20-9.*

SOLUTION: According to the definition for an OR gate, if any of the inputs are 1, the output is a 1. Since some inputs are 1, then M will be a 1.

The Inverter

The *inverter*, Fig. 20-12, is a device that outputs a 1 if its input is a 0 and outputs a 0 if its input is a 1. The mathematical symbol for it is the overscore and it is referred to as not. Thus, $B = \bar{A}$ is pronounced, "B equals A not," and means that the B lead is at a logic level opposite to that of the A lead.

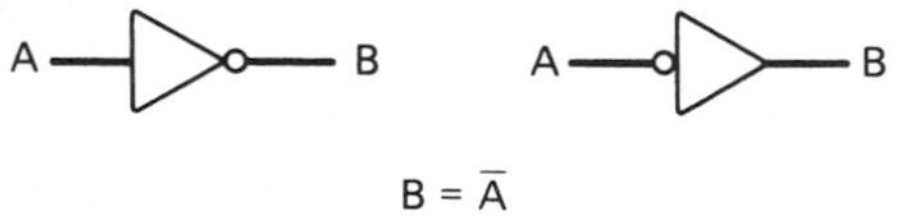

$$B = \bar{A}$$

FIGURE 20-12: *Inverter.*

The actual graphic symbol for the inverter is the circle, with the triangle representing a noninverting amplifier. Thus, the NAND (not AND) gate and the NOR (not OR) gate have the meanings shown in Figs. 20-13 and 20-14.

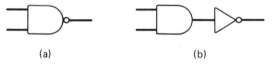

(a)   (b)

FIGURE 20-13: *NAND gate: (a) symbol; (b) meaning.*

(a)   (b)

FIGURE 20-14: *NOR gate: (a) symbol; (b) meaning.*

Flip-Flops

Flip-flops (FF) are the devices used to store numbers—they remember. There are four types available: RS, T, D and JK. The RS FF, Fig. 20-15, has two input leads, S (Set), and R (Reset) and two output leads, Q and $\bar{Q}$. In all FF's when Q is 1, $\bar{Q}$ is 0 and when Q is 0, $\bar{Q}$ is 1. By putting a 1 on the S lead, the Q lead will go to a 1 and stay there. When the S lead returns to 0, the Q will still be a 1. On the other hand, providing a 1 on the R lead will cause Q to go to 0 and stay there, even after R goes to 0. Thus,

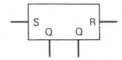

FIGURE 20-15:   *RS flip-flop.*

the FF remembers which lead last had a 1 on it, S or R. The rules of the game do not permit S and R to be 1 simultaneously; when that happens the manufacturer will not guarantee a predictable result.

The type T (toggle) FF, Fig. 20-16, changes its output each time the T lead goes high to low. Thus, it is the basic counting element used in a digital circuit. In the figure note that the output frequency is one-half the input frequency. This type FF is used in many electronic devices for generating a lower frequency from a higher one.

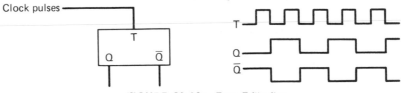

FIGURE 20-16:   *Type-T flip-flop.*

The type D (delay) FF, Fig. 20-17, is used for storing the logic level on a line for future use. When the T lead goes from 1 to 0, the FF samples the D lead. If the D lead is a 1, its Q lead goes to 1 and its $\bar{Q}$ lead to 0. If the D is a 0 at this time, the Q lead goes to 0 and $\bar{Q}$ to 1. it will then stay in this condition until the next high to low transition on the T lead occurs. Thus, using one type D FF on each of 16 wires, a circuit can be developed that will take a snapshot of the logic levels on each wire at any particular time by merely causing all the T leads to go from 1 to 0.

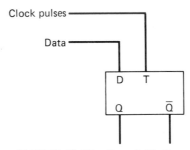

FIGURE 20-17:   *Type-D flip-flop.*

The JK FF is a combination of the type D and T units, Fig. 20-18. It has three inputs: a T clock input and two programming inputs, J and K. By providing different logic levels on J and K, it can be made to operate as a D or T FF. The accompanying operation table shows how the J and K leads affect operation. It should be noted that

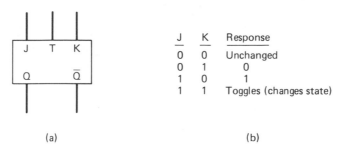

J	K	Response
0	0	Unchanged
0	1	0
1	0	1
1	1	Toggles (changes state)

(a)   (b)

FIGURE 20-18: *JK flip-flop: (a) symbol; (b) operation table.*

nothing at all will happen to the Q and $\bar{Q}$ leads unless the T lead goes from a 1 to a 0. When this occurs:

1. If J and K are both 0s, the Q and $\bar{Q}$ leads will stay as they were before the T transition.

2. If J is zero and K is 1, the Q output will go to the 0 level and $\bar{Q}$ to the 1 level.

3. If J is 1 and K is 0, the Q output will go to the 1 level and the $\bar{Q}$ lead to the 0 level.

4. If J and K are both 1s, the Q and $\bar{Q}$ leads will go the opposite levels they were before the T transition. This is called *toggling* and represents the T mode of operation.

Digital systems are made up of these basic elements: AND gates, OR gates, inverters, and four types of FF's. Each of these is neatly compacted into an integrated-circuit package. However, within the integrated circuit each gate, inverter, or FF is constructed of transistors and diodes (semiconductors) and resistors. We shall, in the next section, introduce these subjects.

20-4. SEMICONDUCTORS

Back in dc we studied wires (conductors) and insulators (nonconductors). In this section we shall investigate devices that become conductors under certain conditions and insulators under other conditions. Thus, they are called *semiconductors*.

The two elements commonly used for semiconductors are germanium and silicon. In their pure form, they exist as crystals, Fig. 20-19(a), with the atoms forming covalent bonds with each other as discussed in Chapter 1. Each atom has a valence of 4. However, when some atoms of arsenic, phosphorus, or antimony having a valence of 5 are introduced into the structure as shown by the black atoms in Fig. 20-19(b), one of the electrons becomes very loosely bound to its maternal atom and can be made to move rather easily by an external voltage. This mixture of material is called *N material* because the charge being moved is an electron (*N* for negative). The process of introducing these pentavalent atoms is called *doping*.

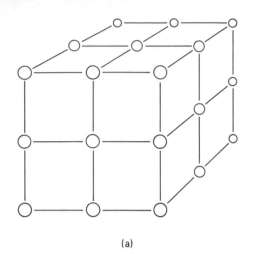

(a)

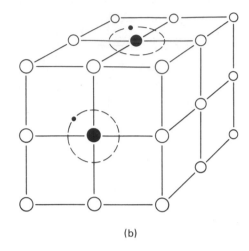

(b)

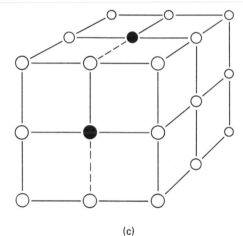

(c)

FIGURE 20-19: *Semiconductor structure: (a) pure crystal; (b) N material, (c) P material.*

In a similar manner, if some trivalent atoms of indium, aluminum, or gallium are introduced into the molten germanium or silicon, when it freezes some valence electrons will be missing from the crystal structure, Fig. 20-19(c). These missing electrons are called *holes*. However, it is possible that under the influence of an external voltage, an adjacent electron could fill hole as illustrated in Fig. 20-20(a). The hole would then appear as in Fig. 20-20(b). But this hole, too, might be filled as in Fig. 20-20(b), resulting in Fig. 20-20(c). This process could easily continue until Fig. 20-20(e) is reached. Note that, although the valence electrons actually did the moving, the hole has effectively been moved four spaces to the left. For this reason we refer to the holes as moving and call the material *P type* (*P* for positive). Thus, the majority current carrier in *N* material is the electron and the majority current carrier in *P* material is the hole.

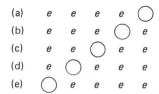

FIGURE 20-20: *Hole flow.*

The Semiconductor Diode

A *semiconductor diode* is formed from a junction of *P* material and *N* material. When the diode is connected as shown in Fig. 20-21(a), the positive end of the battery attracts the electrons and the negative end the holes, resulting in a region around the junction devoid of current carriers; thus, no current flows in the external circuit. However, when the battery is reversed as in Fig. 20-21(b), the current carriers flow toward the junction where the holes combine with the electrons. Meanwhile, new holes are formed at the positive end of the diode and electrons injected into its negative end. Thus, current will flow continuously in the external circuit.

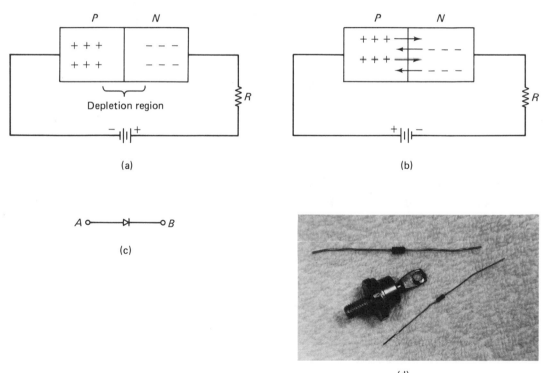

FIGURE 20-21: *Semiconductor diode: (a) reversed-biased; (b) forward-biased; (c) symbol; (d) photograph (Ruple).*

The symbol for the diode is shown in Fig. 20-21(c). If point *A* is more positive that point *B*, current will flow from *B* to *A*. If *A* is more negative than *B*, no current will flow.

The diode is very useful in converting ac to dc, Fig. 20-22. In the half-wave rectifier, Fig. 20-22(a), when point *A* is more positive than *F*, current will flow from *F* to *A*, charging up the capacitor to the peak ac voltage. When point *A* is less than *F*, the diode is back-biased and the load must receive its dc current from the discharging capacitor.

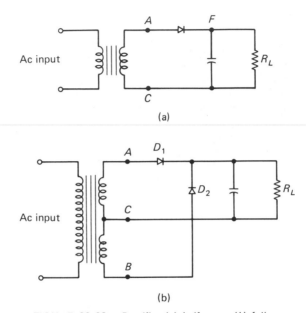

(a)

(b)

FIGURE 20-22:   *Rectifier: (a) half-wave; (b) full wave.*

The full-wave rectifier charges the capacitor on both the positive and negative alternations of the cycle. When *A* is at peak positive voltage with reference to *B*, diode D_1 conducts, charging the capacitor to the peak voltage between *A* and *C*. Meanwhile, D_2 is back-biased and no current flows through it. When, however, *B* is positive with respect to *A*, D_2 conducts and D_1 cuts off.

The Bipolar Transistor

The *bipolar transistor* is formed from three layers of semiconductor material, Fig. 20-23. Both *NPN* and *PNP* devices are available. The emitter, e, is highly doped with impurities, whereas the base, b, is very lightly doped compared to the collector, c. The transistor is commonly connected in a common-emitter configuration, Fig. 20-24, in which the base–emitter junction is forward-biased and the base–

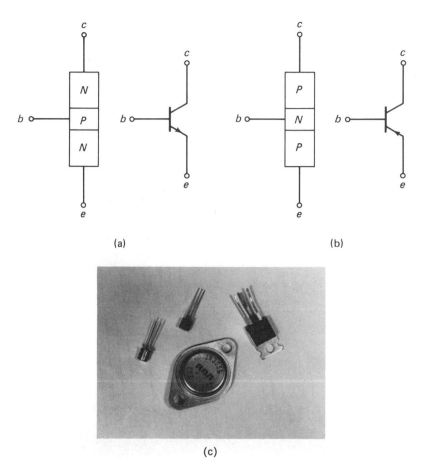

FIGURE 20-23: *Transistor: (a) NPN; (b) PNP; (c) photograph (photo by Ruple).*

collector junction reverse-biased.[1] As electrons from the emitter pass to the base, they find recombination difficult because of the very few holes in this region, and because the base is very thin. Consequently, the highly positive collector voltage sweeps the electrons across the narrow base region into the collector.

The action of the base is to cause more or less emitter electrons to cross from the emitter to the base. If very little base current flows, very little collector current flows. This relationship can be expressed by the equation

$$I_C = \beta I_B$$

---

[1]Although this circuit will work, it is highly dependent upon the β of the transistor. In practical circuits, resistors are added from the base to ground and in series with the emitter to reduce this β dependency.

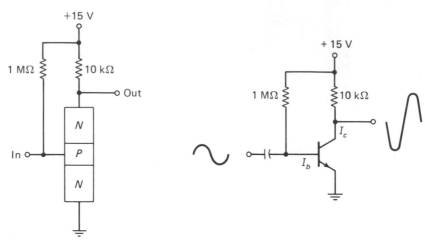

FIGURE 20-24: *Common-emitter amplifier.*

where I_C is the collector current, I_B the base current, and β the current gain of the transistor. Thus, if β were 200, 1 μA of base current would result in 100 μA of collector current, and 5 μA of base current 500 μA of collector current. The transistor is, then, a current amplifier.

If an ac signal of 1 μA p-p is impressed upon the base of a transistor having a β of 100, the collector would have ac current of

$$I_C = \beta I_B = 100 \times 1 = 100 \ \mu A$$

However, if R_L were 10 kΩ, then the voltage between the collector and ground would be

$$V = I_C R_L = 0.1 \times 10 = 1 \text{ V p-p}$$

Thus, our 1-μA signal into the transistor resulted in 5 V p-p at the output of the transistor. Note, however, that the transistor is a current amplifier and that output voltage will depend upon the load resistor.

Field-Effect Transistors

Field-effect transistors are used in a great number of electronic systems due to their extremely low power dissipation and extremely high input impedance. The first field-effect transistors consisted of a bar of substrate material, type N, for example, and two P electrodes on each side, Fig. 20-25(a). When a voltage is applied between the source and the drain, current flows through the N material. However, when a voltage is applied between G_1 and G_2, a depletion region forms around the gates, restricting N-channel current flow to a narrow part of the channel where there are current carriers. Higher gate voltages result in a narrower channel until, if the voltage

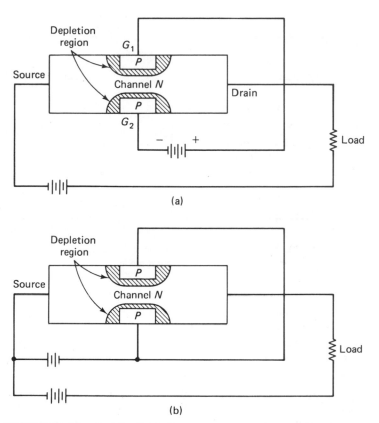

FIGURE 20-25: *Junction field-effect transistor:* (*a*) *dual gates;* (*b*) *single gate.*

is high enough, *N*-channel current is pinched off. Therefore, gate voltage controls drain current.

The next step was to tie both gates together and apply a voltage from the source to the gate, Fig. 20-25(b). This has the same effect, namely, allowing gate voltage to control drain current. This type of transistor is called a *junction field-effect transistor* (JFET). These devices have very high input impedance, typically greater than 1 MΩ. Note that input voltage controls output current, whereas in the bipolar transistor, input current controls output current.

The next stage of development was that of the *metal oxide semiconductor field-effect transistor* (MOSFET). In this device a low-resistivity *N* material is diffused onto a high-resistivity *P* material, called the *substrate*, Fig. 20-26. An insulator, silicon dioxide, is then formed by oxidizing the silicon substrate; the source and drain are etched through this insulator, but the gate is not, forming electrodes for connection to the external circuit.

Applying a positive voltage to the drain results in a very small current flow from the source to the drain, since there are few minority carriers available. (A minority

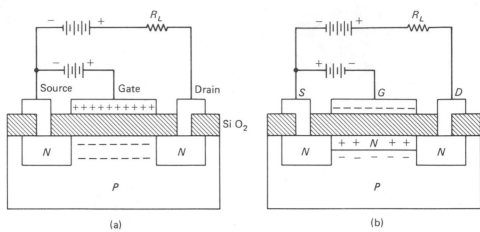

FIGURE 20-26: *Metal oxide semiconductor field-effect transistor (MOSFET):*
(a) enhancement mode; (b) depletion mode.

carrier is an electron in *P* material and a hole in *N* material.) Note that this is the same as applying a voltage from collector to emitter of a bipolar transistor with the base open. However, we shall now apply a positive voltage to the gate. This will attract many of the minority electrons to the channel beneath the gate, providing more carriers for the source-drain current flow. With this new channel, electron flow is enhanced—hence the name *enhancement mode*.

A thin bar of material can be added to the structure so that source-drain current will flow when the gate is at 0 V, Fig. 20-26(b). This current can be decreased by driving the gate negative, forcing the *N*-bar electrons into the *P* substrate and forming a deple-tion region around the gate. This type of device is called a *depletion-mode field-effect transistor*.

As can be seen, the input impedance of a MOSFET is extremely high, typically $10^{12}\ \Omega$. Additionally, its low source-drain voltage when it is fully on and its extremely low power requirements make it a very desirable element.

20-5. VACUUM TUBES

The vacuum tube was the first amplifying device but has, in all but special-purpose applications, been replaced with the transistor and the integrated circuit. The vacuum-tube *diode* consists of two elements, a cathode and a plate, and a filament wire called a *heater*, Fig. 20-27. These are then enclosed in an evacuated glass or metal case and the leads brought out to the external world. As voltage is applied to the heater, it begins to glow, heating the cathode, resulting in electrons literally boiling off the cathode surface into the space immediately around the cathode. If a positive voltage is applied to the plate, these electrons then are pulled into the plate and current flows in the external circuit, Fig. 20-27(b). If the battery shown as V_B is removed and the

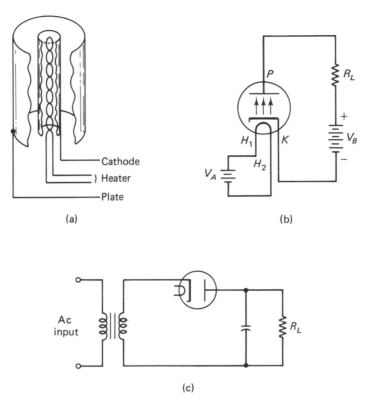

FIGURE 20-27: *Vacuum-tube diode: (a) physical construction; (b) electron flow; (c) rectifier circuit.*

secondary of a transformer inserted, the result is a rectifier, changing ac to dc, Fig. 20-27(c).

The Triode

The *triode* vacuum tube can be used as an amplifier and has a third element called a *control grid*, Fig. 20-28(a). Electrons must pass from the cathode through the helical control grid to reach the plate. If, however, the grid has a highly negative voltage with respect to the cathode, electrons will be repelled back to the cathode instead of passing to the plate. Thus, the grid voltage controls cathode to plate current.

The triode is usually connected as shown in Fig. 20-28(b). Current flowing through the tube causes a voltage drop across R_k, making the cathode a few volts above the grid. This grid to cathode voltage is called *bias*, and permits some current flow through the tube. If an ac signal is presented to the control grid, the current passing from cathode to plate is varied, varying the current through R_L and, consequently, the voltage across R_L. This much-amplified voltage can then be coupled to the outside world through C_2.

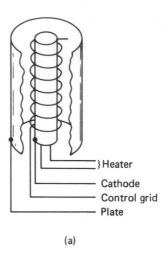

(a)

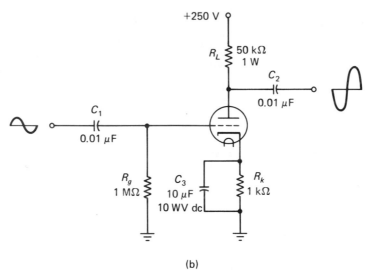

(b)

FIGURE 20-28:   *Vacuum-tube triode: (a) physical construction; (b) amplifier circuit.*

Multigrid Vacuum Tubes

Some tubes have two grids and are called *tetrodes*. The second grid, called a *screen grid*, is used to reduce the effect of the capacitance between the control grid and the plate. A *pentode* has a third grid called a *suppressor grid* which prevents electrons that have been bounced out of the plate from being drawn to the screen grid. It does this by repelling them back into the plate. Figure 20-29 illustrates these two tubes.

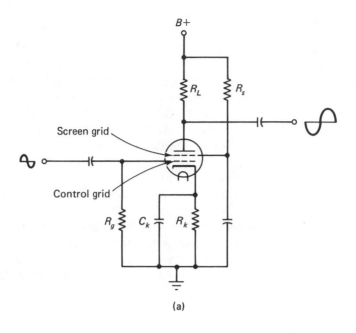

Screen grid

Control grid

R_L R_s

$B+$

R_g C_k R_k

(a)

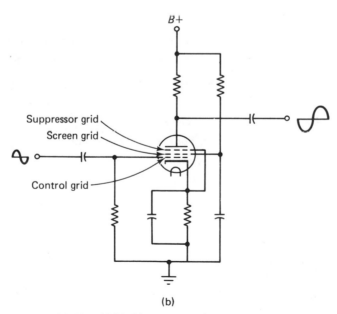

Suppressor grid

Screen grid

Control grid

$B+$

(b)

FIGURE 20-29: *Multigrid vacuum tubes: (a) tetrode amplifier; (b) pentode amplifier.*

20-6. SUMMARY

An amplifier receives a small signal and outputs a large signal. There are three types: voltage amplifiers, current amplifiers, and power amplifiers. Gain of each can be expressed in absolute or decibel quantities. Operational amplifiers have an assumed gain of infinity, an assumed input impedance of infinity, an assumed output impedance of zero ohms, and an input considered at virtual ground. The gain of the op amp is determined by the ratio of its feedback resistor to its input resistor.

Digital circuits consist of AND gates, OR gates, inverters, and flip-flops. An AND gate must have all its inputs at a logical 1 in order to output a 1. The OR gate outputs a 1 any time any of its inputs are 1. The inverter outputs a 1 with a 0 input and outputs a 0 with a 1 input. There are four types of flip-flops, each used for storing a 1 or a 0: RS, T, D, and JK.

Semiconductor diodes are constructed of a layer of N material and a layer of P material and may be used as rectifiers. Bipolar transistors consist of three layers of material, either NPN or PNP, and are used as amplifiers. Field-effect transistors have very high input impedances. Vacuum-tube diodes may be used as rectifiers, and triodes, tetrodes, and pentodes as amplifiers.

20-7. REVIEW QUESTIONS

20-1. What is the gain of an amplifier?

20-2. What are the equations for finding decibels of current? Voltage? Power?

20-3. How can an amplifier have a power gain with a less-than-unity current gain?

20-4. What are the four basic assumed specifications of an operational amplifier?

20-5. What determines the gain of an op amp when it is in a circuit?

20-6. Name the four building blocks used in digital circuits.

20-7. Under what conditions does an AND gate output a 1?

20-8. Under what conditions does an OR gate output a 1?

20-9. A NAND gate is composed of what two building blocks?

20-10. A NOR gate is composed of what two building blocks?

20-11. A one on the S lead and a zero on the R lead of an RS FF results in what logic level on the $\bar{Q}$ lead?

20-12. A pulse train having a frequency rate of F_r is applied to the T lead of a type T FF. What is the frequency of the waveform on the Q lead?

20-13. A type D FF has a frequency of 1 kHz on its D lead and a logical one on its T lead. What will appear at the Q output?

20-14. State the four possible actions of a JK FF as its J and K leads change while receiving clock pulses.

20-15. How many valence electrons do the impurity atoms of P material have?

436

20-16. How many valence electrons does silicon have? Germanium?

20-17. What is a hole?

20-18. The negative side of a battery is connected to the *P* side of a diode and the positive side to the *N* side (through a resistor). What will happen?

20-19. Why don't electrons and holes simply combine in the base of an *NPN* transistor instead of the electrons passing to the collector?

20-20. What are the relative input impedances of bipolar and field-effect transistors?

20-21. What is the difference between the enhancement and depletion mode MOSFETs?

20-22. Name the four connecting pins to a vacuum-tube diode.

20-23. What is the purpose of a rectifier?

20-24. What is the purpose of a control grid in a vacuum-tube triode?

20-25. What is the purpose of a screen grid? A suppressor grid?

20-8. PROBLEMS

20-1. Compute the gain of a current amplifier in which an input of 2.5 mA causes an output of 45 A. Show your answer in both absolute and decibel units.

20-2. Compute the gain for a voltage amplifier in which an input of 12 mV causes an output of 45 V. Use both absolute and decibel units.

20-3. An amplifier has an input of 2.5 mV at 2.5 mA which causes an output of 20 V at 35 mA. Compute its voltage gain, current gain, and power gain in both absolute and decibel units.

20-4. An amplifier with a gain of 45 dB must supply 45 W at its output. What power must be provided its input?

20-5. Design an op-amp circuit with a gain of 20.

20-6. A three-input AND gate has zeros on all its inputs. What is its output?

20-7. A four-input OR gate has a 1 on one input lead and 0s on the other three. What is its output?

20-8. A type-T FF has a frequency of 30 kHz on its input. What is the frequency of its output?

20-9. A half-wave rectifier has 15-V rms ac applied to it. What will its output voltage be under no load?

20-10. A transistor with a β of 100 has 2.8 μA applied to its base. What will its collector current be?

20-9. PROJECTS

20-1. Build an amplifier with a gain of 10 using a 741 op amp.

20-2. Using a 7410 digital IC, go through all possible combinations of inputs and verify its proper functioning.

20-3.   Connect a 1-kHz, 0- to 5-V square wave to the T lead of a 7473 digital IC. What is its output frequency? (*Note:* The square wave can be obtained by the circuit shown in Fig. 20-30.)

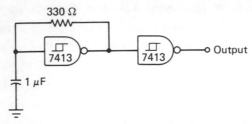

FIGURE 20-30:   *1-kHz square-wave generator.*

20-4.   Connect a 1N4001 diode in series with a 100-Ω resistor and a voltage source. Vary the voltage source and observe the voltage across and current through the diode. Next, reverse the leads of the diode and repeat the procedure.

20-5.   Connect a 2N2222 as shown in Fig. 20-31. Vary R_1 and observe both I_b and I_c. From this, compute the β of the transistor.

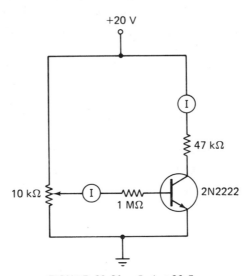

FIGURE 20-31:   *Project 20-5.*

20-6.   Connect a 6SL7 triode as shown in Fig. 20-28(b). Measure its voltage gain.

20-7.   Write a short essay on the relative importance of the discoveries of the vacuum tube and the transistor.

20-8.   What are some of the important uses for vacuum tubes that cannot be done using transistors?

20-9.   Describe the SCR.

APPENDIX

DETERMINANTS

Although there are several ways in which simultaneous equations may be solved, the advent of the hand calculator has made the method of determinants even more attractive. We shall discuss only two and three variable equations here, for this is usually sufficient for electronics. It should be pointed out that manipulations beyond those presented here will be required for those problems of four variables and beyond.

TWO VARIABLES

To solve a two-variable problem, the equations must first be put in proper form such that the variables and their coefficients are on the left side of the equals sign and the constants on the right. Additionally, the X terms and Y terms must be lined up. For example, the following are in proper form:

$$2X - 3Y = -13$$
$$4X + 5Y = 7$$

The next step is to form the denominator determinant by placing all the variable coefficients in a 2×2 format. Note that the signs are retained:

$$\begin{vmatrix} 2 & -3 \\ 4 & 5 \end{vmatrix}$$

To solve for X, form a numerator determinant by substituting the constants on the right for the X coefficients in the denominator determinant; divide this by the denominator determinant:

$$X = \frac{\begin{vmatrix} -13 & -3 \\ 7 & 5 \end{vmatrix}}{\begin{vmatrix} 2 & -3 \\ 4 & 5 \end{vmatrix}}$$

We are now ready to evaluate the determinant according to the equation

$$\begin{vmatrix} a & c \\ b & d \end{vmatrix} = ad - bc$$

Thus, solve for X:

$$X = \frac{\begin{vmatrix} -13 & -3 \\ 7 & 5 \end{vmatrix}}{\begin{vmatrix} 2 & -3 \\ 4 & 5 \end{vmatrix}} = \frac{(-13 \times 5) - (-3 \times 7)}{(2 \times 5) - (-3 \times 4)} = \frac{-44}{22} = -2$$

In a similar manner, we can solve for Y by forming its numerator determinant and dividing by the denominator determinant. The numerator is formed by substituting the constants on the right of the equation for the Y coefficients:

$$Y = \frac{\begin{vmatrix} 2 & -13 \\ 4 & 7 \end{vmatrix}}{\begin{vmatrix} 2 & -3 \\ 4 & 5 \end{vmatrix}} = \frac{(2 \times 7) - (-13 \times 4)}{(2 \times 5) - (-3 \times 4)} = \frac{66}{22} = 3$$

EXAMPLE I-1: Solve the equations:

$$4V_1 = 10V_2 - 16$$
$$7V_2 - 5V_1 - 12 = 0$$

SOLUTION: Put them into standard form:

$$4V_1 - 10V_2 = -16$$
$$-5V_1 + 7V_2 = 12$$

Solve for V_1:

$$V_1 = \frac{\begin{vmatrix} -16 & -10 \\ 12 & 7 \end{vmatrix}}{\begin{vmatrix} 4 & -10 \\ -5 & 7 \end{vmatrix}} = \frac{(-16 \times 7) - (-10 \times 12)}{(4 \times 7) - [-10 \times (-5)]} = \frac{8}{-22} = -0.3636 \text{ V}$$

Solve for V_2:

$$V_2 = \frac{\begin{vmatrix} 4 & -16 \\ -5 & 12 \end{vmatrix}}{-22} = \frac{(4 \times 12) - [-5 \times (-16)]}{-22} = \frac{-32}{-22} = 1.455 \text{ V}$$

Three-variable determinants are formed the same way as the two variables:

1. Put the equations in standard form.
2. Form a 3 × 3 denominator determinant from the coefficients of the variables.
3. Form numerator determinants by substituting the constants on the right of the equation for the coefficients of the variable under consideration.

Thus, assume the following equations:

$$3X + 1 - Z = 0$$
$$2X = 14 + 4Y - Z$$
$$-3X - Y = -1$$

Rearrange into standard form:

$$3X + 0Y - Z = -1$$
$$2X - 4Y + Z = 14$$
$$-3X - 1Y + 0Z = -1$$

Form determinants:

$$X = \frac{\begin{vmatrix} -1 & 0 & -1 \\ 14 & -4 & 1 \\ -1 & -1 & 0 \end{vmatrix}}{\begin{vmatrix} 3 & 0 & -1 \\ 2 & -4 & 1 \\ -3 & -1 & 0 \end{vmatrix}} \qquad Y = \frac{\begin{vmatrix} 3 & -1 & -1 \\ 2 & 14 & 1 \\ -3 & -1 & 0 \end{vmatrix}}{\begin{vmatrix} 3 & 0 & -1 \\ 2 & -4 & 1 \\ -3 & -1 & 0 \end{vmatrix}} \qquad Z = \frac{\begin{vmatrix} 3 & 0 & -1 \\ 2 & -4 & 14 \\ -3 & -1 & -1 \end{vmatrix}}{\begin{vmatrix} 3 & 0 & -1 \\ 2 & -4 & 1 \\ -3 & -1 & 0 \end{vmatrix}}$$

We must next evaluate each determinant. Using the denominator determinant as an example, copy down the first two rows again and form three positive diagonal products and three negative diagonal products:

Negative Products

$$D_E = \begin{vmatrix} 3 & 0 & -1 \\ 2 & -4 & 1 \\ -3 & -1 & 0 \end{vmatrix} \begin{matrix} 3 & 0 \\ 2 & -4 \\ -3 & -1 \end{matrix}$$

Positive Products

$$D_E = +(3)(-4)(0) + (0)(1)(-3) + (-1)(2)(-1)$$
$$- (-3)(-4)(-1) - (-1)(1)(3) - (0)(2)(0)$$
$$= 17$$

441

Apply this principle to the solutions of the variables:

$$X = \frac{\begin{vmatrix} -1 & 0 & -1 \\ 14 & -4 & 1 \\ -1 & -1 & 0 \\ 3 & 0 & -1 \\ 2 & -4 & 1 \\ -3 & -1 & 0 \end{vmatrix} \begin{matrix} -1 & 0 \\ 14 & -4 \\ -1 & -1 \\ 3 & 0 \\ 2 & -4 \\ -3 & -1 \end{matrix}}{} = \frac{\begin{matrix} +(-1)(-4)(0) + (0)(1)(-1) \\ +(-1)(14)(-1) - (-1)(-4)(-1) \\ -(-1)(1)(-1) - (0)(14)(0) \end{matrix}}{17}$$

$$= \frac{17}{17} = 1$$

$$Y = \frac{\begin{vmatrix} 3 & -1 & -1 \\ 2 & 14 & 1 \\ -3 & -1 & 0 \end{vmatrix} \begin{matrix} 3 & -1 \\ 2 & 14 \\ -3 & -1 \end{matrix}}{17} = \frac{\begin{matrix} +(3)(14)(0) + (-1)(1)(-3) \\ +(-1)(2)(-1) - (-3)(14)(-1) \\ -(-1)(1)(3) - (0)(2)(-1) \end{matrix}}{17}$$

$$= \frac{-34}{17} = -2$$

$$Z = \frac{\begin{vmatrix} 3 & 0 & -1 \\ 2 & -4 & 14 \\ -3 & -1 & -1 \end{vmatrix} \begin{matrix} 3 & 0 \\ 2 & -4 \\ -3 & -1 \end{matrix}}{17} = \frac{\begin{matrix} +(3)(-4)(-1) + (0)(14)(-3) \\ +(-1)(2)(-1) - (-3)(-4)(-1) \\ -(-1)(14)(3) - (-1)(2)(0) \end{matrix}}{17}$$

$$= \frac{68}{17} = 4$$

EXAMPLE I-2: Solve the following equations:

$$3I_1 = 2I_2 + 6I_3$$
$$4I_1 - I_3 = 6$$
$$7I_1 + 6I_2 - I_3 - 7 = 0$$

SOLUTION: Rearrange into standard form:

$$3I_1 - 2I_2 - 6I_3 = 0$$
$$4I_1 + 0I_2 - 1I_3 = 6$$
$$7I_1 + 6I_2 - 1I_3 = 7$$

Solve for each of the variables:

$$I_1 = \frac{\begin{vmatrix} 0 & -2 & -6 \\ 6 & 0 & -1 \\ 7 & 6 & -1 \\ 3 & -2 & -6 \\ 4 & 0 & -1 \\ 7 & 6 & -1 \end{vmatrix}}{} = \frac{\begin{matrix} +(0)(0)(-1) + (-2)(-1)(7) \\ +(-6)(6)(6) - (7)(0)(-6) \\ -(6)(-1)(0) - (-1)(6)(-2) \\ \hline +(3)(0)(-1) + (-2)(-1)(7) \\ +(-6)(4)(6) - (7)(0)(-6) \\ -(6)(-1)(3) - (-1)(4)(-2) \end{matrix}}{} = \frac{-214}{-120} = 1.783 \text{ A}$$

$$I_2 = \frac{\begin{vmatrix} 3 & 0 & -6 \\ 4 & 6 & -1 \\ 7 & 7 & -1 \end{vmatrix}}{-120} = \frac{\begin{array}{l} +(3)(6)(-1) + (0)(-1)(7) \\ +(-6)(4)(7) - (7)(6)(-6) \\ -(7)(-1)(3) - (-1)(4)(0) \end{array}}{-120} = \frac{87}{-120} = -0.7250 \text{ A}$$

$$I_3 = \frac{\begin{vmatrix} 3 & -2 & 0 \\ 4 & 0 & 6 \\ 7 & 6 & 7 \end{vmatrix}}{-120} = \frac{\begin{array}{l} +(3)(0)(7) + (-2)(6)(7) \\ +(0)(4)(6) - (7)(0)(0) \\ -(6)(6)(3) - (7)(4)(-2) \end{array}}{-120} = \frac{-136}{-120} = 1.133 \text{ A}$$

ANSWERS TO
ODD-NUMBERED PROBLEMS

CHAPTER 1

1-1. 29 electrons, 29 protons, 35 neutrons

CHAPTER 2

2-1. -3.000 C

2-3. 4.000 A

2-5. 9.339 ms

2-7. 3.000 V

2-9. **(a)** Nominal $= 2.2$ kΩ, 20%
Maximum $= 2.640$ kΩ
Minimum $= 1.760$ kΩ
(c) Nominal $= 750$ kΩ, 5%, 0.1% rel.
Minimum $= 712.5$ kΩ
Maximum $= 787.5$ kΩ
(b) Nominal $= 10.0\,\Omega$, 10%, 1% rel.
Minimum $= 9.000\,\Omega$
Maximum $= 11.00\,\Omega$
(d) Nominal $= 5.6\,\Omega$, 5%,
0.001% rel.
Minimum $= 5.320\,\Omega$
Maximum $= 5.880\,\Omega$

2-11. **(a)** Brown, red, orange, silver
(c) Brown, black, black
(b) Orange, blue, red, gold
(d) Blue, red, silver, gold

2-13. **(a)** 13 kΩ **(b)** 820 kΩ **(c)** 470 Ω **(d)** 2.4 Ω

2-15. **(a)** 4.56 mV **(b)** 23.643 kΩ **(c)** 67.450 μV **(d)** 67.985 pA

3-1. 700 Ω   **3-11.** 600.0 mW, 666.7 Ω

3-3. 111.6 mA   **3-13.** 410.0 mA, 73.17 Ω

3-5. 16.45 V   **3-15.** 10.00 mA, 300.0 mW

3-7. 10.00 mA   **3-17.** 707.1 V, 70.71 mA

3-9. 416.7 mA, 288.0 Ω   **3-19.** 2.4 kΩ, 10 W

CHAPTER 4

4-1. $V_1 = 2$ V, $V_2 = 5$ V, $V_3 = 6$ V, $V_4 = 8$ V; CCW more positive.

4-3. 21.00 kΩ   **4-5.** $V_4 = 15.00$ V

4-7. $I = -126.8\ \mu$A, $V_1 = -8.624$ V (left side more positive)

4-9. $V_c = -525$ V, $V_1 = 164$ V, $V_2 = 240$ V, $V_3 = 136$ V

4-11. $I = 3.070\ \mu$A, $V_1 = 3.070$ V, $V_2 = 20.88$ V, $V_3 = 11.05$ V

4-13. $I = 4.286$ mA, $V_A = 164.3$ V, $V_2 = 25.71$ V, $V_3 = 8.571$ V

4-15. 1.080 W   **4-17.** 107.5 μW

4-19.   **4-21.**

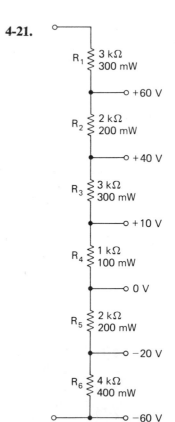

5-1. 16 mA

5-3. 92.92 Ω

5-5. 3.333 kΩ

5-7. 15.67 kΩ

5-9. 63.60 W

5-11. $P_t = 200$ mW, $R_t = 500$ Ω, $R_1 = 1$ kΩ, $R_2 = 10$k Ω, $R_3 = 5$ kΩ,
$R_4 = 1.429$ kΩ, $P_1 = 100$ mW, $P_2 = 10$ mW, $P_3 = 20$ mW, $P_4 = 70$ mW,
$I_4 = 7$ mA

5-13. $I_t = 431.7$ mA, $P_t = 8.633$ W, $R_t = 46.33$ Ω, $R_3 = 266.7$ Ω,
$R_4 = 222.2$ Ω, $P_1 = 4.000$ W, $P_2 = 1.333$ W, $P_4 = 1.800$ W,
$I_1 = 200$ mA, $I_2 = 66.67$ mA, $I_3 = 75.00$ mA

5-15. $V_s = 100$ V, $I_t = 20.00$ mA, $P_t = 2.000$ W, $R_4 = 7.500$ kΩ,
$P_1 = 100$ mW, $P_2 = 66.67$ mW, $P_3 = 500$ mW, $P_4 = 1.333$ W,
$I_2 = 666.7$ μA, $I_3 = 5.000$ mA, $I_4 = 13.33$ mA

CHAPTER 6

6-1. 14.97 kΩ

6-3. $I_t = 1.069$ mA, $V_1 = 2.731$ V, $V_2 = 7.269$ V, $I_1 = 487.8$ μA,
$I_2 = 581.2$ μA

6-5. $V_s = 71.60$ V, $I_t = 7$ mA, $R_2 = 6$ kΩ, $V_1 = 24$ V, $V_2 = 47.60$ V

6-7. 12.50 mA **6-9.** 2.703 V **6-11.** 55.11 mW **6-13.** 11,830 ft or 2.241 mi

6-15.

R_1 50.00 Ω 180 mW
+15 V
R_2 111.1 Ω 225 mW
+10 V
R_3 62.50 Ω 100 mW
+7.5 V
R_4 750 Ω 75.00 mW
0 V

6-17. $R_1 = 363.6$ Ω,
$P_1 = 44.00$ mW
$R_2 = 1.571$ kΩ,
$P_2 = 77.00$ mW
$R_3 = 1.000$ kΩ,
$P_3 = 25.00$ mW
$R_6 = 20.00$ kΩ,
$P_6 = 20.00$ mW

CHAPTER 7

7-1. 14.40 V

7-3. 10.43 A

7-5. 10,000 cmil

7-7. 160.8 Ω

7-9. 10.27 Ω

7-11. AWG 6

7-13. 344.8 mV

7-15. $P_{\text{wire}} = 154.1$ W, $P_{\text{heater}} = 1172$ W

CHAPTER 8

8-1. $I_1 = 10.00$ μA, right to left

8-3. $V_{75} = 128.0$ V, $V_{10} = 97.53$ V

8-5. 6.780 V, left side more positive

8-7. 438.2 μA

8-9. 57.14 V

8-11. $V_{\text{th}} = 13.33$ V, $R_{\text{th}} = 16.67$ kΩ

8-13. 678.0 μA

8-15. 181.8 μA

8-17. $I_{R1} = 1.636$ mA, $I_{R2} = 2.727$ mA, $I_{R3} = 1.091$ mA

8-19. $V_{20} = 78.51$ V, $V_{150} = 107.5$ V

8-21. 15.44 kΩ

CHAPTER 9

9-1. $R = 4.900$ kΩ

9-3. **(a)** 1 kΩ/V   **(b)** 10 kΩ/V

9-5. Yes

9-9. $R = 52.632$ Ω

9-7.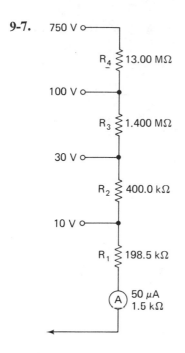

9-11.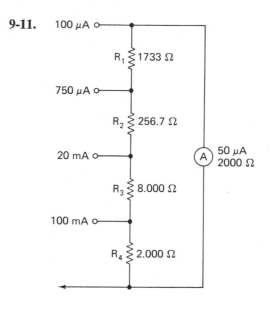

9-13. Approx. 1.1%

CHAPTER 10

10-1. 20 A

10-3. 50,000 A/Wb

10-5. 10.61 T

10-7. 80 A/m

10-9. 200

CHAPTER 11

11-1. 10.998

11-3. $f = 17.94$ Hz, $p = 55.75$ ms

11-5. **(a)** 0.4520 rad   **(b)** 1.868 rad   **(c)** -1.134 rad

11-7. **(a)** 65   **(b)** 4567 rad/s   **(c)** 726.9 Hz   **(d)** 1.376 ms
     (e) 23°   **(f)**

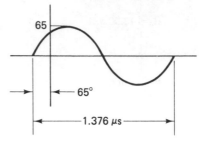

11-9. **(a)** 0.023 A   **(b)** 45,000 rad/s   **(c)** 7162 Hz   **(d)** 139.6 μs
     (e) 65°   **(f)**

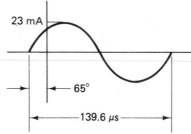

11-11. 10.51 V

11-13. **(a)** 4 MHz   **(b)** 17 MHz

11-15. 1.546 m

11-17. Av = 6.667, rms = 7.071

11-19. Av = 214.9 V, rms = 238.6 V

CHAPTER 12

12-1. **(a)** $-j$ **(b)** j **(c)** $-j$

12-3. $162.2 + j147.0$

12-5. **(a)** $1 - j16$ **(b)** $-26 - j4$ **(c)** $-9 - j1369.3$

12-7. **(a)** $-0.6800 - j1.760$ **(b)** $-0.500 - j1.500$ **(c)** $0.7071 - j0.3987$

12-9. **(a)** $7.616/\!-66.80°$ **(b)** $9.849/66.04°$ **(c)** $82.57/\!-62.29°$

12-11. **(a)** $49/69°$ **(b)** $339.6/\!-79.66°$ **(c)** $43.35/6.463°$

12-13. **(a)** $1.876/0°$ **(b)** $0.6241/60.30°$ **(c)** $3.098/95.50°$

CHAPTER 13

13-1. 20 V

13-3. 295 mH

13-5. 2.609 H

13-7. 556.8 mH

13-9. 11.5.1 mH

13-11. 59.26 V

13-13. 11.85 mA

13-15. $I = 1.389$ A, $V = 14.40$ V

13-17. 79.06

13-19. $V_L = 30.12$ V, $V_R = 69.88$ V

13-21. $V_L = 13.075$ mV, $V_R = 6.925$ mV, $I = 135.8$ nA

CHAPTER 14

14-1. **(a)** 6.283 Ω   **(b)** 62.83 Ω   **(c)** 628.3 Ω

14-3. 318.3 Hz

14-5. 34.56 kΩ

14-7. 2565 Ω

14-9. 678.6 kΩ

14-11. 3034 Ω

14-13. $3.016\underline{/70°}$ V

14-15. $I = 83.77\underline{/-90°}$ μA, $V_{150} = 157.9\underline{/0°}$ mV, $V_{200} = 210.5\underline{/0°}$ mV, $V_{600} = 631.6\underline{/0°}$ mV

14-17. $12.72\underline{/38.15°}$ kΩ

14-19. $3900\underline{/75.14°}$ Ω

14-21. $V_{1.5} = 3.190\underline{/-39.40°}$ V, $V_{40} = 1.604\underline{/50.60°}$ V, $V_{55} = 2.205\underline{/50.60°}$ V, $V_{680} = 1.446\underline{/-39.40°}$ V

14-23. $3.917\underline{/65.38°}$ V

CHAPTER 15

15-1. 3 V

15-3. 50 nA

15-5. 75,000 V/m

15-7. 22.125×10^{-12} F/m

15-9. 5 μF

15-11. 26.552 μF

15-13. 576.9 pF

15-15. 6.480 V

15-17. 13.952 kΩ

15-19. 10.865 V

CHAPTER 16

16-1. **(a)** 3.386 MΩ **(b)** 338.6 kΩ **(c)** 33.86 kΩ **(d)** 3.386 kΩ

16-3. $X_C = 723.4\ \Omega$, $I = 2.765\underline{/90°}$ mA

16-5. $212.2\underline{/-90°}$ mV

16-7. 677.3 Hz

16-9. 424.6 Ω

16-11. 1573 Ω

16-13. 673.7 mΩ

16-15. $5.399\underline{/-79.33°}$ kΩ

16-17. $5.305\underline{/89.99°}$ MΩ

16-19. $V_R = 51.44\underline{/46.70°}$ V, $V_C = 54.58\underline{/-43.30°}$ V

16-21. 78.00 V

CHAPTER 17

17-1. 20.48 mW

17-3. 156.3 kvar

17-5. $S = 281.3$ VA, $Q = 144.9$ var, $P = 241.1$ W, $PF = 0.8572$

17-7. $Q = 196.2$ var, $P = 405.0$ W

17-9. $27.59\underline{/43.53°}$ kΩ

17-11. $4.762\underline{/-9.276°}$ kΩ

17-13. $I = 651.3\underline{/-12.31°}$ μA, $V_L = 8.185\underline{/77.69°}$ V, $V_C = 6.479\underline{/-102.3°}$ V, $V_R = 7.816\underline{/-12.31°}$ V

17-15. **(a)** $7.407\underline{/9.043°}$ kΩ **(b)** $3.990\underline{/+57.86°}$ kΩ

17-17. $3.136\underline{/-18.13°}$ kΩ

17-19. $I_L = 6.366\underline{/-90°}$ mA, $I_C = 4.021\underline{/90°}$ mA, $I_R = 1.600\underline{/0°}$ mA, $I_t = 2.838\underline{/-55.69°}$ mA

17-21. $I_{C1} = 8.618\underline{/93.90°}$ mA, $I_{C2} = 9.449\underline{/88.23°}$ mA

17-23. $2.618\underline{/43.89°}$ kΩ

17-25. $I_{R1} = 36.03\underline{/21.22°}$ mA flowing north, $I_{R2} = 18.80\underline{/156.21}$ mA flowing south

CHAPTER 18

18-1. 2771 Hz

18-3. 810.6 pF

18-5. 480 Hz

18-7. $L = 42.44$ mH, $C = 1842$ pF

18-9. $f_r = 5.365$ kHz, $Q = 17.98$, BW $= 298.4$ Hz

18-11. 26.70 Ω

18-13. $I_L = 1.414\underline{/-90°}$ mA, $I_C = 1.414\underline{/90°}$ mA, $I_R = 100\underline{/0°}$ μA, $I_t = 100\underline{/0°}$ μA, $Z = 10\underline{/0°}$ kΩ

18-15. $f_{ar} = 522.8$ Hz, $f_r = 538.0$ Hz

18-17. $L = 1.492$ mH, $C = 27,162$ pF

CHAPTER 19

19-1. 21.24 dB

19-3. 910.5 pF

19-5. $C = 0.01492$ μF in parallel with the load

19-7. $L = 568.4$ mH in parallel with the load

19-9. $C_1 = C_2 = 10,610$ pF, $L = 1.9098$ mH

19-11. $L_1 = L_2 = 954.9$ μH, $L_3 = 11.46$ mH, $C_1 = 5305$ pF

CHAPTER 20

20-1. $G_i = 18,000 = 85.11$ dB

20-3. $G_v = 8000 = 78.06$ dB, $G_i = 14.00 = 22.92$ dB, $G_p = 112,000 = 50.49$ dB

20-5. The ratio of R_f to R_i must be 20. One such answer is $R_f = 200$ kΩ and $R_i = 10$ kΩ.

20-7. One

20-9. 21.21 V

INDEX